SCIENCE to GCSE

Stephen Pople
Michael Williams

OXFORD
UNIVERSITY PRESS

OXFORD
UNIVERSITY PRESS

Great Clarendon Street, Oxford OX2 6DP

Oxford University Press is a department of the University of Oxford.
It furthers the University's objective of excellence in research,
scholarship, and education by publishing worldwide in

Oxford New York

Auckland Cape Town Dar es Salaam Hong Kong Karachi
Kuala Lumpur Madrid Melbourne Mexico City Nairobi
New Delhi Shanghai Taipei Toronto

With offices in

Argentina Austria Brazil Chile Czech Republic France Greece
Guatemala Hungary Italy Japan Poland Portugal Singapore
South Korea Switzerland Thailand Turkey Ukraine Vietnam

Oxford is a registered trade mark of Oxford University Press
in the UK and in certain other countries

First published 1995. Second edition 2002

British Library Cataloguing in Publication Data

Data available

ISBN 978 0 19 914831 8
10 9 8 7 6 5

Printed in China by Printplus

Paper used in the production of this book is a natural, recyclable product made
from wood grown in sustainable forests. The manufacturing process conforms to
the environmental regulations of the country of origin.

Acknowledgements

The Publisher would like to thank the following for their kind permission to reproduce copyright material:

p18 J Durham/SPL (left), BSIP Laurent H Americain/SPL (right); **p19** N Cobbing/Still Pictures (left), R Giling/Still Pictures (right); **p20** C Nuridsany & M Perennou/SPL; **p23**. T Bucholz/Bruce Coleman Photo Library ; **p27** N Cattlin/Holt Studios; **p32** SPL; **p34** CNRI/SPL; **p38** D Scharf/SPL; **p43** D Reed/Corbis; **p44** H Morgan/SPL; **p46** Dr G Settles/SPL; **p47** CNRI/SPL; **p48** D Sharf/SPL; **p50** OUP; **p51** A Walsh/Oxford Scientific Films; **p55** N Cattlin/Holt Studios; **p56** G Kidd/OSF; **p59** P Parks/OSF (both); **p60** A & H Michler/SPL (top), Bios/Still Pictures (bottom); **p61** H Smith/alamy.com; **p63** S Rowner/OSF; **p64** A Hay/OSF (left); **p64** M Hamblin/OSF (right); **p66** M St. Maur Sheil/Corbis; **p67** Hart Davis/SPL (left), V Fleming/SPL (right); **p69** OUP; **p70** OUP; **p71** B Kent/OSF; **p72** M Howes/alamy.com; **p73** S Jonasson/Nordicphotos/alamy.com; **p75** Geoscience Features; **p76** Peter Gould; **p78** R Sanford/SPL (top), Somboon-Unep/Still Pictures (bottom); **p80** ZEFA (top), Peter Gould (bottom); **p83** Silkeborg Museum, Denmark/Munoz-yague/SPL; **p84** S Bloom/TCL/Getty Images; **p86** OUP; **p88** OUP; **p89** C D Winters/SPL; **p90** Dr B Booth/Geoscience Features (both); **p96** OUP; **p101** OUP (left), N Aberman/Hutchison Library (top right), Gusto Productions/SPL (bottom right); **p104** OUP (top right); **p104** N Cattlin/Holt Studios; **p105** H Rogers/TRIP; **p106** Synetix; **p107** Dinodia/TRIP; **p111** OUP (top), Lambert/TRIP (bottom); **p112** P Crapet/Stone/Getty Images; **p115** Winters/SPL (top), BMW (bottom); **p118** Geoscience Features (top), **p118** H Rogers/TRIP (bottom); **p120** Dr B Booth/Geoscience Features(top), B Johnson/SPL (bottom); **p122** Alcan Aluminium (top), B Heinsohn/Stone/Getty Images (bottom); **p123** H Rogers/TRIP; **p124**. Winsford Rock Salt Mine/Salt Union Ltd; **p126** OUP (both); **p127** W Lesch/Image Bank/Getty Images; **p128** Grantpix/SPL; **p129** OUP; **p130** Image State; **p131** Paul Glendale/Still Pictures; **p132** OUP; **p133** Tony Waltham/Geophotos; **p134** SPL; **p135** The Flight Library; **p138** H Schneebeli/SPL; **p140** OUP; **p144** OUP; **p148** OUP; **p149** OUP; **p151** I West/OSF; **p152** Allsport; **p159** J Allan Cash; **p162** T Craddock/SPL; **p164** J Allan Cash; **p166** Colorsport; **p168** OUP (top), Photo Library International/SPL (bottom); **p169** U Walz/OSF; **p170** J Amos/SPL; **p171** OUP; **p172** J Allan Cash; **p178** J Watts/OSF; **p181** Spectrum Colour Library; **p183** T Tilford/OSF (top), **p183** SPL (bottom); **p186** OUP; **p189** K & H Benser/Powerstock Zefa; **p191** D Parker/SPL; **p192** Rex Features; **p195** P Plailly/SPL (top), **p195** M Bond/SPL (bottom); **p197** M Dohrn/SPL (left), **p197** D Parker/SPL (right); **p198** NASA/SPL; **p199** T Hallas/SPL; **p200** D Parker/SPL (left), Astro Space/SPL (top right), NASA/SPL (bottom right); **p202** Celestial Image Co./SPL.

The illustrations are by Chris Duggan, Jones Sewell, Patricia Moffett, Pat Murray, Mike Ogden, Oxford Designers & Illustrators, Steve Scanlan, Pat Thorne, Borin Van Loon, and Pamela Venus.

Cover photograph by Images Colour Library

Introduction

If you are working towards GCSE Science (double or single award), then this book is designed for you. It explains the science concepts that you will meet and helps you find what you need to know. The topics are covered in double-page units which we have called ***spreads***.

Contents Here you can see how the spreads are arranged.

Test and check Try answering these questions when you revise. At the end of each question there is a number telling you which spread to look up if you need to check information or find out more.

Spreads 1.01 to 1.04 The first two spreads tell you how to plan and carry out investigations, draw conclusions, and evaluate what you have done. The next two spreads deal with scientific ideas, how they can change, and some of the issues they raise.

Spreads 2.01 to 4.35 These are grouped into three sections, matching Attainment Targets 2, 3, and 4 of the National Curriculum.

Summaries These tell you the main points covered in each section and the particular spreads which deal with them.

Exam-style questions These are similar to the questions that you might meet in your GCSE examinations. A single question may cover topics from more than one area of science.

Answers Here you will find brief answers to the questions in the spreads and to the exam-style questions. But try the questions before you look at the answers!

Units and symbols Here you will find the main units of measurement, their abbreviations, and the prefixes (such as 'kilo' and 'milli') that are used to form bigger or smaller units.

Index Use this if you have a particular scientific word or term that you need to look up.

Periodic table This is on the last page of the book. It shows all the elements.

To be a good scientist, you need to carry out investigations. This book should help you to understand the scientific ideas that support your investigations. We hope that you will find it useful.

Stephen Pople

Contents

Your examination specification may not include the topics covered in spreads 4.14, 4.15, and 4.16. Please check with your teacher.

Test and check

Can you answer the following? The spread number in brackets tells you where to find the information.

1 What features do animals and plants have in common? *(2.01)*
2 Where is a cell's 'control centre'? *(2.01)*
3 How do plant cells and animal cells differ? *(2.01)*
4 How do plants obtain their food? What is the process called? *(2.02)*
5 How do plants and animals get their energy? What is the process called? *(2.02)*
6 On what factors does the rate of photosynthesis depend? What is meant by a limiting factor? *(2.03)*
7 What do guard cells do? *(2.03)*
8 Why do plants need minerals? How do these get into the roots and up through the plant? *(2.03)*
9 Where do plants store the food they make? *(2.03)*
10 Where in a flower are the male sex cells and the female sex cells? *(2.04)*
11 In a plant, what does auxin do? *(2.04)*
12 What are artificial plant hormones used for? *(2.04)*
13 What can happen to a plant if minerals are missing from the soil? *(2.04)*
14 In your body, how do food, water, and oxygen get to your cells? *(2.05)*
15 What are the main organs of the body? *(2.05)*
16 What happens to food in the gut? *(2.06)*
17 What are the five main types of substance in food? *(2.06)*
18 How would you test for glucose, starch, protein, and fat? *(2.07)*
19 How is digested food absorbed by your blood? *(2.07)*
20 What unit is used to measure amounts of food energy? *(2.07)*
21 What is the difference between aerobic and anaerobic respiration? *(2.07)*
22 Why do your cells make ATP? *(2.07)*
23 Can you describe some of the jobs done by enzymes in the digestive system? *(2.07)*
24 Can you describe five important jobs done by the blood? *(2.08)*
25 What is the difference between an artery and a vein? *(2.08)*
26 Why does the heart need to work as two separate pumps? *(2.08)*
27 What job is done by the lungs? *(2.09)*
28 What are alveoli? Why are they surrounded by blood capillaries? *(2.09)*
29 What do the kidneys do? *(2.09)*
30 In human females, where are ova (female sex cells) stored? *(2.10)*
31 In human males, where are sperms (male sex cells) stored? *(2.10)*
32 What is the central nervous system? What does it do? *(2.11)*
33 What is the difference between a sensory neurone and a motor neurone? *(2.11)*
34 What is a reflex action? Can you give an example? *(2.11)*
35 In the human eye, what jobs are done by the iris, lens, retina, and optic nerve? *(2.11)*
36 In the body, where are hormones made? What do they do? *(2.12)*
37 How do hormones control the menstrual cycle? *(2.12)*
38 Can you give three examples of hormones being taken as drugs? Are any of these harmful *(2.12)*
39 What is homeostasis? Can you give three examples of homeostasis in the human body? *(2.13)*
40 Where are nephrons and what do they do? *(2.13)*
41 What effect does have insulin on the liver? *(2.13)*
42 Can you describe three ways in which the body can control its temperature? *(2.13)*
43 What are germs? How can they spread from one person to another? *(2.14)*
44 How does your body defend itself against germs? *(2.14)*
45 How do antibiotics and vaccines work? *(2.14)*
46 Can you describe the structure of a typical bacterial cell? *(2.14)*
47 How does your skin help protect you against disease? *(2.15)*
48 What does the mucus in your nose and windpipe do? *(2.15)*
49 What are your white blood cells for? (2.15)
50 Where are antibodies made? What do they do? *(2.15)*

Test and check

51 What diseases can smoking cause? *(2.15)*
52 Can you give two examples of inherited characteristics? *(2.16)*
53 Where is information about your inherited characteristics stored? *(2.16)*
54 Can you give an example of how the environment can affect a characteristic? *(2.16)*
55 Can you give an example of selective breeding? *(2.16)*
56 Can you give an example of genetic modification? *(2.16)*
57 What are alleles? *(2.17)*
58 If an allele is recessive, what does this mean? *(2.17)*
59 What is the difference between a genotype and a phenotype? *(2.17)*
60 What are gametes? How does the number of chromosomes in a gamete compare with that in other cells? *(2.17)*
61 What is meiosis? What does it produce? *(2.17)*
62 If one person has XX chromosomes in their cells and another XY, what are their sexes? *(2.17)*
63 What is mitosis? What does it produce? *(2.18)*
64 What is DNA? Where is it found, and what job does it do? *(2.18)*
65 How does DNA make copies of itself? *(2.18)*
66 What is the difference between sexual and asexual reproduction? *(2.18)*
67 What is the difference between being homozygous for a characteristic and being heterozygous? *(2.19)*
68 What is the difference between the F_1 generation and the F_2 generation? *(2.19)*
69 What is cloning? For plants, how is it done, and what is it used for? *(2.19)*
70 Can you give an example of a genetic disease? *(2.19)*
71 Why is the theory of natural selection, sometimes called the survival of the fittest? *(2.20)*
72 Can you give an example of how a plant or animal is adapted to its environment? *(2.20)*
73 What are mutations? How can they be caused? *(2.20)*
74 What are fossils? How are they formed? *(2.21)*
75 What branch of science was started by Mendel? *(2.21)*
76 Can you name one living and one non-living factor that can affect a population of animals or plants? *(2.22)*
77 Can you give an example of how an animal or plant is adapted to its environment? *(2.22)*
78 Can you explain why a change in the size of one population of animals or plants may affect the size of another population? *(2.22)*
79 What factors limit the size of a population of animals or plants? *(2.23)*
80 What is meant by sustainable development? *(2.23)*
81 Can you describe five different types of pollution caused by humans? *(2.23)*
82 What is global warming, and what is its most likely cause? *(2.24)*
83 Why is the ozone layer important, and how has it been damaged? *(2.24)*
84 What are the possible causes and effects of acid rain? *(2.24)*
85 What problems are caused by forest destruction? *(2.24)*
86 Can you give an example of a food chain? *(2.25)*
87 What is a food web? *(2.25)*
88 What is a pyramid of biomass? *(2.25)*
89 What are decomposers? What do they do? *(2.25)*
90 How can carbon atoms in the atmosphere end up in the body of an animal? How are they returned to the atmosphere? *(2.26)*
91 How can nitrogen atoms in the atmosphere end up in the body of an animal? How are they returned to the atmosphere? *(2.26)*
92 What is a trophic level? *(2.27)*
93 Why are there only three or four trophic levels in most food chains? *(2.27)*
94 When plant or animal matter decomposes, what factors affect its rate of decay? *(2.27)*
95 Why do polluting chemicals become more concentrated as they pass along a food chain? *(2.27)*
96 If food production is increased, how can this cause environmental damage? *(2.27)*

SECTION 3

Test and check

Can you answer the following? The spread number in brackets tells you where to find the information.

1 What is the smallest bit of an element called? *(3.01)*
2 What are molecules? *(3.01)*
3 What does the formula CO_2 tell you? *(3.01)*
4 What is the difference between an element and a compound? *(3.01)*
5 What is the difference between a compound and a mixture? *(3.01)*
6 How do the particles behave in a solid, in a liquid, and in a gas? *(3.02)*
7 What is diffusion? *(3.02)*
8 What is meant by a change of state? *(3.02)*
9 When ice melts, what happens to its particles? *(3.02)*
10 When water evaporates, what happens to its particles? *(3.02)*
11 How is density calculated? *(3.03)*
12 What is a solute? What is a solvent? What is a solution? *(3.03)*
13 How would you separate copper(II) sulphate from water? *(3.03)*
14 Can you give an example of fractional distillation? *(3.03)*
15 Can you give two features of a physical change? *(3.04)*
16 Can you give an example of a physical change? *(3.04)*
17 Can you give two features of a chemical change? *(3.04)*
18 Can you give an example of a chemical change? *(3.04)*
19 What three types of particle make up an atom? *(3.05)*
20 What is meant by atomic number and mass number? *(3.05)*
21 What are isotopes? *(3.05)*
22 What is meant by relative atomic mass? *(3.05)*
23 In the periodic table, in what ways are elements in the same group similar? *(3.06)*
24 What properties (features) do the noble gases have in common? *(3.06)*
25 What are electron shells? *(3.06)*
26 What is an ion? *(3.07)*
27 What is ionic bonding? *(3.07)*
28 Can you give an example of a giant ionic lattice? *(3.07)*
29 Can you give an example of a compound ion? *(3.07)*
30 What is covalent bonding? *(3.08)*
31 Can you give an example of a double bond? *(3.08)*
32 What makes the molecules stick together in water? *(3.08)*
33 What do you understand by an ionic crystal, molecular crystal, and giant molecule? *(3.09)*
34 What is the structure of graphite? Why does graphite conduct electricity? *(3.09)*
35 What are metallic bonds? *(3.09)*
36 Why are alloys harder than pure metals? *(3.09)*
37 Why must a chemical equation balance? *(3.10)*
38 What is meant by the relative molecular mass (M_r) of a compound? *(3.10)*
39 What is an ionic equation? *(3.10)*
40 In chemistry, what is meant by a mole? *(3.10)*
41 Can you calculate the masses and/or volumes of products of a given reaction? *(3.10)*
42 Can you use ideas about bond breaking and making to describe what an exothermic reaction is? *(3.11)*
43 Can you use ideas about bond breaking and making to describe what an endothermic reaction is? *(3.11)*
44 Knowing the bond energies of the reactants and products, can you calculate the energy changes which occur during reactions? *(3.11)*
45 Can you give an example of a synthesis reaction? *(3.12)*
46 What is thermal decomposition? *(3.12)*
47 What is precipitation *(3.12)*
48 What happens in a redox reaction? *(3.12)*
49 What gas is given off when an acid reacts with a metal? *(3.13)*
50 How does an acid affect litmus indicator? *(3.13)*
51 How does an alkali affect litmus indicator? *(3.13)*
52 If a solution has a pH of 1, what does this tell you about it? *(3.13)*
53 If a solution has a pH of 7, what does this tell you about it? *(3.13)*
54 If a base neutralizes an acid, what is produced? *(3.14)*
55 If a carbonate neutralizes an acid, what is produced? *(3.14)*

Test and check

56 Can you give two practical examples of the use of neutralization? *(3.14)*
57 What factors affect the rate of a chemical reaction? *(3.15)*
58 What are catalysts? *(3.15)*
59 What are enzymes? *(3.16)*
60 Can you give an example of the use of enzymes in the food production? *(3.16)*
61 What happens during fermentation? *(3.16)*
62 Can you describe some uses of biotechnology? *(3.16)*
63 What is a reversible reaction? *(3.17)*
64 If a reversible reaction is in equilibrium, what does this mean? *(3.17)*
65 When ammonia is produced in a chemical plant, what factors affect the yield? *(3.17)*
66 How is sulphuric acid produced in a chemical plant? *(3.17)*
67 How is ammonia made by the Haber process? Can you write the equation for the reaction? *(3.18)*
68 How is ammonia used to produce fertilizer? *(3.18)*
69 As you move down a group in the periodic table, how do the chemical properties of the elements tend to change? *(3.19)*
70 How do the alkali metals react with air, with water, and with chlorine? *(3.19)*
71 What are the halogens? *(3.20)*
72 Why do the halogens form negative ions? *(3.20)*
73 Can you give some uses of halogens? *(3.20)*
74 What are the transitions metals? Can you list some of their properties? *(3.21)*
75 Can you give a feature that many compounds of transition metals have in common? *(3.21)*
76 Can you give examples of transition metals being used as catalysts? *(3.21)*
77 What is the reactivity series? *(3.22)*
78 Can you name a metal that will not react with acids? *(3.22)*
79 What happens when a metal corrodes? *(3.22)*
80 What conditions are necessary for rusting? *(3.22)*
81 What happens in a displacement reaction? *(3.23)*
82 Why do the legs of oil rigs have blocks of magnesium or zinc wired to them? *(3.23)*
83 When might two metals compete for oxygen? *(3.23)*
84 Why is gold found in the ground as a pure metal, while iron is only found in compounds? *(3.24)*
85 Why are some metals more difficult to separate from their ores than others? *(3.24)*
86 How is iron produced from iron ore? *(3.24)*
87 What is steel? How is it made? *(3.24)*
88 How is aluminium separated from its ore? *(3.25)*
89 How is aluminium anodized? *(3.25)*
90 How can copper be purified by electrolysis? *(3.25)*
91 Can you describe some of the uses and products of sodium chloride? *(3.26)*
92 Can you describe the reactions that take place at the anode and cathode during the electrolysis of brine? *(3.26)*
93 Can you describe some of the uses and products of limestone? *(3.26)*
94 What are hydrocarbons? *(3.27)*
95 What are alkanes? Can you give an example, along with its molecular formula? *(3.27)*
96 What are the main fractions in crude oil? What are they used for? *(3.27)*
97 How are the fractions in crude oil separated? *(3.27)*
98 What is formed when a large hydrocarbon molecule is cracked? *(3.28)*
99 What is an unsaturated hydrocarbon? *(3.28)*
100 What are polymers? *(3.28)*
101 Can you give an example of a polymer? *(3.28)*
102 How are addition polymers produced? *(3.28)*
103 Why is the sea salty? *(3.29)*
104 Where did the gases in the Earth's early atmosphere come from? *(3.29)*
105 Which gas was most abundant in the Earth's early atmosphere? Why is there much less of this gas in the present atmosphere? *(3.29)*
106 How are igneous rocks formed? *(3.30)*
107 When molten rock cools and solidifies, what effect does the cooling rate have? *(3.30)*
108 How are sedimentary rocks formed? *(3.30)*
109 How are metamorphic rocks formed? *(3.30)*
110 What is the rock cycle? *(3.30)*

SECTION 4

Test and check

Can you answer the following? The spread number in brackets tells you where to find the information.

1 Where does electric charge come from? *(4.01)*
2 Which materials are the best conductors of electricity? *(4.01)*
3 What happens when like charges are brought close? *(4.01)*
4 What devices make use of the force between charges? *(4.01)*
5 How are current and voltage measured? *(4.02)*
6 Can you draw a circuit with a battery and two bulbs **(a)** in series **(b)** in parallel? What are the advantages of the parallel arrangement? *(4.02)*
7 Can you draw circuits using symbols? *(4.02)*
8 Why do heating elements get hot, while copper connecting wires do not? *(4.03)*
9 Could you use a variable resistor to control the brightness of a bulb? *(4.03)*
10 What is measured in 'kW h'? *(4.03)*
11 How is the cost of electricity calculated? *(4.03)*
12 How is resistance measured? *(4.04)*
13 What equation links voltage, current, and resistance? *(4.04)*
14 What are the properties of thermistors, light-dependent resistors, and diodes? *(4.04)*
15 Can you use the equation $V = IR$? *(4.05)*
16 How is electrical power calculated? *(4.05)*
17 What is the link between charge, current, and time? *(4.05)*
18 What is the link between energy, charge, and voltage? *(4.05)*
19 What happens when like poles of two magnets are brought close? *(4.06)*
20 Why would the core of an electromagnet be made of iron rather than steel? *(4.06)*
21 How do a magnetic relay and a circuit breaker work? *(4.06)*
22 Why does the coil in an electric motor turn when a current flows through it? *(4.07)*
23 What is electromagnetic induction? *(4.07)*
24 What is an alternator? How does it work? *(4.07)*
25 What is the difference between AC and DC? *(4.08)*
26 With mains circuits, why are switches and fuses fitted in the live wire? *(4.08)*
27 What colours are used for the live, neutral, and earth wires in a mains plug? *(4.08)*
28 What do transformers do? *(4.09)*
29 What equation links the output and input voltages of a transformer? *(4.09)*
30 Why is mains power transmitted across country at very high voltage? *(4.09)*
31 Can you give examples of when friction is a nuisance and when it is useful? *(4.10)*
32 What factors affect a car's stopping distance when the brakes are applied? *(4.10)*
33 How is speed calculated? *(4.11)*
34 What is the difference between speed and velocity? *(4.11)*
35 How is acceleration calculated? *(4.11)*
36 Can you interpret distance–time and speed–time graphs? *(4.11)*
37 How is force measured? *(4.12)*
38 What is weight? *(4.12)*
39 What is the link between force, mass, and acceleration? *(4.12)*
40 What does g stand for? *(4.12)*
41 How are action and reaction forces linked? *(4.12)*
42 What is meant by a resultant force? *(4.13)*
43 How will an object move if the forces on it are balanced? *(4.13)*
44 How will an object move if the forces on it are unbalanced? *(4.13)*
45 What is terminal velocity? *(4.13)*
46 How is pressure calculated? *(4.14)*
47 How do hydraulic machines work? *(4.14)*
48 What is Hooke's law? *(4.15)*
49 What is meant by the elastic limit of a material? *(4.15)*
50 What is Boyle's law? *(4.15)*
51 How do you calculate the moment of a force? *(4.16)*
52 What is the law of moments? *(4.16)*
53 What is an object's centre of gravity? *(4.16)*
54 What makes an object stable? *(4.16)*
55 Can you give some examples of different forms of energy *(4.17)*
56 What is the law of conservation of energy? *(4.17)*
57 What is the difference between heat and temperature? *(4.18)*

Test and check

58 What is the difference between conduction and convection? *(4.18)*
59 Why are wool and fur good insulators? *(4.18)*
60 How does energy reach us from the Sun? *(4.19)*
61 Which type of surface is the best emitter of thermal radiation? *(4.19)*
62 Which type of surface is the best absorber of thermal radiation? *(4.19)*
63 How does a fuel-burning power station produce electricity? *(4.20)*
64 If a power station has an efficiency of 35%, what does this mean? *(4.20)*
65 Can you give examples of renewable and non-renewable energy sources? *(4.20)*
66 Can you explain how the energy in petrol originally came from the Sun? *(4.21)*
67 What types of power station do not burn fuel? *(4.20 and 4.21)*
68 How do you calculate **(a)** work **(b)** power? *(4.22)*
69 How do you calculate **(a)** gravitational potential energy **(b)** kinetic energy? *(4.22)*
70 What is the difference between longitudinal waves and transverse waves? Can you give an example of each? *(4.23)*
71 What equation links the speed, frequency, and wavelength of waves? *(4.23)*
72 When waves are refracted, what happens to their speed? *(4.23)*
73 What is diffraction? What factors affect it? *(4.23)*
74 What causes sound waves? *(4.24)*
75 Why cannot sound travel through a vacuum? *(4.24)*
76 How could an echo be used to calculate the speed of sound? *(4.24)*
77 Comparing sound waves, how are loud sounds different from quiet sounds? How are high sounds different from low sounds? *(4.25)*
78 What is ultrasound? What is it used for? *(4.25)*
79 What is magma? *(4.26)*
80 What are tectonic plates? In what ways can they move at boundaries? *(4.26)*
81 What are P-waves and S-waves? What evidence do they give about the Earth's interior? *(4.26)*
82 How does a flat mirror form an image? *(4.27)*
83 How do light rays bend when they enter glass? What is the effect called? *(4.27)*
84 What is total internal reflection? *(4.27)*
85 What is meant by the critical angle? *(4.27)*
86 What happens to white light when it passes through a prism? *(4.28)*
87 Can you list the different types of wave in the electromagnetic spectrum? Can you describe some uses of these different types? *(4.28)*
88 What is the difference between analogue and digital signals? *(4.29)*
89 What are the advantages of digital transmission? *(4.29)*
90 How do optical fibres work? What are they used for? *(4.29)*
91 What are the three main types of nuclear radiation? How are they different? *(4.30)*
92 Which type of radiation is most ionizing? *(4.30)*
93 Why is nuclear radiation dangerous? *(4.30)*
94 What happens during radioactive decay? *(4.31)*
95 What is meant by half-life? *(4.31)*
96 What is **(a)** fission? **(b)** a chain reaction? *(4.31)*
97 How can radioactivity be used to date rocks? *(4.31)*
98 What are radioactivite tracers? *(4.31)*
99 Why do we get day and night? *(4.32)*
100 How can you tell a planet from a star? *(4.32)*
101 What keeps the Earth in orbit around the Sun? *(4.32)*
102 What is the Solar System? *(4.33)*
103 How does the time for a planet's orbit vary with its distance from the Sun? *(4.33)*
104 What is a galaxy? Why are humans never likely to visit other galaxies? *(4.33)*
105 Can you describe some of the jobs satellites are used for? *(4.34)*
106 What is a geostationary orbit? *(4.34)*
107 What is the orbit of a comet like? *(4.34)*
108 How were the Sun and planets formed? *(4.35)*
109 What will happen to the Sun when all its hydrogen fuel has been used up? *(4.35)*
110 What evidence is there for the big bang and an expanding Universe? *(4.35)*

1.01 Doing an investigation

This spread should help you to:
- *plan an experimental procedure*
- *obtain and present your evidence*
- *consider your evidence*
- *evaluate your investigation.*

The right-hand side of each of these pages shows one student's thoughts about her investigation.

Planning

- **Decide on a problem to investigate**
  In the example shown on the right, the student has decided to investigate how quickly sugar dissolves in water.
- **Write down your prediction**
  You may already have an idea of what you expect to happen in your investigation. This is your ***prediction***. It may not be right! The aim of your investigation is to test it.
- **Decide what the key factors are**
  The ***key factors*** are those things which can affect the outcome of your experiment. For example, in sugar-dissolving tests, two of the key factors are how much water you use and whether you stir the mixture or not. In your experiment, you must decide what the key factors are and which ones you can control. Of those, which will you keep fixed and which will you change?
- **Check that your tests will be fair**
  To make sense of your results, you need to change just one factor at a time and find out how it affects one other. Otherwise, it will not be a ***fair test***. For example, if you want to find out how the crystal size affects dissolving, it wouldn't be fair to compare big, brown sugar crystals in hot water with small, white ones in cold water.
- **Decide what equipment you need and in what order you will do things**
  To help in your planning, you may need to carry out a trial run of the experiment.
- **Prepare tables for your results**
  First, decide what readings you will take, how many of them, and over what range.

I will investigate how quickly sugar dissolves in water.

Sugar is made up of tiny crystals. In icing sugar, the crystals are very small. In caster sugar, they are a bit bigger. In ordinary sugar, they are a bit bigger again.

If the crystals are small, the water can get in contact with all the sugar more quickly. So I think that small crystals will dissolve more quickly than bigger ones.

I also think that sugar particles will dissolve more quickly in hot water...

Three of the key factors are:
time for the sugar to dissolve
particle size
temperature

Other key factors are the amount of sugar, the amount of water, the type of sugar (brown or white), and whether I stir it or not. I will use the same amounts of sugar and water each time. I will use white sugar, and I will stir it gently to separate the crystals. If the crystals are in a heap, the water cannot reach them properly.

Items needed:
ordinary sugar, caster sugar, icing sugar, beaker, thermometer...

Obtaining and presenting evidence

- **Make your measurements and record them**
  For greater accuracy, repeat measurements.

- **Present your results**
  Decide whether it would be useful to show them in the form of a chart or graph (see below).

Considering your evidence

- **Look for patterns in your results**
  A graph shows how one factor changes with another. Some measurements are difficult to make accurately, so the points on your graph may be uneven. Draw a ***line of best fit*** through them. This is the straight line or smooth curve which passes close to as many points as possible.

- **Present your conclusions**
  What links did you find between the factors you measured? Can you explain these links? Did your results match your prediction?

Evaluating your investigation

- **Evaluate your evidence**
  Did you get enough data to reach firm conclusions? Did any of your measurements seem to be 'wrong'? If so, can you explain why?

- **Evaluate your methods**
  Could your experiment be improved in any way?

First, I will find out how temperature affects dissolving. I will use ordinary sugar each time, so that the crystal size is fixed. I will dissolve the sugar in water at different temperatures. I will measure the temperature with a thermometer, and the time for dissolving with a stopwatch.

Next, I will see how crystal size affects dissolving. I might be able to measure the size of the sugar crystals with a microscope. If not, I will just call them 'small', 'medium', and 'large'.

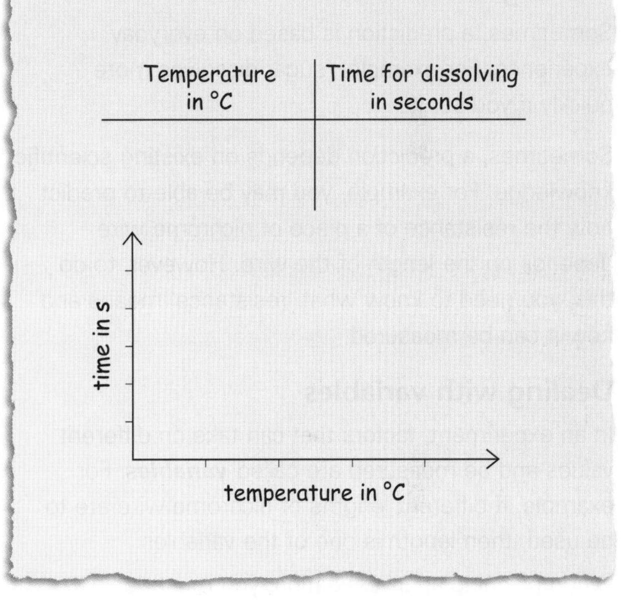

Temperature in °C	Time for dissolving in seconds

There seems to be a link between time for dissolving and temperature. As the temperature gets higher, the time for dissolving gets...

This is why I think there is a link. When the temperature rises...

Measuring the time accurately was difficult. It was hard to tell when all the sugar had dissolved. Also, I would like to measure the crystal size more accurately.

1.02 Investigating further

This spread should help you to:
- *carry out investigations in which you use your scientific knowledge*
- *identify and measure variables*
- *use graphs to help you reach conclusions*
- *assess the reliability of your findings.*

The left-hand side of each of these pages describes some of the aspects you should consider in a more advanced investigation. The right-hand side shows one student's thoughts about her investigation.

Starting with science

Sometimes, a prediction is based on everyday experience. For example: 'sugar dissolves more quickly if you stir it'.

Sometimes, a prediction depends on existing scientific knowledge. For example, you may be able to predict how the resistance of a piece of nichrome wire depends on the length of the wire. However, to do this, you need to know what 'resistance' means and how it can be measured.

Dealing with variables

In an experiment, factors that can take on different values and be measured are called ***variables***. For example, if different lengths of nichrome wire are to be used, then length is one of the variables.

When planning an investigation, you must decide what the key variables are, how to measure them, and over what range. In the nichrome wire experiment:

- What should the highest voltage and current values be? (Safety is a factor here.)
- What lengths of wire will you use?

As well as the key variables, there may be other variables to consider. You may not need to measure them, but you might have to control them. In the nichrome wire experiment, you might want to keep the temperature of the wire steady.

Some factors are difficult to control. For example, in growth experiments, some plants may be in more shade than others. You must take factors like this into account when deciding how reliable your results are.

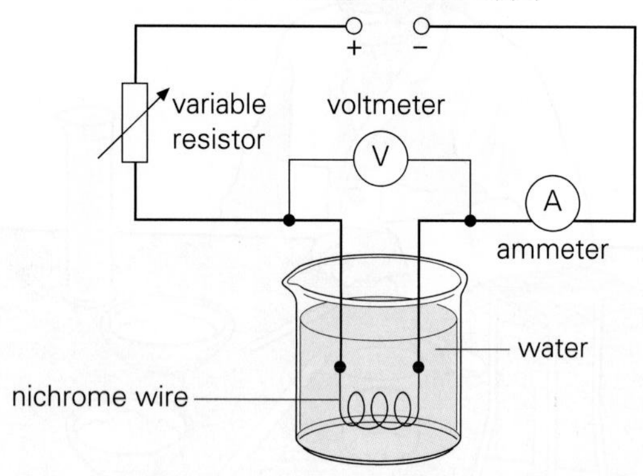

I will investigate how the resistance of a piece of nichrome wire depends on its length. I know from earlier work that resistance can be calculated with this equation:

$$\text{resistance (in } \Omega) = \frac{\text{voltage (in V)}}{\text{current (in A)}}$$

I think I can predict how the resistance will vary with temperature. If the length of wire is doubled, the current (flow of electrons) has to be pushed between twice as many atoms. So I would expect the resistance to double as well.

In my circuit, I shall start with 50 cm of thin nichrome wire, put a voltage of 6 V across it, and measure the current through it. (I will do a trial run first, using an ammeter that can measure several amperes. I may be able to change to a more sensitive meter for my main experiment.)

From my voltmeter and ammeter readings, I will calculate the resistance. Then, I will take more sets of readings, shortening the wire by 5 cm each time until it is only 10 cm long.

In my experiment, the key variables will be voltage, current, and length. From data books, I know that the resistance of nichrome changes with temperature, so temperature is another variable which I must control. A large beaker of cold water should keep the temperature of the nichrome steady.

Uncertainties

No measurement is exact. There is always some ***uncertainty*** about it. For example, you may only be able to read a voltmeter to the nearest 0.1 V.

Say that you measure a voltage of 3.3 V and a current of 1.3 A. To work out the resistance, you divide the voltage by the current on a calculator and get...

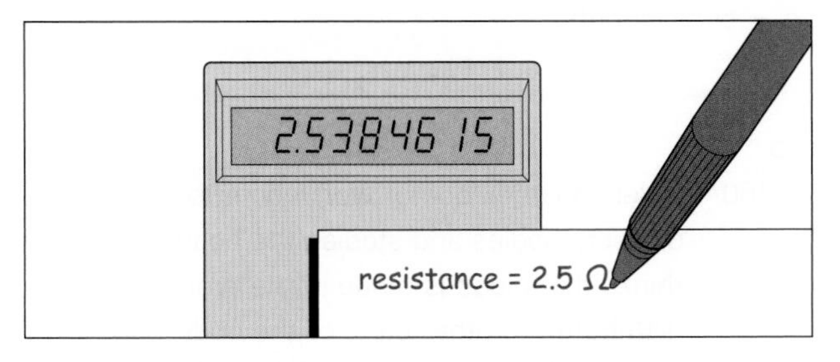

This should be recorded as 2.5 Ω. With the uncertainties in your voltage and current readings, you cannot justify any more figures than this.

Uncertainties mean that the points on a graph will be scattered. So you need to draw a straight or smoothly curved line of best fit. Before you do this:

- Decide whether the line goes through the origin.
- Decide whether any readings should be rejected. Some may be so far out that they are probably due to mistakes rather than uncertainties. See if you can find out why they occurred.

From the way points scatter about a line of best fit, you can see how reliable your readings are. But for this, you need plenty of points on the graph.

Trends and conclusions

Graphs help you see trends or patterns in your data. The simplest form of graph is a straight line through the origin. A graph of resistance against wire length might be like this. If so, it means that if the length doubles, the resistance doubles...and so on. Resistance and length are in ***direct proportion***.

If you think that your graph supports your original prediction, then explain your reasons.

In reaching your conclusions, remember that there are uncertainties in your readings, and variables which you may not have allowed for. So your results can never prove your prediction. You must decide how far they support it. Having done this, see if you can suggest other ways of getting more evidence.

I have calculated the resistance of each length of nichrome wire. Now I shall use these values to plot a graph of resistance against length.

Length goes along the bottom axis because it is the independent variable – it is the variable which I chose to change. Resistance goes up the side axis because it is the dependent variable. Its value depends on the length of wire chosen.

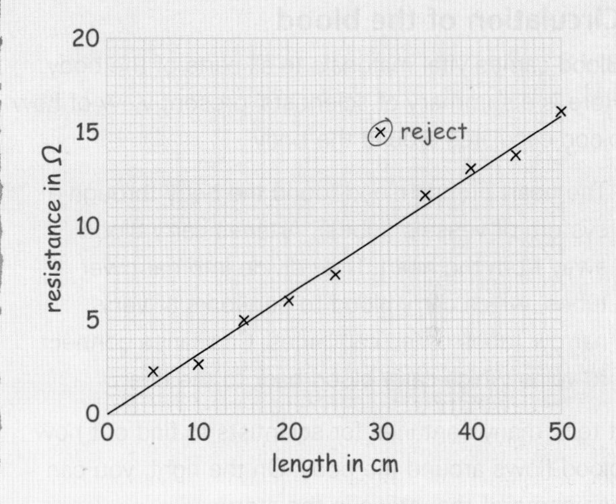

The points on my graph are a little scattered, but I think that the line of best fit is a straight line.

This line ought to go through the origin. If the wire has no length, its resistance should be zero.

I have rejected one point on my graph. In my table, the current reading for that point seems far too low. I probably misread the ammeter.

As the graph is a straight line through the origin, the resistance of the nichrome wire is in direct proportion to its length. This agrees with my original prediction because...

I would like more evidence that the graph really is really straight line. To improve my investigation, I need a more accurate method of measuring resistance. I might get some ideas from reference books...

1.03 Changing ideas

By the end of this spread, you should be able to:
- *explain how scientific ideas develop, and how they may be affected by social and other factors.*

For more on the topics covered here, see:
Spread 2.08 (circulation of the blood)
Spreads 4.33 and 4.35 (planets, stars, Universe)

Scientific ideas have changed over the centuries as people have come to rely more and more on experimental evidence. Here are two examples.

Circulation of the blood

Blood carries vital materials to all parts of the body. Here is a summary of scientists' present view of how blood circulates around the body.

The heart pumps blood round the body through a system of vessels (tubes). Arteries carry blood away from the heart. They divide into narrower tubes, which carry blood to networks of very narrow tubes called capillaries. Capillaries connect to veins which carry blood back to the heart.

It took many centuries for scientists to find out how blood flows around the body. On the right, you can see some of the steps in the story.

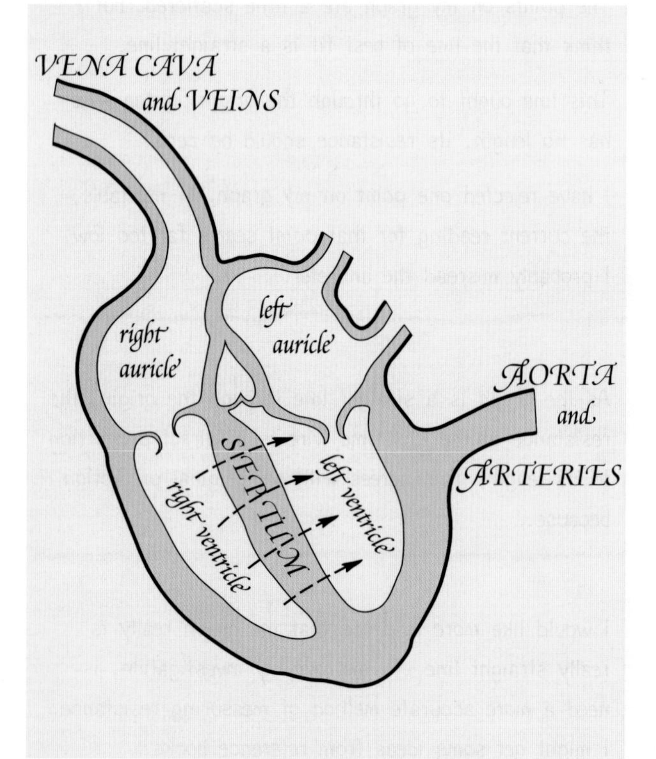

The inside of the heart according to Galen

c. = circa (about)

AD

c. 150	Galen, a Greek doctor working in Rome, dissects bodies and studies the heart. He thinks that blood is made in the liver, distributed to other parts of the body, and used up. He discovers two types of blood vessel (arteries and veins) and believes that blood passes from one to the other through the septum in the heart. His ideas stand unchallenged for 1400 years.
1531	Vesalius, in Padua, Italy, does more dissections and starts to question some of Galen's teachings about blood flow.
c. 1550	Columbo finds that blood from the lungs flows back to the heart through a vein.
1603	Fabricius discovers that some veins have valves in them which let blood flow in one direction only.
1628	William Harvey, in London, publishes the results of his many experiments on blood flow. He suggests that blood cannot possibly be made at the rate required by Galen's theory. Instead it circulates around the body, leaving the heart through arteries and returning through veins. He deduces that there must be very narrow tubes linking arteries to veins but is unable to see them with his magnifying glass.
1661	Microscopes are more widely available. Using one, Malpighi is able to identify the capillaries linking arteries to veins.
Today	Although doctors have much more detailed information about the blood and its circulation, the basic principles established by Harvey still stand.

Earth and Universe

Here is a summary of scientists' present view of the Universe and the Earth's place within it:

The Earth is a planet, one of several orbiting the Sun. The Sun is a star. It belongs to a huge group of billions of stars called a galaxy. There are billions of galaxies in the Universe. They appear to be moving apart. Their matter was created in a gigantic burst of energy called the big bang which happened about 13 billion years ago.

No one can be sure that the above description is correct in every detail. It is just a useful picture – scientists call it a model. It matches the observations and measurements that we now have. It is very different from earlier models. Here is its story:

350 BC Aristotle believes that the Earth is at the centre of the Universe. The Sun, Moon, planets, and stars lie on crystal spheres and move about the Earth in perfect circles. This model is based on Ancient Greek beliefs about the perfection of the heavens, rather than on observations.

AD c. 150 Ptolemy realizes that the motion of a planet can appear to wobble. He does not want to abandon the idea of perfect circles, so he develops a complicated model of orbits with little circles added to bigger ones. He keeps the Earth at the centre.

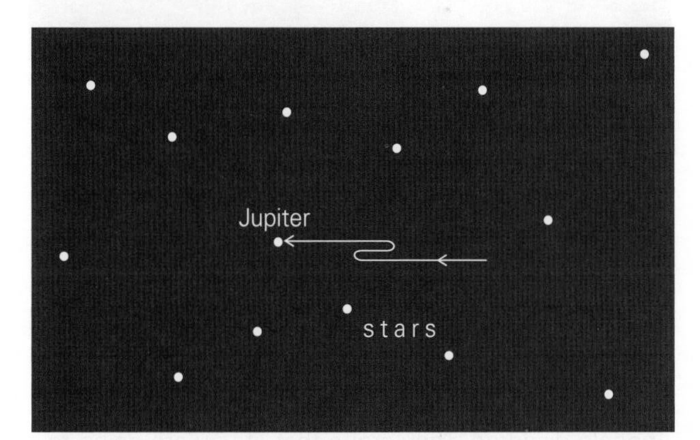

Viewed from Earth, the planet Jupiter shows a wobble in its motion. It doesn't really go backwards. It only appears to do so because our viewpoint changes as the Earth moves round the Sun.

1543 Copernicus suggests a new model to explain the observed motion of the planets. The Earth is a planet. It orbits the Sun with other planets. His ideas are rejected.

1608 The telescope is invented. Later, Galileo makes observations which indicate that the Earth is not at the centre of the Universe. However, this brings him into conflict with the Church. Under threat of torture, Galileo is forced to renounce his ideas

1609 From astronomical measurements made by Brahe, Kepler deduces that planets must orbit the Sun in ellipses or circles.

1687 Newton publishes his theory of gravitation and laws of motion. He is able to explain why planets orbit in the way they do.

c. 1790 Herschel makes a careful survey of the stars. He concludes that they are arranged in a huge group which he calls the Galaxy. He estimates that the Sun, which is also a star, is at the centre of the Galaxy.

1918 By now, telescopes are much improved. Shapley maps the positions of the stars in the Galaxy and concludes that the Sun is not at its centre after all.

1923 Hubble discovers that there are many more galaxies besides our own.

1929 Hubble analyses the light from distant galaxies and concludes that the galaxies are moving apart.

1946 Gamow revives a theory from the 1920s. This is the big bang theory. At some moment in the distant past, the whole of space exploded from a tiny concentration of matter and energy. This created the matter from which the galaxies would form. They have been moving apart ever since.

1965 Using a radio telescope, astronomers detect microwave radiation from all directions in space. It may be an 'echo' of the big bang.

Today Although gravitational attraction should be slowing down the Universe's rate of expansion, observations show that it may be speeding up. The story is far from over.

1.04 Rights and wrongs

By the end of this spread, you should be able to:
- *explain how scientific research raises ethical issues (issues of right and wrong) and cannot provide definite answers to all questions.*

New technologies come from scientific discoveries. Here are some examples. Are they helpful to humans and other living things, or harmful? Science cannot answer that question. You must form your own opinion.

Antibiotics

Diseases such as sore throats, typhoid, dysentry, and pneumonia are caused by microscopic living things known as bacteria (or germs). Antibiotics are medicines which can damage or kill bacteria.

Penicillin was the first major antibiotic. It was discovered in 1928, by Alexander Fleming. Today, there are hundreds of antibiotics. They have revolutionized the treatment of bacterial diseases.

Helpful or harmful? Antibiotics have saved many thousands of lives. But they have been over-used. When antibiotics kill off germs, the rarer and more resistant forms of the germ survive and evolve. As a result, dangerous and previously unknown 'superbugs' are starting to appear, and these may be untreatable.

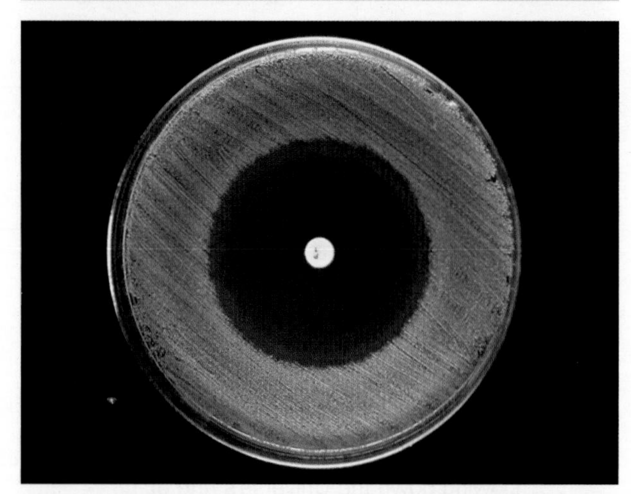

Bacteria (light brown) grow on a jelly in a dish. The dark area shows where bacteria have been killed by a pellet of penicillin (white) placed in the middle.

Lasers

The physicist Theodore Maiman built the first laser in 1960. He found a way of energizing a specially cut crystal so that it produced an intense, narrow beam of light. Here are some of the uses of lasers:

Bar code readers scan labels with a tiny laser beam. Pulses of reflected light are picked up and processed by a computer, which displays the price. CD players process light pulses reflected from a spinning disc.

Optical fibre cables carry digital telephone signals in the form of pulses of laser light. They are thinner and lighter than electric cables and can carry more signals with less loss of power.

Laser scalpels are used by surgeons to make very fine cuts which also seal the tissue:

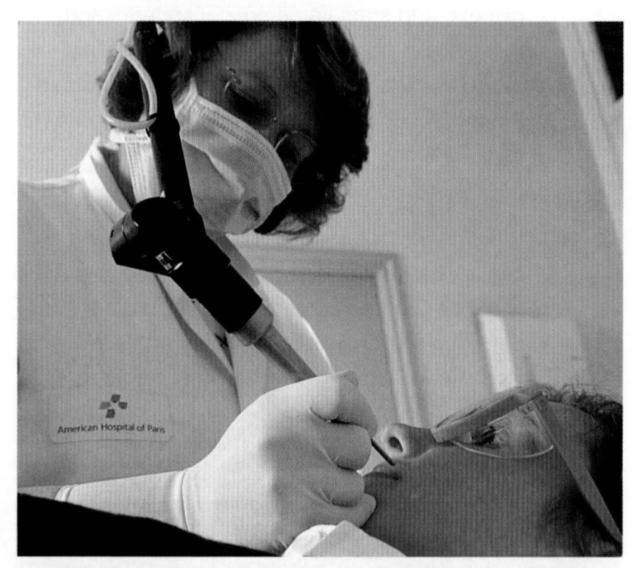

Laser-guided weapons can be rifles, bombs, or missiles. They use a laser beam to 'designate' (mark) the target, which could be a person, tank, or building. A missile, for example, detects the laser spot, locks on to it, and flies an accurate course towards it.

Helpful or harmful? Despite the benefits of lasers, many think that laser-guided weapons, like all weapons, should never have been invented. Military people disagree. They argue that, if you are going to have weapons, it is better to have accurate ones, rather than those which might cause unintended damage.

GM foods

Living things contain chemical instructions called genes. These control their growth and life processes. GM (genetically modified) foods are made from plants which have had genes altered to produce some special feature. Here is an example:

Soya is used in many processed food. When ordinary soya is grown, the soil is sprayed with weedkiller, but not during the growing season, otherwise the soya plants would be killed as well as the weeds. In GM soya, a gene has been altered to make the plant resistant to the weedkiller. This means that the weeds can be controlled after the young plants have started to grow, and less weedkiller is needed.

In genetic modification, the altered gene can come from a different plant or from an animal. For example, an 'antifreeze' gene that allows some types of fish to survive in polar water can be added to plants so that their fruit is not damaged by frost.

Helpful or harmful? Critics argue that genetic modification is unnatural; that new genes may have unpredictable effects; that GM foods might contain undetected poisons; that genes spread by pollen might create indestructible 'superweeds'. Supporters argue that GM is a controlled and carefully tested process while normal plant-breeding mixes genes in a much more random way; that GM crops need fewer chemicals sprayed on them, give better quality, and higher yields; that GM could prevent famines because crops can be modified to grow in near-drought conditions.

GM plants. The source of 'Frankenstein foods'?

In answering these questions, try to think of new examples, rather than those given in the spread:

1 Think of a scientific discovery or invention that has had both helpful and harmful effects for humans or other living things. List these helpful and harmful effects.

2 Think of a question that might be answered by scientific research, but only after evidence has been collected for several years.

3 Think of a question that scientific research would never be able to answer with 100% certainty.

Uncertain answers

Scientific research cannot provide instant answers to all questions. Sometimes, it cannot provide firm answers at all. Here are some examples.

Is smoking safe? For most people, no. But to produce that answer, the health records of smokers and non-smokers had to be compared over many years.

Is smoking safe for you? Genetic (inherited) factors affect your level of risk. But, as yet, there is no readily available test to tell you what it is.

Are GM foods safe to eat? This question will never be answered with 100% certainty. Scientific research cannot prove that all foods of a particular type are safe under all circumstances. But that applies as much to non-GM foods as to GM ones.

...or the solution to problems like this?

2.01 Animals, plants, and cells

By the end of this spread, you should be able to:
- *explain what cells are*
- *describe the similarities and differences between animals and plants, and between their cells.*

Animals and plants are ***organisms*** (living things with organized parts). They have these features in common:

Nutrition (food) Animals and plants need food. Animals must take in food. Plants take in simple materials and make their own (see the next spread).

Respiration (releasing energy from food) Animals and plants need energy to stay alive, grow, and move. They get it by respiring (see the next spread).

Excretion (getting rid of waste products) Animals and plants make waste materials which they must get rid of. You excrete when you go to the toilet.

Growth Animals and plants grow bigger. They may also grow new parts to replace old or damaged ones.

Movement Animals usually make bigger and faster movements than plants.

Reproduction Animals and plants can produce more of their own kind. For example, humans have children.

Sensitivity Animals and plants react to the outside world. For example, they may be sensitive to light. Animals usually react more quickly than plants.

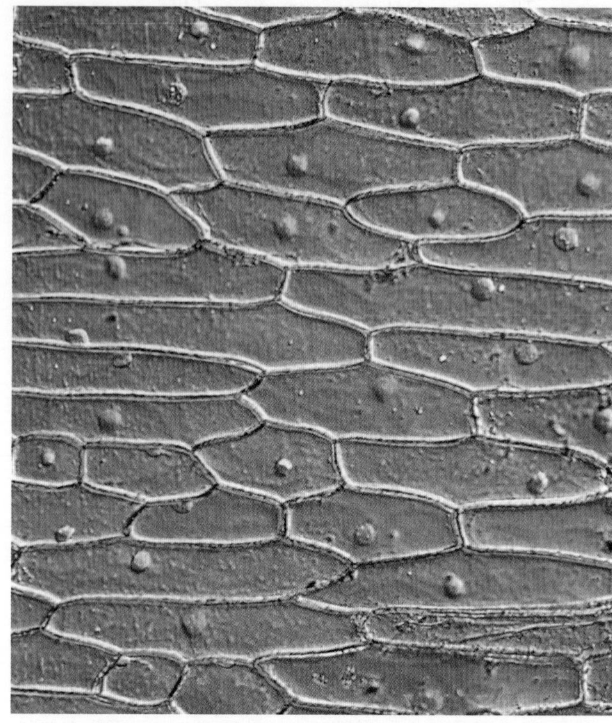

Onion cells: magnification ×500

Animals and plants are made from ***cells***. These are tiny chemical factories where the vital processes of life take place. They take in substances, make new ones, release energy from food, and give out waste.

Animals and plants grow by ***cell division***. A cell splits to form two new cells...and so on. You started as a single cell, but there are billions in your body now.

Animal cells

Animal cells exist in many shapes and sizes. But they all have several features in common:

Nucleus This controls all the chemical reactions that take place in the cell. It contains, thread-like ***chromosomes*** which store the chemical instructions needed to build the cell and make it do its job.

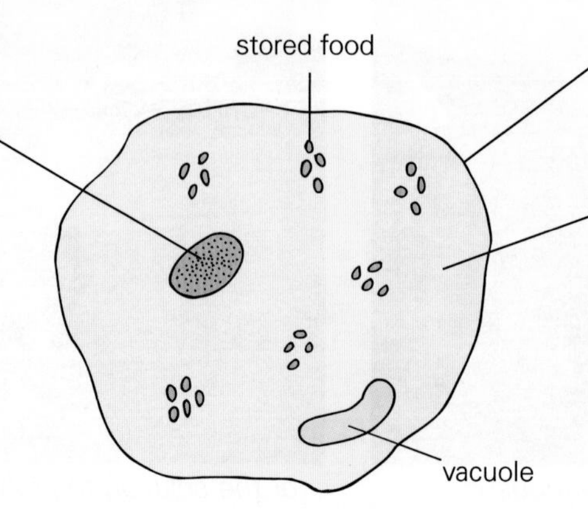

Cell membrane This is a thin skin which controls the movement of materials in and out of the cell.

Cytoplasm In this jelly, the cell's vital chemical reactions take place. New substances are made, and energy is released and stored. Sometimes, cytoplasm contains tiny droplets of liquid called ***vacuoles***.

Plant cells

Plant cells also have a ***cell membrane***, ***cytoplasm***, and ***nucleus***. But they have some features which make them different from animal cells:

Cell wall Plant cells are surrounded by a firm wall made of ***cellulose***. This holds plant cells together and gives them much of their strength. For example, wood is mainly cellulose.

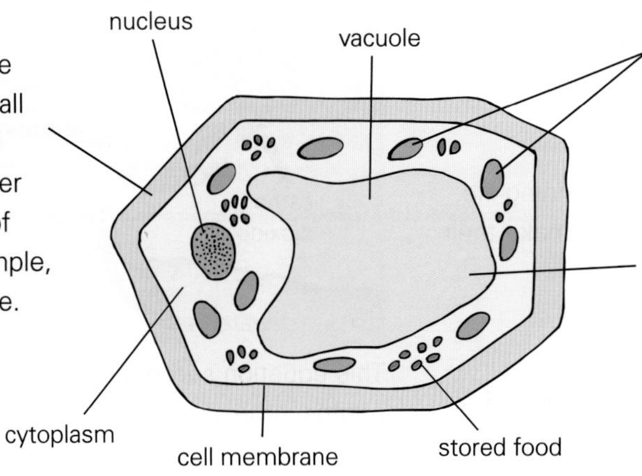

Chloroplasts These contain a green substance called ***chlorophyll***. This absorbs the energy in sunlight. Plants need the energy to make their food.

Cell sap This is a watery liquid in a large vacuole. Pressure from the liquid keeps the cell firm, rather like a tiny balloon. If a plant loses too much liquid from its cells, the pressure falls and the plant wilts.

Groups of cells

In animals and plants, different groups of cells have different jobs to do. Groups of similar cells are called ***tissue***. A collection of tissues doing a particular job is an ***organ***. Eyes are organs, so are muscles.

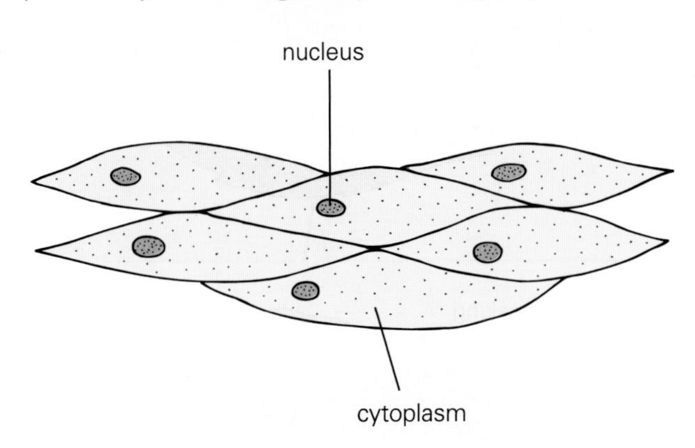

The muscles in your arms and legs are made from cells like this. The cells can shorten so that the muscle contracts.

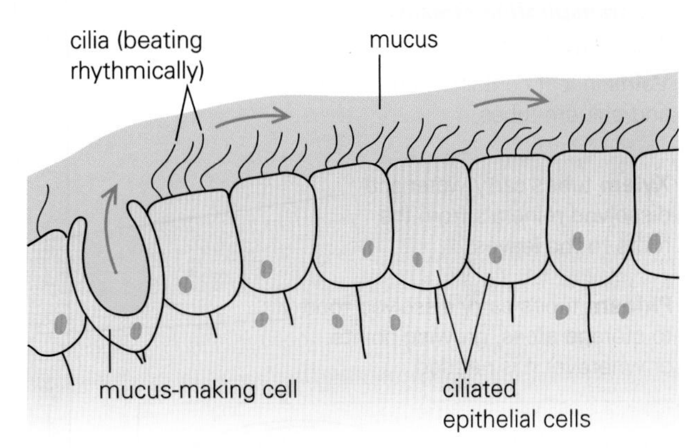

Your nose is lined with cells like this to 'clean' the air. The mucus traps dust and germs. The tiny cilia (hairs) push the mucus along to your throat.

1 *nucleus* *cellulose* *membrane* *chloroplast* *cytoplasm*
Which of the above matches each of these?
(a) Absorbs the energy in sunlight.
(b) The walls of plant cells are made of this.
(c) The part of a cell where new substances are made and energy is stored and released.
(d) A thin skin around the cytoplasm.
(e) A cell's control centre.

2 In what ways are living things different from non-living things?

3 Why do animals and plants need food?

4 **(a)** Give *three* ways in which plant cells are similar to animal cells.
(b) Give *three* ways in which plant cells are different from animal cells.

5 **(a)** What is tissue? **(b)** What word means a collection of tissues doing one particular job?

2.02 Making and using food

By the end of this spread, you should be able to:
- *explain how plants make their food*
- *explain why animals and plants respire*
- *name the gases involved in making and using food.*

Living things need food. It supplies them with materials for growth, and energy for maintaining life. Animals have to find their food. But plants make their own.

Photosynthesis

Plants take in carbon dioxide gas from the air, and water from the soil. They use the energy in sunlight to turn these into food such as glucose sugar. The process is called ***photosynthesis***.

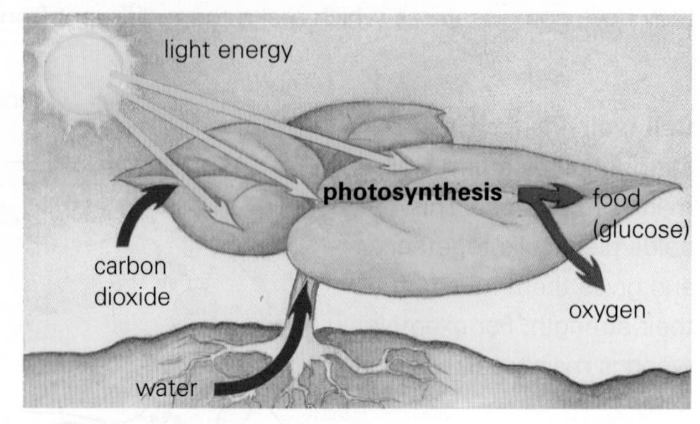

This equation summarizes what happens during photosynthesis:

$$\text{carbon dioxide + water} \xrightarrow{\text{light energy}} \text{glucose + oxygen}$$

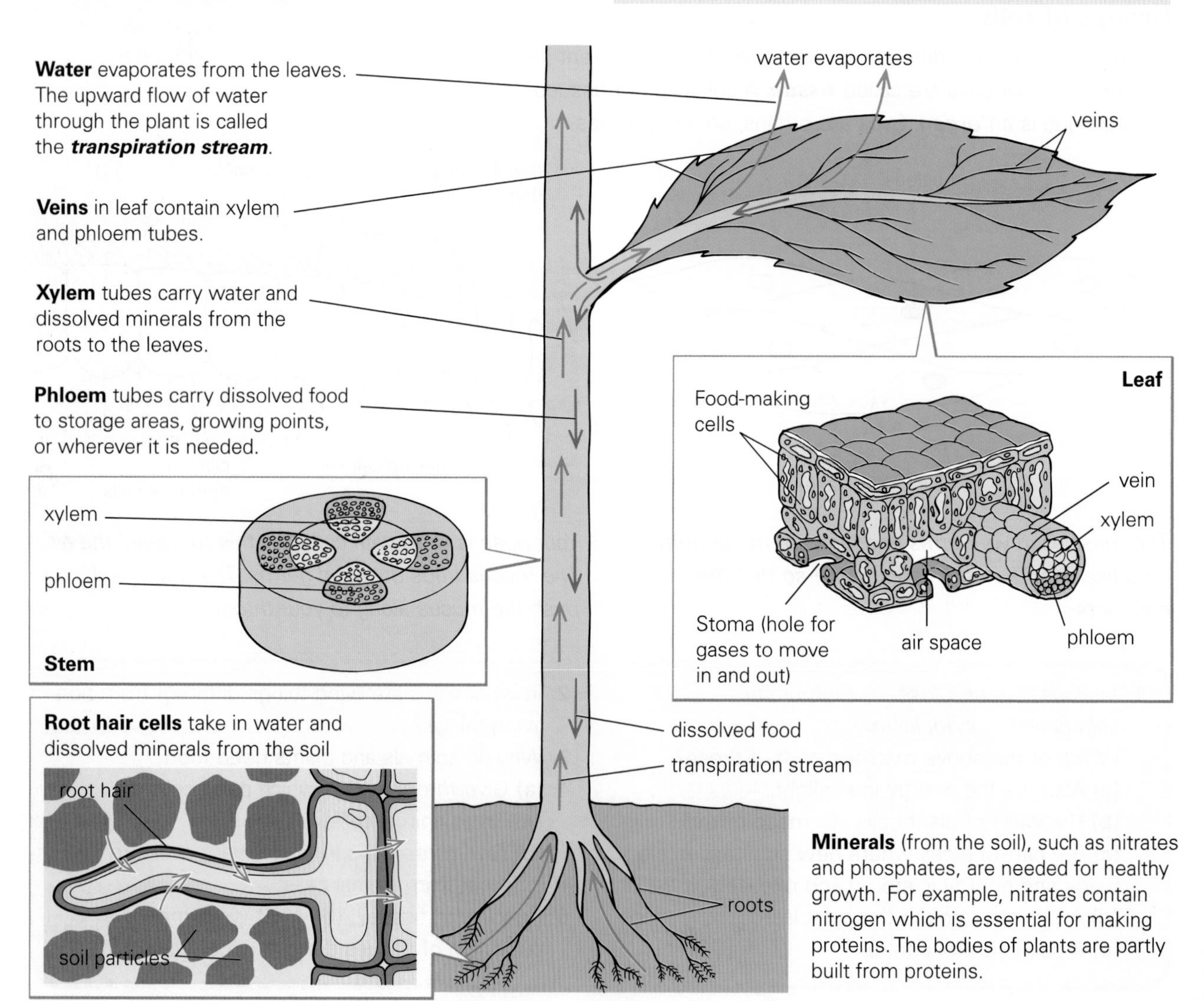

Minerals (from the soil), such as nitrates and phosphates, are needed for healthy growth. For example, nitrates contain nitrogen which is essential for making proteins. The bodies of plants are partly built from proteins.

To absorb the energy in sunlight, plants have a green chemical called chlorophyll in their leaves.

During photosynthesis, plants make oxygen as well as food. They need some of this oxygen. But the rest comes out of their leaves through tiny holes called ***stomata*** (each hole is called a ***stoma***). The same holes are also used for taking in carbon dioxide.

When plants make their food, they can store it in their leaves and roots to be used later on. Some is stored in the form of starch. By eating plants, animals can use this stored food.

Respiration

Plants and animals get energy from their food by a chemical process called ***respiration***. It is rather like burning, but without any flames. Usually, the food is combined with oxygen:

food + oxygen → carbon dioxide + water + *energy*

Carbon dioxide and water are the waste products. For example, the air you breathe out contains extra carbon dioxide and water vapour produced by respiration.

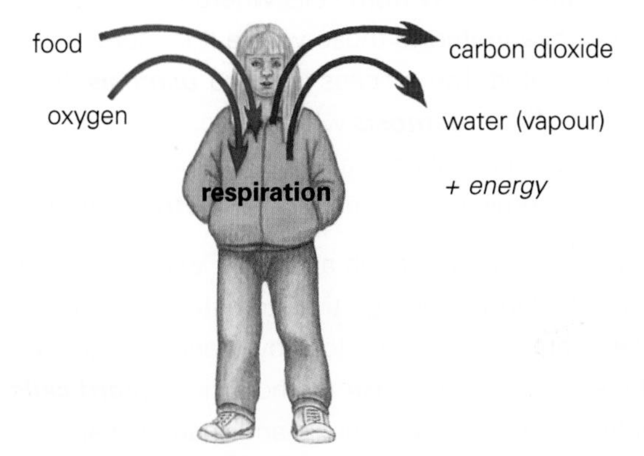

Gases in balance

Respiration takes place all the time. So plants and animals need a steady supply of oxygen.

During daylight hours, plants make oxygen by photosynthesis. They make more than they need for respiration, so they put their spare oxygen into the atmosphere.

At night, photosynthesis stops. So plants must take in oxygen – just like animals. However, they use less oxygen during the night than they give out during the day.

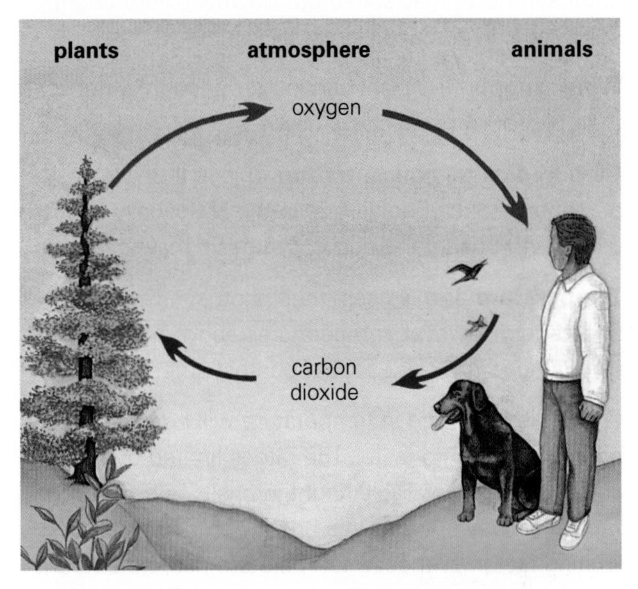

Overall, plants take in carbon dioxide and give out oxygen, while animals take in oxygen and give out carbon dioxide. Between them, they keep the gases in the atmosphere in balance.

1 During photosynthesis, what gas is **(a)** used up **(b)** made?
2 What else do plants make during photosynthesis?
3 Why does photosynthesis usually stop at night?
4 Why do plants need minerals from the soil?
5 Where do gases enter and leave a plant?
6 What is a transpiration stream?
7 During respiration, what gas is usually **(a)** used up **(b)** made?
8 Animals are using up oxygen all the time. Why does the amount of oxygen in the atmosphere not go down?

2.03 Plant processes

By the end of this spread, you should be able to:
- *explain how the rate of photosynthesis can change and what happens to the substances made*
- *explain how materials are transported in a plant*

Photosynthesis factors

The rate of photosynthesis depends on these factors:

Light intensity If the light is brighter, then photosynthesis may speed up. However, very bright sunshine can damage plants.

Water supply If a plant cannot get enough water, then photosynthesis slows down.

Carbon dioxide concentration This is the percentage of carbon dioxide in the air (usually about 0.03%). If it rises, then photosynthesis may speed up.

Temperature If this rises, then photosynthesis may speed up. (However, temperatures above about 40 °C can damage the plant so that photosynthesis stops.)

On a dull day, a rise in temperature will not make photosynthesis go faster. The rate is limited by the amount of light available. Light intensity is a ***limiting factor***. However, when it is bright, temperature is a limiting factor. Without a temperature rise, no amount of extra light can speed up photosynthesis. The graph below illustrates these results. It shows how light intensity affects the rate of photosynthesis at two different temperatures.

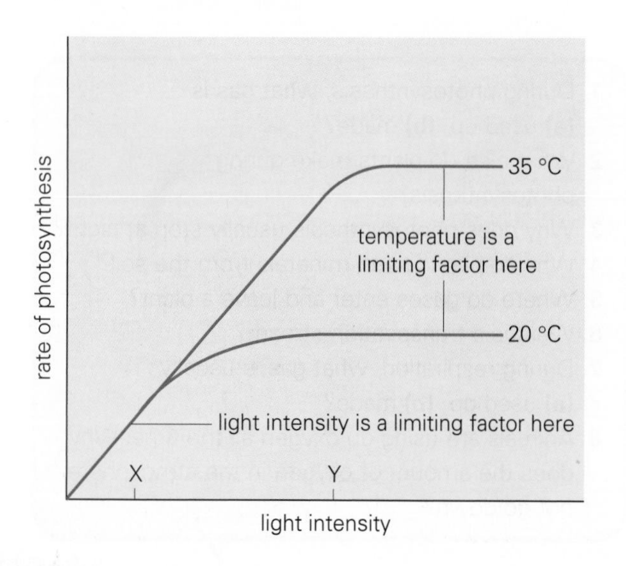

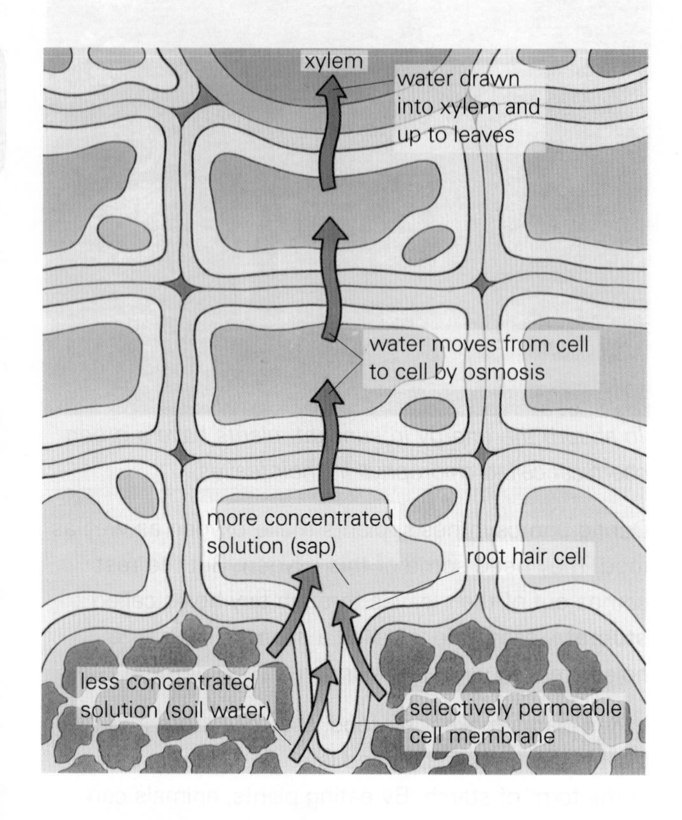

Materials on the move

Dissolved minerals in the soil enter roots by a process called ***diffusion*** (see Spread 3.02). The same effect makes dyes spread when dropped into water.

In a plant, the mixture of water and dissolved minerals is called ***sap***. If the amount of dissolved material is increased, then the solution is more concentrated.

In roots, sap passes from cell to cell as follows. Cell membranes are ***selectively permeable*** – they allow smaller molecules through but not bigger ones. Water passes through, but some dissolved materials do not. This causes a flow from cells where the solution is less concentrated to those where it is more concentrated. The process is called ***osmosis***. (It is pressure from ***osmosis*** which keeps cells firm.) Dissolved minerals can also be pumped from cell to cell by chemical reactions – called ***active transport***.

Sap is drawn up through a plant as water evaporates from the leaves through the stomata (tiny holes). The plant controls the water loss by changing the size of the stomata. To increase the hole size, ***guard cells*** either side swell by osmosis and change shape.

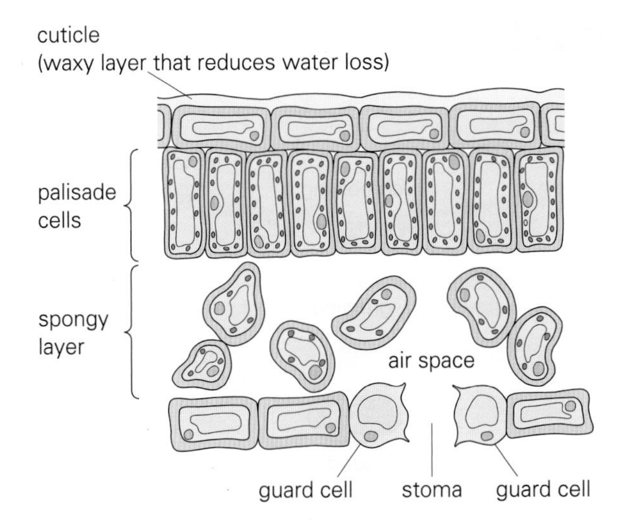

Photosynthesis mainly takes place in the ***palisade cells*** of a leaf. These are near the upper surface, where sunlight can reach them. Beneath them, there are air spaces so that carbon dioxide, oxygen, and water vapour can pass between cells and air by diffusing across the cell membranes. The guard cells control the flow of gases in or out of the leaf.

Food in store

During daylight hours, when a leaf makes plenty of glucose, some is turned into starch and stored in the leaf. Later, it is turned back into glucose and carried (in water) to other parts of the plant. There, it can be stored as starch or used to make other materials.

Many plants need to build up a store of food so that they can survive the winter. Often, we eat their food stores. For example, carrots are swollen roots.

Like other seeds, the broad bean seed below contains a store of food. It must rely on this until its shoot grows out of the ground and can make more food by photosynthesis.

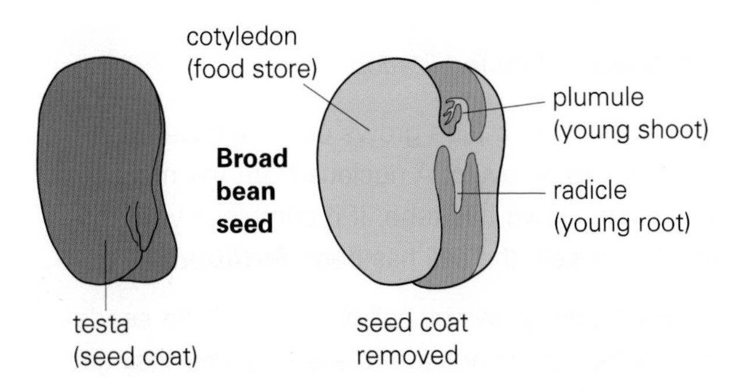

Making new materials

Plants are complicated chemical factories. They turn incoming materials into a whole range of new materials which are needed for food or new growth. Photosynthesis is the first step in the process. It produces glucose and other types of sugar. Sugars are made from carbon, hydrogen, and oxygen. Substances like this are called ***carbohydrates***.

The chart shows some of the materials made in a plant and how minerals are vital for many of them.

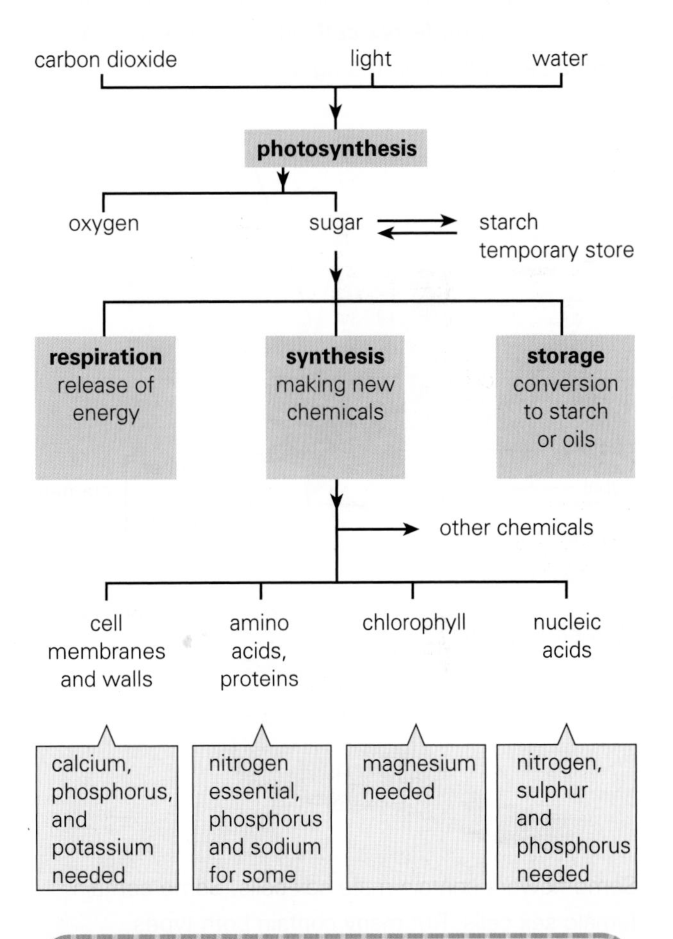

1 Look at the graph on the opposite page. **(a)** If the plant is at 20 °C and the light intensity is increased beyond Y, what will happen to the rate of photosynthesis? **(b)** At X, light intensity is a limiting factor. What does this mean?

2 By what process does water move from one cell to another in the roots of a plant?

3 What do guard cells do? How do they work?

4 Describe *three* things that can happen to sugar made in a leaf by photosynthesis.

2.04 Cells, seeds, and growth

By the end of this spread, you should be able to:

- explain how seeds are produced
- describe what plant hormones do, and how commercial growers make use of them.

Flowers

Many plants have flowers. Flowers produce the seeds which will grow into new plants. But before a seed can develop, a ***male sex cell*** must combine with a ***female sex cell***, as on the right.

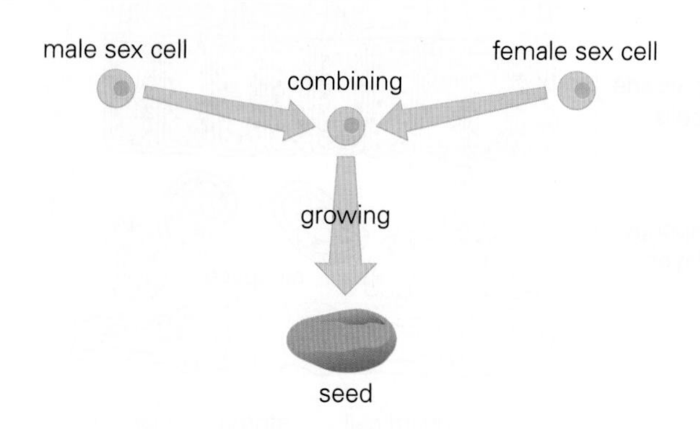

Inside a flower

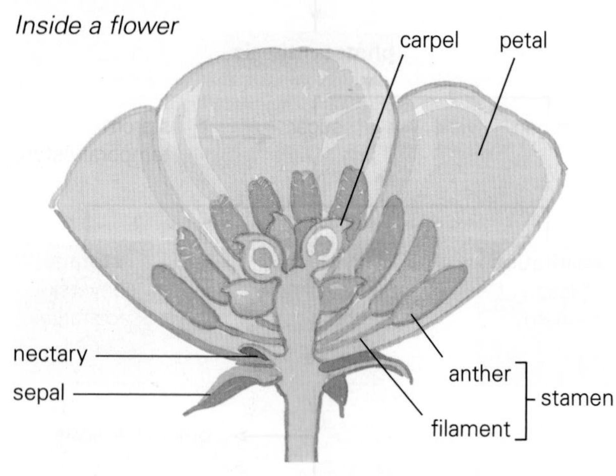

Anther: lower half

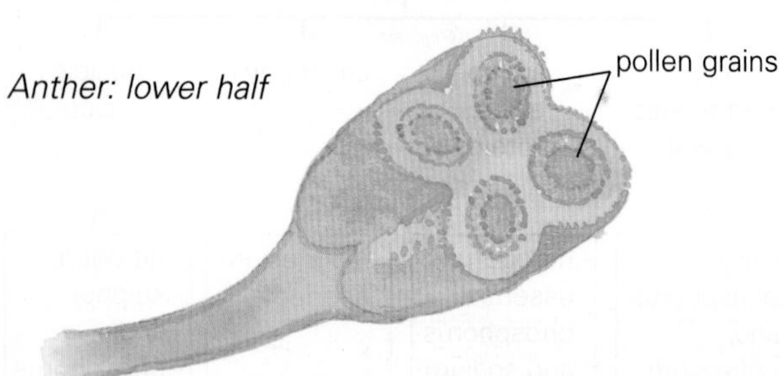

Some flowers contain male sex cells. Some contain female sex cells. But many contain both types.

Stamens hold the grains of ***pollen*** which contain the male sex cells. Pollen in released when the ***anther*** ripens and splits open.

Carpels have a space inside called an ***ovary***. In the ovary are tiny ***ovules***, each containing a female sex cell. Carpels also have a sticky tip called a ***stigma***. Pollen can stick to this.

Not all plants come from seeds. To find out about another method of plant reproduction, see 2.18.

Pollination

Before male and female sex cells can combine, pollen must get across to a stigma and stick to it. This is called ***pollination***. Some flowers are pollinated by wind. Others are pollinated by insects which fly from flower to flower, covered in pollen.

Self-pollination means that pollen is transferred to a stigma in the same flower.

Cross-pollination means that pollen is transferred to a different flower of the same type.

Fertilization...and after

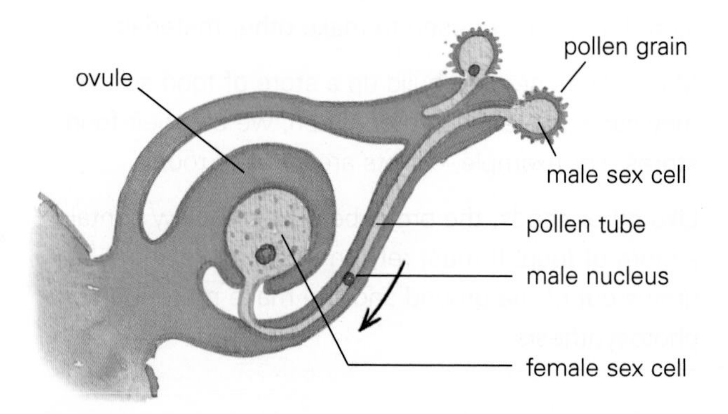

Carpel just before fertilization

After pollination, a tube grows out of a pollen grain and down to an ovule. A nucleus from the male sex cell passes down this tube. If it combines with the female sex cell, the cell has been ***fertilized***.

Fertilized cells grows by cell division to form seeds. Later, when the seeds are released, some may get into the soil and grow into new plants.

Plant hormones

Plant contain natural chemicals called ***hormones***. These control how different parts of the plant work and grow. The hormones are carried with the dissolved food in the phloem tubes.

Here is an example of a hormone in action:

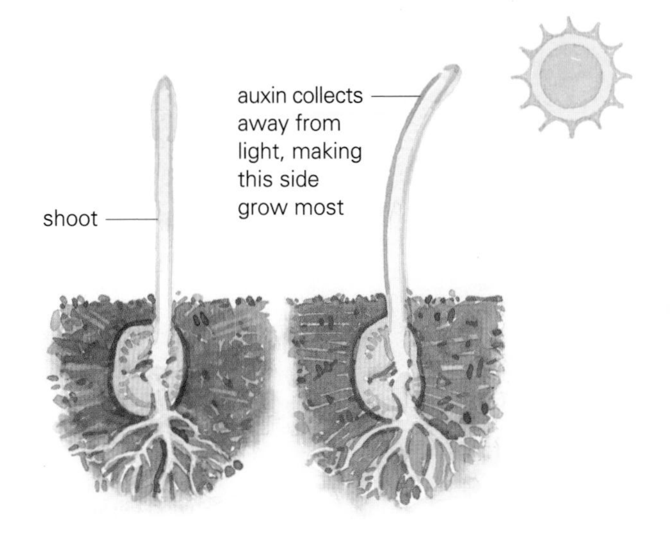

Auxin is a light-sensitive hormone which stimulates plant growth. The diagram above shows how auxin may make a shoot grow towards the light. Auxin tends to collect on the side of the shoot away from the light. This side grows more than the other, so the shoot starts to bend over. Other hormones make roots grow downwards or towards water.

Using plant hormones

Gardeners and commercial growers use hormones to control the growth and development of plants. Here are some examples:

Rooting powder When gardeners take cuttings, they sometimes dip them in hormone powder to stimulate the growth of new roots.

Control of ripening Hormones can be used to speed up or slow down ripening. For example, tomatoes are usually picked before they are ripe. Just before they are delivered to the shops, they are exposed to ethene gas, a hormone which triggers ripening.

Weedkillers Some weedkillers contain artificial auxins. These make weeds grow too quickly so that they run out of food and die.

Missing minerals

For healthy growth, plants need minerals from the soil. The minerals are present in the form of charged atoms (or groups of atoms) called ***ions***: see Spread 3.07. Below are some examples of how plant growth may be affected if mineral ions are missing:

Shortage of...	Effect on plant
nitrate ions	stunted growth; older leaves turn yellow
phosphate ions	poor root growth; younger leaves turn purple
potassium ions	leaves turn yellow, with dead spots

The result of a shortage of potassium

1. Which part of a flower contains **(a)** the female sex cells **(b)** the male sex cells?
2. If a female sex cell has become fertilized, what has happened to it?
3. What is the difference between self-pollination and cross-pollination?
4. How does the hormone auxin make the shoot of a plant grow towards the light?
5. **(a)** Why are hormones used in some weedkillers?
   (b) Give one other way in which plant hormones are used by gardeners or growers.
6. Give *two* reasons why the leaves of a plant might turn yellow.

2.05 Action in the body

By the end of this spread, you should be able to:
- *explain why the body needs to take in some materials and get rid of others*
- *describe what different organs of the body do.*

An ***organ*** is any part of the body with a special job to do. The next page shows some of the main organs of the human body. The organs are made of cells.

Heart and blood

Your heart pumps blood round the body through tubes called ***blood vessels***. Blood leaves the heart through ***arteries*** and returns through ***veins***. These are linked by networks of very narrow tubes called ***capillaries***. Every cell in the body is close to a capillary, so blood can bring new materials to the cell and take waste products away. (See Spread 2.08 for full details.)

Carried in the blood

Your body takes in oxygen, water, and food (which it changes into liquid form). The blood carries these to cells in all your organs. Each cell is a tiny chemical factory. It uses some incoming materials to make new substances for growth and repair, and others to get energy by respiration (see Spread 2.07).

Respiration produces carbon dioxide and water. The blood carries these and other waste products to the organs that get rid of them. Your body also has other waste – unused food that goes right through you.

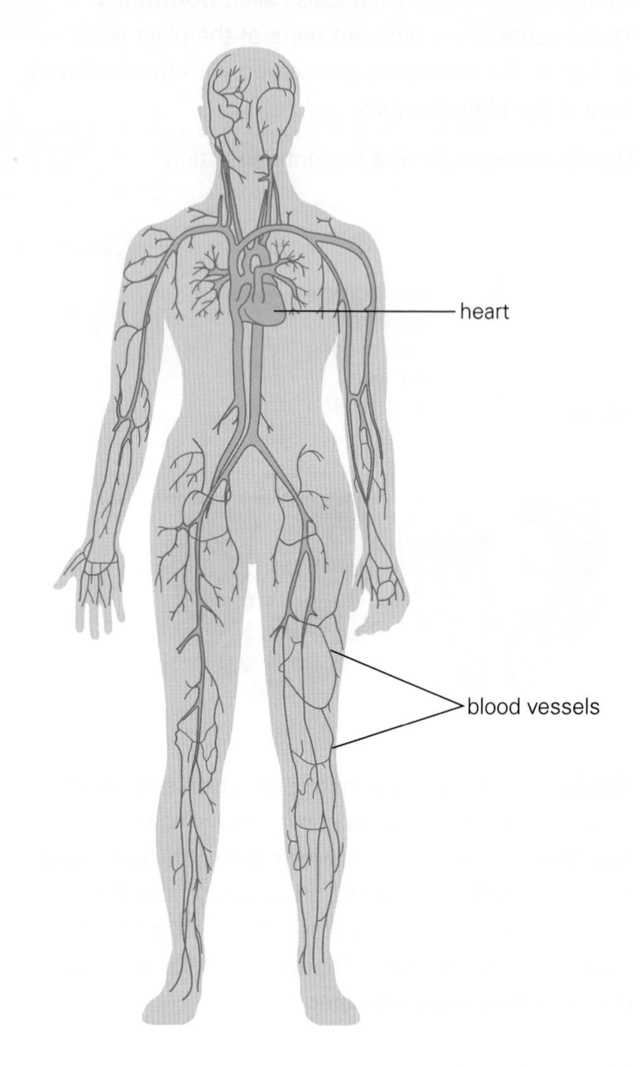

Some of the blood vessels in the human body

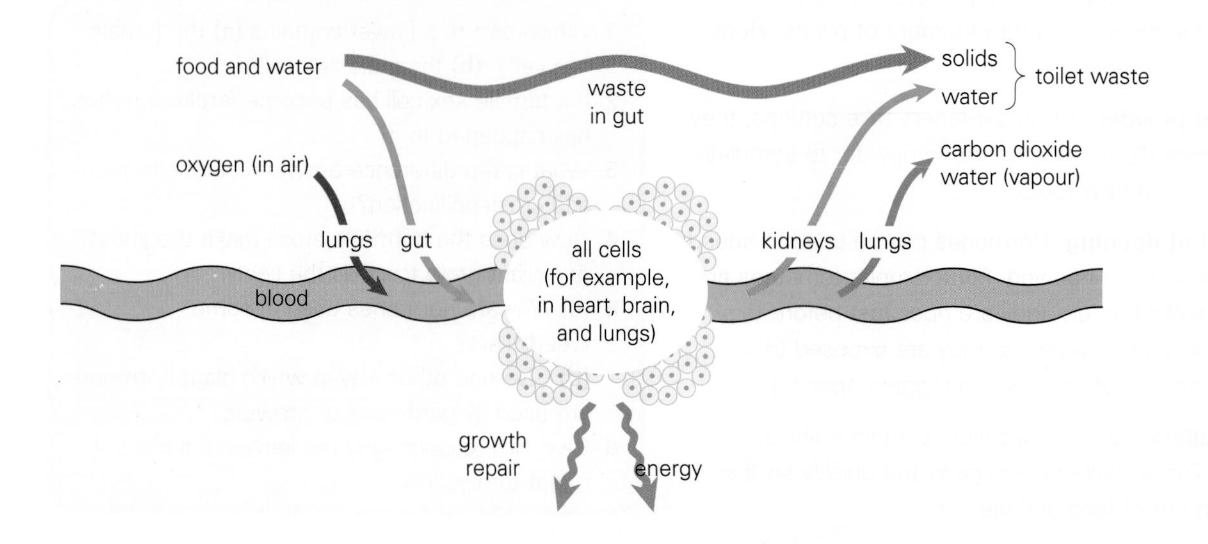

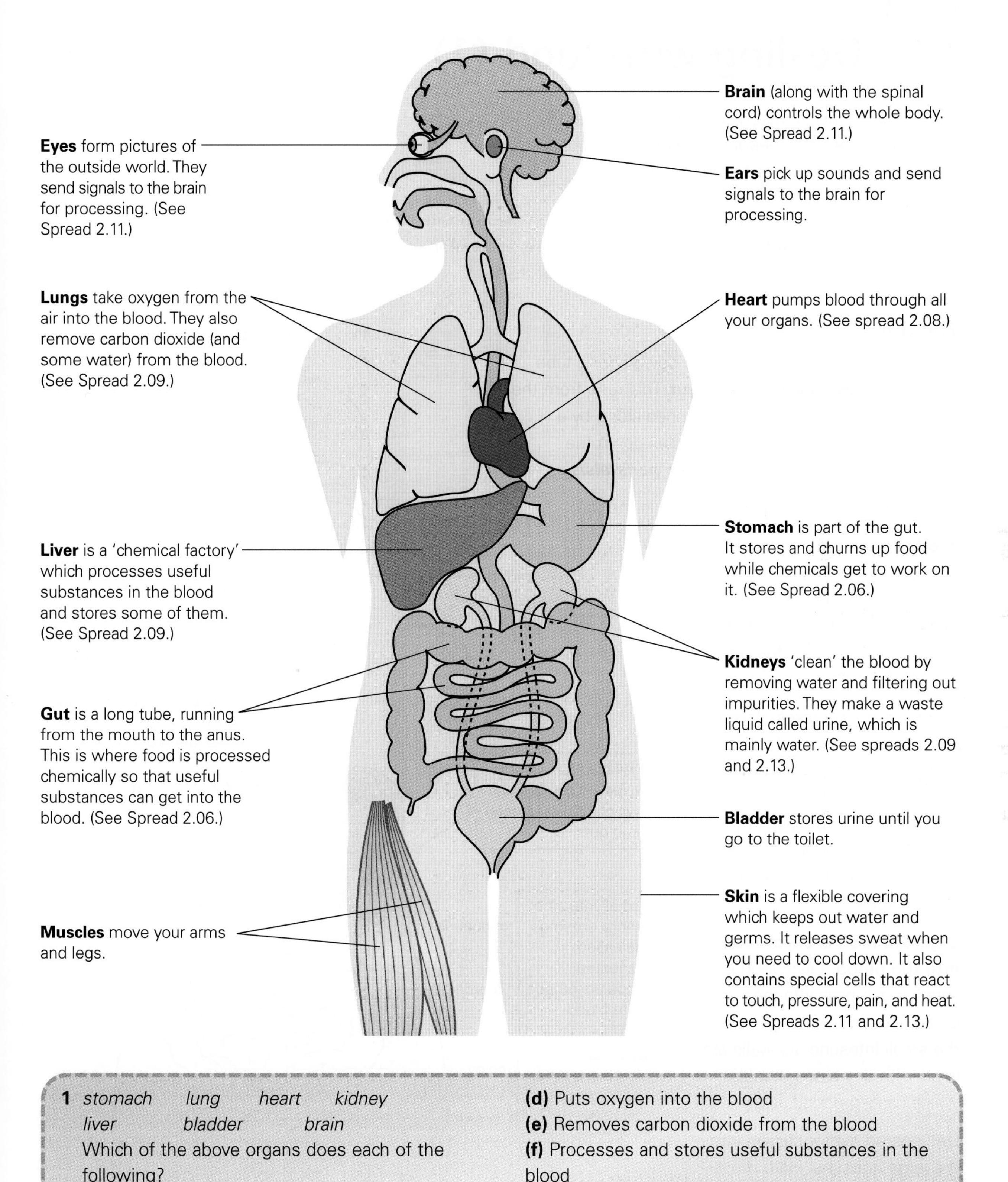

1 *stomach* *lung* *heart* *kidney* *liver* *bladder* *brain*

Which of the above organs does each of the following?

(a) Controls the whole body

(b) Pumps blood to all the organs

(c) Stores and churns up food so that chemicals can get to work on it

(d) Puts oxygen into the blood

(e) Removes carbon dioxide from the blood

(f) Processes and stores useful substances in the blood

(g) Cleans the blood by making urine.

2 Write down *three* things that the body must take in.

3 Write down *two* ways in which the body can get rid of water.

2.06 Dealing with food (1)

By the end of this spread, you should be able to:
- *describes the foods needed for a balanced diet*
- *describe what happens to food in the body.*

Food is a mixture of the useful substances shown on the opposite page: carbohydrates, fats, proteins, minerals, vitamins, fibre, and water. A balanced diet gives you the right amounts of all of these.

The gut

When you swallow food, it moves down a long tube called the ***alimentary canal***, or ***gut***. This runs from the mouth to the anus. The food is pushed along by a squeezing action that moves in ripples down the sides of the gut. This action is called ***peristalsis***.

Two important things happen to food in the gut:

Food is digested This means that it is changed into a simpler, liquid form. The chemicals that do this are called ***enzymes***. For more about enzymes and their effects, see the next spread, 2.07.

Digestion mainly takes place in the stomach and small intestine. But it begins in the mouth. As you chew, an enzyme called ***amylase*** in your saliva starts to break down any starch into liquid glucose, a type of sugar.

Digested food is absorbed into the blood Once food is liquid, it can pass into the blood. This mainly happens in the small intestine. Its walls are lined with tiny blood vessels which carry the food away.

Undigested matter passes into the large intestine. Here most of its water is reabsorbed by the body. This leaves a semi-solid waste (faeces) which comes out of the anus when you go to the toilet.

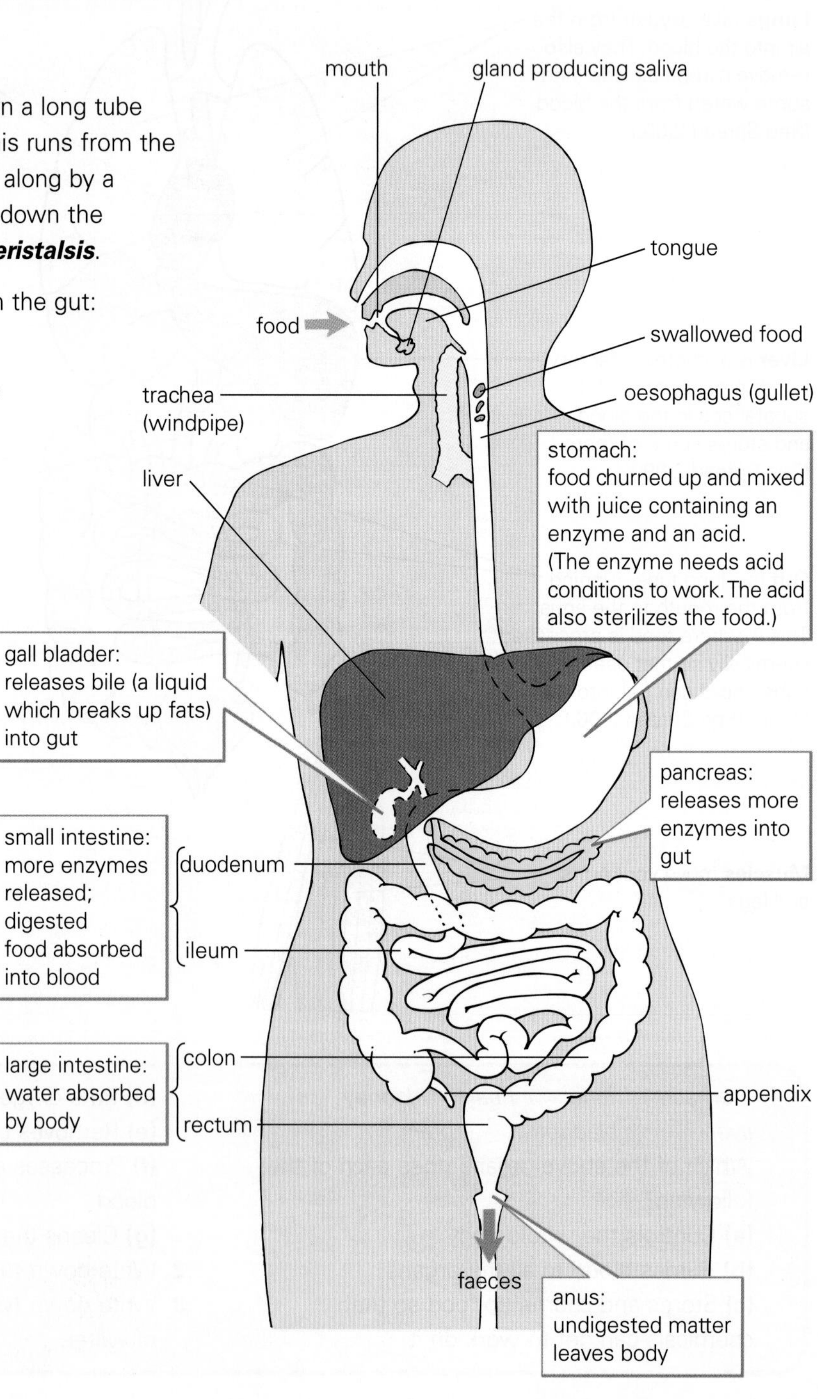

	Examples
Carbohydrates These supply about 50% of your energy. The body breaks down sugars and starches into simple sugars like glucose. Some may also be converted into fat for storage.	**Sugars** in... jams, cakes, sweets, glucose, sweet fruit **Starches** in... potatoes, rice, bread, flour
Fats (including oils) These supply about 40% of your energy. Fats are rich in energy. They can be stored by the body as a reserve supply of food.	Butter, margarine, vegetable oil, lard, meat, cheese
Proteins Needed for growth and for replacing dead cells. The body breaks down proteins into amino acids, which it can use to build new body tissues.	Meat, eggs, fish, milk, cheese, bread
Minerals Small amounts of minerals are needed for some body tissues and for some of the body's chemical reactions.	**Calcium** (for making bones and teeth) in cheese, milk, vegetables **Iron** (for making blood) in liver, eggs, bread **Sodium** (for muscle movements) in salt
Vitamins Small amounts of vitamins are needed to speed up some of the body's chemical reactions.	**Vitamin A** in margarine, butter, liver, carrots, green vegetables, fish oil **Vitamin B_1** in yeast, bread, meat, potatoes, milk **Vitamin B_2** in milk, liver, eggs, cheese **Vitamin C** in blackcurrants, green vegetables, oranges **Vitamin D** from margarine, eggs, fish oil
Fibre This is the cellulose from plants. You cannot digest it, but it provides bulk. It helps food pass through the system more easily.	Vegetables, cereals, bread
Water You need about a litre of water every day – more if it is hot or you are very active.	Drinks, fruit, and other food with water in

Food tests

Foods can be tested for the following substances:

Glucose (e.g. in grape juice) Add Benedict's solution to liquid. If glucose is present, it turns reddy brown.

Starch (e.g. in potato) Put drops of iodine solution on food. If starch is present, it turns blue-black.

Protein (e.g. in milk) Mix liquid with sodium hydroxide solution. Add a few drops of copper sulphate solution. If protein is present, it turns violet.

Fat (e.g. in cooking oil) Add ethanol to liquid. Shake. Add water. If fat is present, liquid turns cloudy white.

1 **(a)** During digestion, what do enzymes do?
(b) What happens to food after it is digested?
(c) What happens to undigested food?

2 Why does the body need **(a)** proteins **(b)** calcium? Name some foods which can suppy a) and b).

3 The body cannot digest fibre (cellulose). Why is it still important in our diet?

4 *carbohydrates fats proteins minerals vitamins fibre water*
(a) Which of the above are the body's main sources of energy?
(b) Make a table to show which of the above are in each of these foods: eggs, cheese, milk, bread.

2.07 Dealing with food (2)

By the end of this spread, you should be able to:
- *describe how food is digested and absorbed*
- *explain where respiration takes place, and how aerobic and anaerobic respiration differ*

Digestion and absorption

During digestion, food is broken down into substances which will dissolve. These are then absorbed by the blood. This mainly happens in the small intestine.

The inner surface of the small intestine is covered with million of tiny bumps called ***villi***. These contain a network of blood capillaries. Together, the villi have a huge surface area for absorbing food into the blood. The blood flows to the liver where there may be further processing of the food.

Villi also absorb undigested droplets of fat. These pass into tiny tubes containing a liquid called ***lymph***. Lymph drains into the blood elsewhere in the body.

Energy values

Some foods give us more energy than others. Scientists normally measure amounts of food energy in ***kilojoules (kJ)*** (see Spread 4.17). The energy available is the same as if the food were burnt. The chart below gives some typical energy values. Fats are the most concentrated source of energy.

About half the energy from your food is needed to drive essential body processes – like circulating the blood and keeping you warm. The rest is needed for moving your muscles. Different people have different energy requirements. An active 18-year-old uses 50% more energy per day than an inactive 50-year-old.

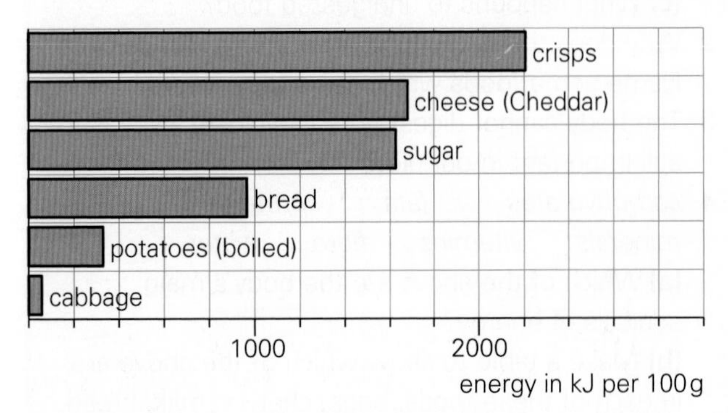

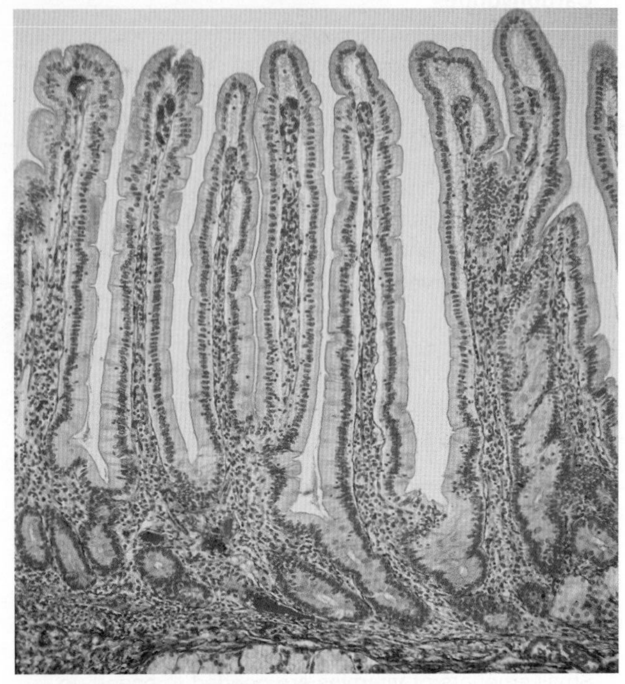

Section through villi: magnification ×80

Aerobic and anaerobic respiration

The body's main 'fuel' is glucose sugar, produced by digestion. Its energy is released during respiration. This equation summarizes what normally happens:

glucose + oxygen → carbon dioxide + water + *energy*

This type of respiration is called ***aerobic respiration*** ('aerobic' means 'using oxygen').

During vigorous exercise, the lungs, heart, and blood system cannot always deliver oxygen fast enough for aerobic respiration. Fortunately, your muscles can work for short periods without oxygen. They can still release some energy from glucose, but the chemical reactions are incomplete and the glucose is turned into ***lactic acid***:

glucose → lactic acid + *energy*

This is an example of ***anaerobic respiration*** ('anaerobic' means 'without oxygen'). In humans, it is only possible for a minute or so because the build up of lactic acid soon stops the muscles working. However, after a rest, when extra oxygen has been breathed in, the lactic acid is changed into glucose, carbon dioxide, and water.

Energy from cells

Respiration takes place in the cells of your body. But it involves a long and complicated series of chemical reactions. If, say, your muscle cells need energy quickly to produce movement, the respiration reactions would be too slow. To overcome this problem, cells do not use the energy from respiration directly. Instead, they use it to make a substance called ***ATP*** (adenosine triphosphate). ATP acts as a temporary energy store. It can release that energy very quickly whenever it is needed.

Cells contain tiny structures called ***mitochondria***. Respiration takes place and ATP is made inside these.

Cell
(animal)

mitochondria: a cell can contain 50–1000 of these

stored food

More about enzymes and digestion

A ***catalyst*** is a substance which speeds up a chemical reaction without being used up itself. Enzymes are natural catalysts, made in the body. Without them, the vital chemical reactions needed for life would work too slowly or not at all.

Different enzyme have different jobs to do. The table on the right shows some of the enzymes used by your digestive system. Other liquids are also needed to create the right conditions for enzymes to work.

Like most other substances, food is made up of molecules (see Spread 3.01). During digestion, large molecules are broken down into smaller, simpler ones. In the cells of your body, these can then be combined to make new substances. For example:

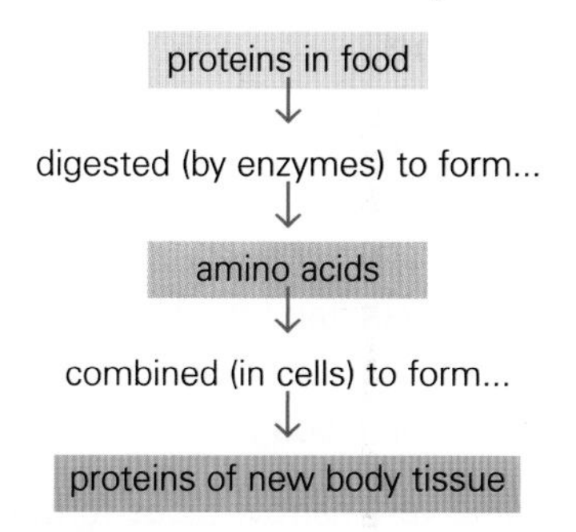

Type of enzyme	Where made	Used for
amylase	salivary glands pancreas small intestine	breaking down starch into sugars
protease	stomach pancreas small intestine	breaking down proteins into amino acids
lipase	pancreas small intestine	breaking down fats into fatty acids and glycerol

Liquid	Where made	Used for
hydro-chloric acid	stomach	killing bacteria; creating acid conditions needed by enzymes in stomach
bile	liver (stored in gall bladder)	neutralizing stomach acid so that enzymes in small intestine can work; breaking up fat into smaller droplets for enzymes to digest

1 Use the chart on the opposite page to estimate how much energy is stored in a 100 g of **(a)** crisps **(b)** bread **(c)** Cheddar cheese.

2 Crisps give you far more energy than boiled potatoes. Why is it not a good idea to eat crisps instead of boiled potatoes with your meals?

3 What are *villi* and what do they do?

4 Why is your respiration sometimes anaerobic? How is this different from aerobic respiration?

5 What do enzymes do?

6 Why does your stomach need acid in it?

7 Where is bile made, and what does it do?

2.08 The blood system

By the end of this spread, you should be able to:
- *explain what blood is and what it does*
- *describe how blood is pumped round the body.*

Blood

Blood is a mixture of ***red cells, white cells***, and ***platelets***, in a watery liquid called ***plasma***.

There are hundreds of times more red cells than white. They give blood its red colour.

Jobs done by the blood

- Bringing oxygen, water, and digested food (such as glucose) to the cells of the body.
- Taking carbon dioxide and other waste products away from the cells.
- Distributing heat to all parts of the body.
- Carrying ***hormones*** (chemicals which control how different organs work: see 2.12)
- Carrying substances which help fight disease.

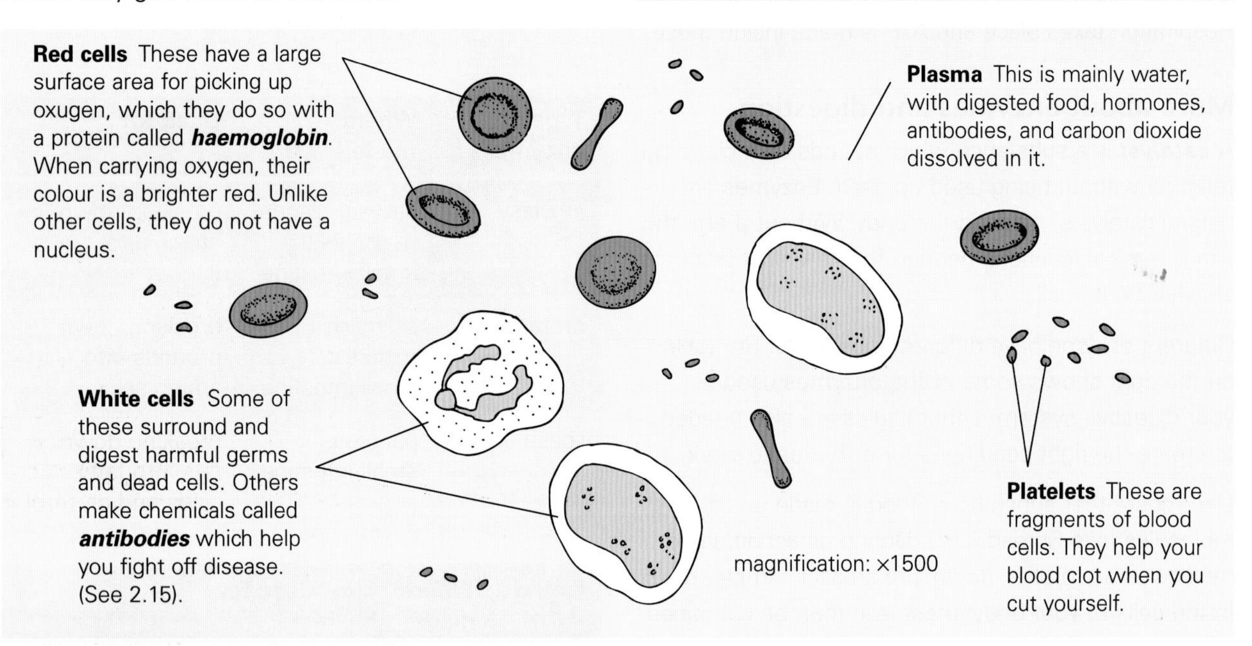

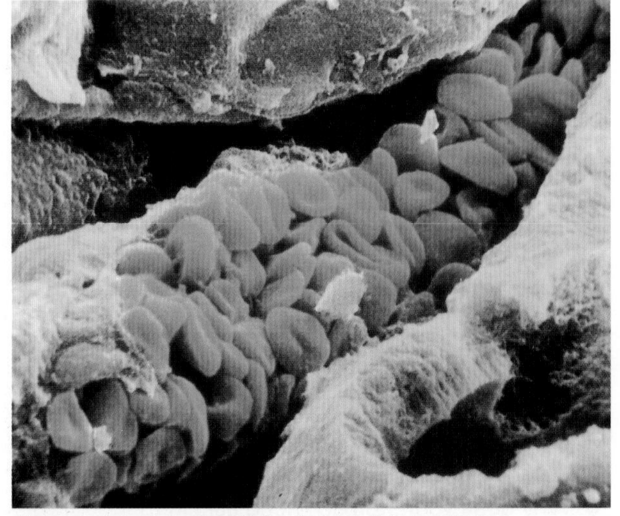

False-colour photograph of red blood cells in a capillary: magnification ×1800

Circulating blood

The heart pumps blood round the body through a system of blood vessels (tubes): see also Spread 2.05.

Arteries These carry blood away from the heart. They have thick walls to withstand the high pressure of the heart's pumping action. They divide into narrower tubes, which carry the blood to...

Capillaries These very narrow tubes form networks throughout the body. Every living cell in the body is close to a capillary. Water and dissolved substances can pass between cells and capillaries through the thin capillary walls. Blood from capillaries drains into...

Veins These carry blood back to the heart at low pressure. Some veins have one-way valves in them to stop the blood flowing in the wrong direction.

The heart

The heart is really two separate pumps in one. One pump sends blood through capillaries in the lungs, where it absorbs oxygen. The other pump takes in this oxygen-carrying blood and pumps it round the rest of the body. The diagram on the right shows the principle, but not the actual layout.

Heart

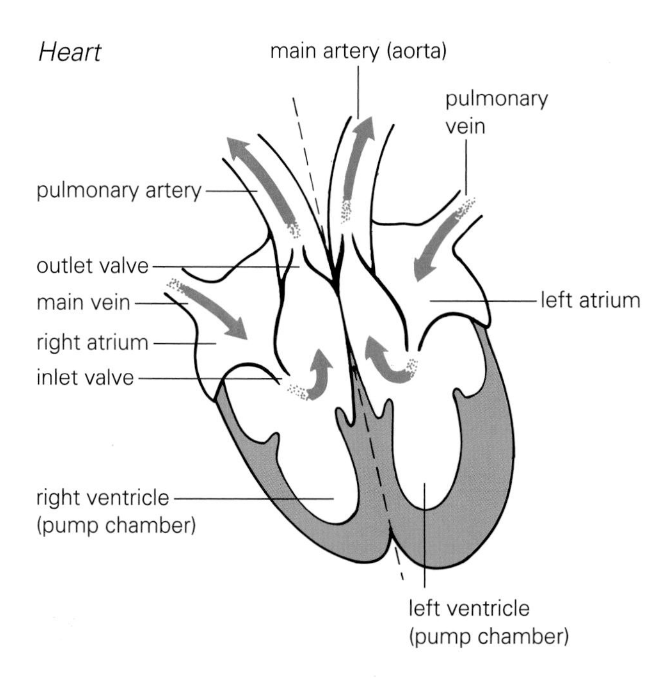

Each pump has two valves. These let blood through in one direction only. Between the valves is a chamber called a ***ventricle***. When muscles around it contract, it gets smaller, and blood is pushed out through the outlet valve. When the muscles relax again, more blood flows into the ventricle through the inlet valve.

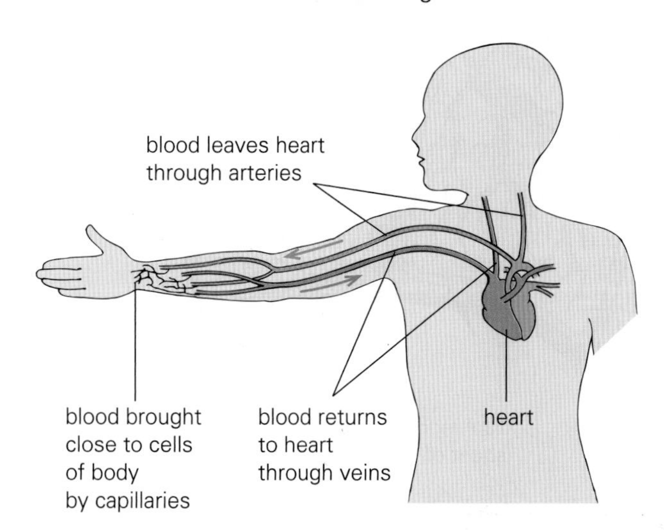

Circulation of the blood

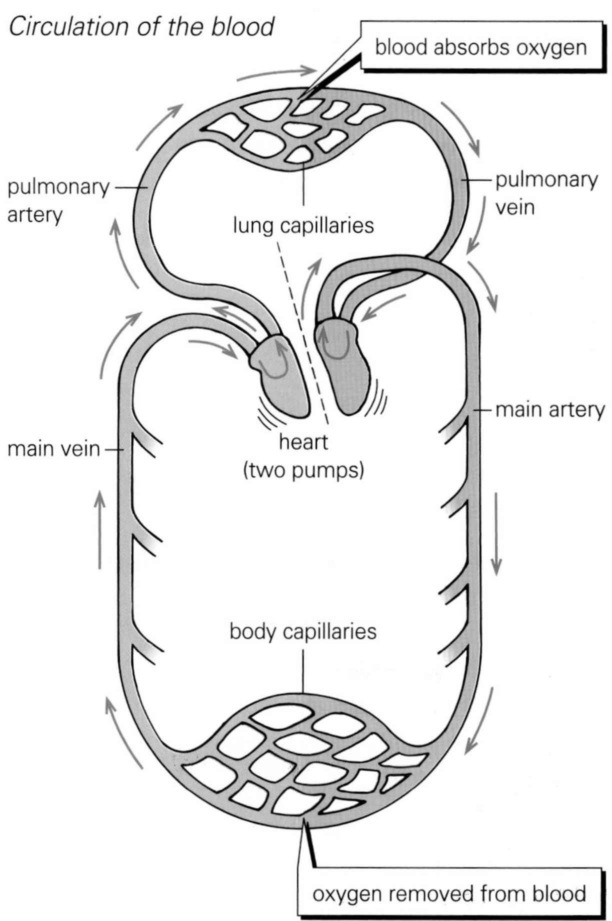

The muscle contractions are called ***beats***. They are set off by nerve impulses produced in the heart itself. However, the beat rate can be changed by the nervous system. If you are exercising, your heart beats faster so that your muscles get oxygen more quickly.

1 Veins and arteries carry blood. Which of these
(a) carry blood away from the heart?
(b) carry blood back to the heart?
2 What are capillaries? What job do they do?
3 Why do the cells of the body need to be close to a supply of blood?
4 What is the liquid part of blood called?
5 Which blood cells help your body fight disease?
6 **(a)** Which blood cells carry oxygen?
(b) What substance do they use to carry oxygen?
7 Where does blood absorb oxygen?
8 Why does the heart need valves?
9 Describe what happens to blood as it leaves the pulmonary artery and circulates round the body, back to where it started.

2.09 Lungs, liver, and kidneys

By the end of this spread, you should be able to:
- *describe how gas exchange occurs in the lungs*
- *describe how the liver and kidneys affect the blood.*

The lungs, liver, and kidneys help to keep substances in the blood in the right proportions.

The lungs

During respiration, the cells of your body use up oxygen and make carbon dioxide and water. The job of the lungs is to put oxygen into the blood and remove the unwanted carbon dioxide.

The lungs are two spongy bags of tissue. They are filled with millions of tiny air spaces, called ***alveoli***. These have very thin walls and are surrounded by a network of blood capillaries. Oxygen in the air can seep through these walls and into the blood. Also, carbon dioxide (and some water) can seep from the blood into the air.

As you breathe in and out, some of the old air in your lungs is replaced by new, and the ***exchange*** of oxygen and carbon dioxide takes place.

Breathing in

ribs pulled upwards and outwards by muscles

lungs fill with air

diaphragm pulled downwards by muscles

Breathing out

air pushed out from lungs

diaphragm relaxes

Typical percentages in air…

	…breathed in	…breathed out
oxygen	21%	16%
carbon dioxide	0.04%	4%

Your breathing rate is controlled by the amount of carbon dioxide in your blood

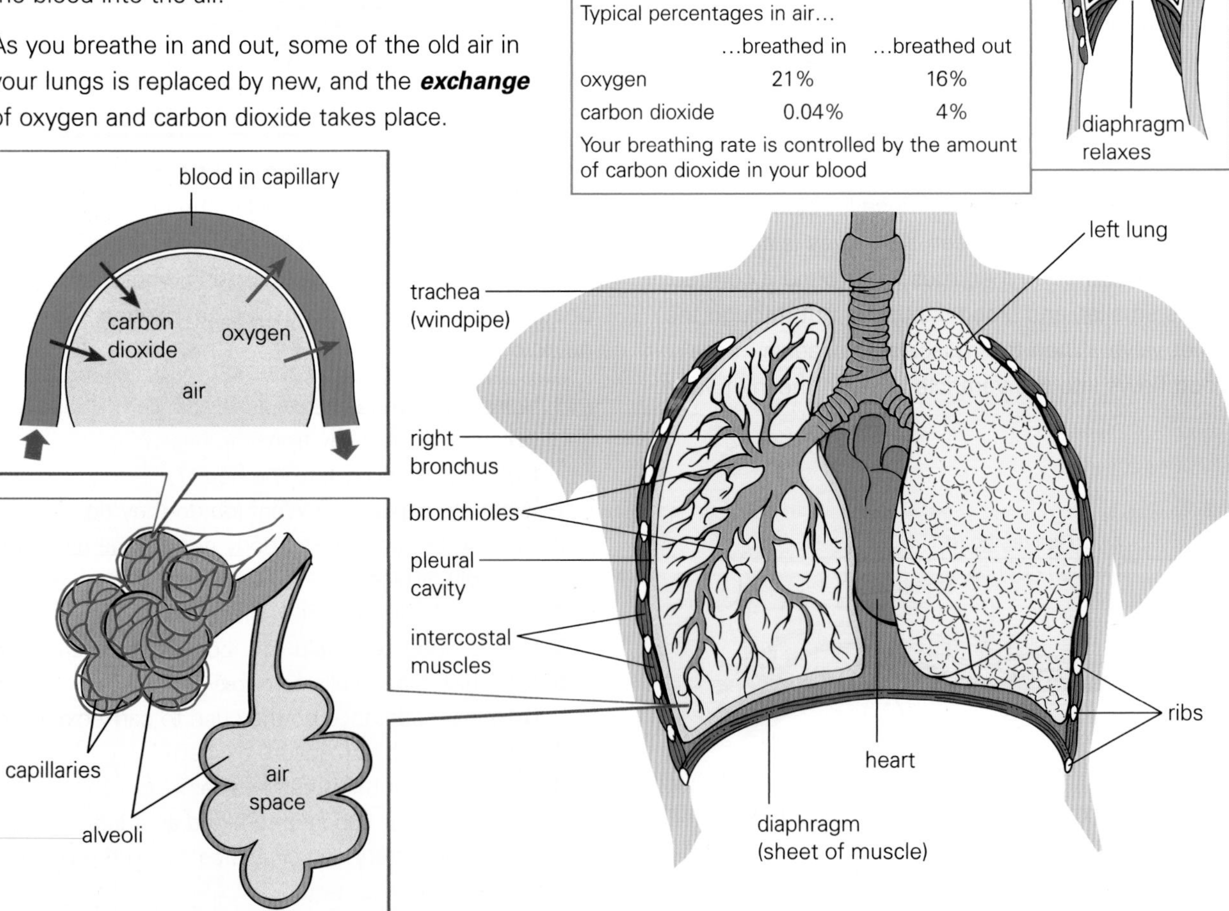

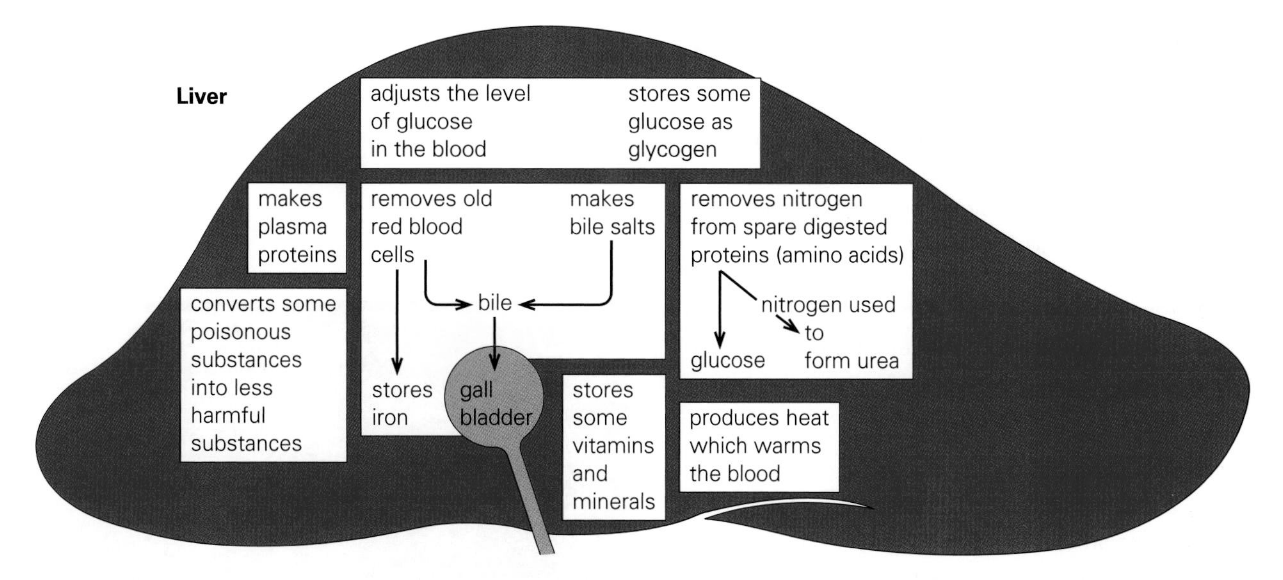

The liver

The liver is the largest organ in the body. It is a complicated chemical factory with many jobs to do. Some of its jobs are shown in the diagram above.

The kidneys

The kidneys 'clean' your blood by filtering it. First, they remove water and other substances. Then they put some of these back so that the proportions are correct. Unwanted water and other substances collect in your bladder as ***urine***.

The kidneys are organs of ***excretion***: they remove unwanted substances from your body. The lungs partly work as organs of excretion as well. For more on how the kidneys work, see Spread 2.13

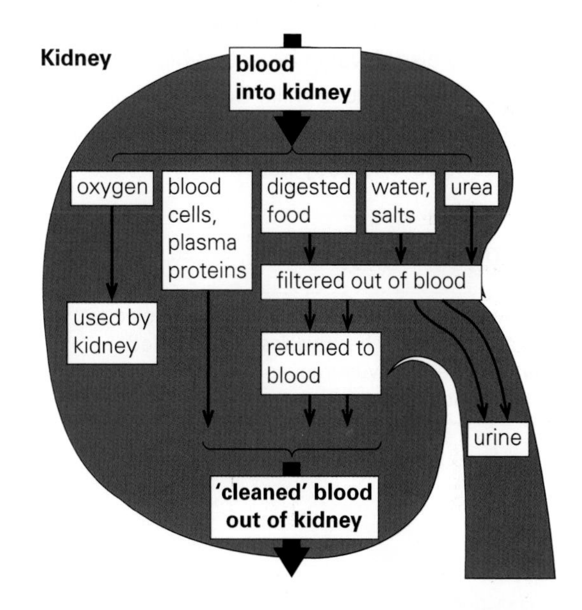

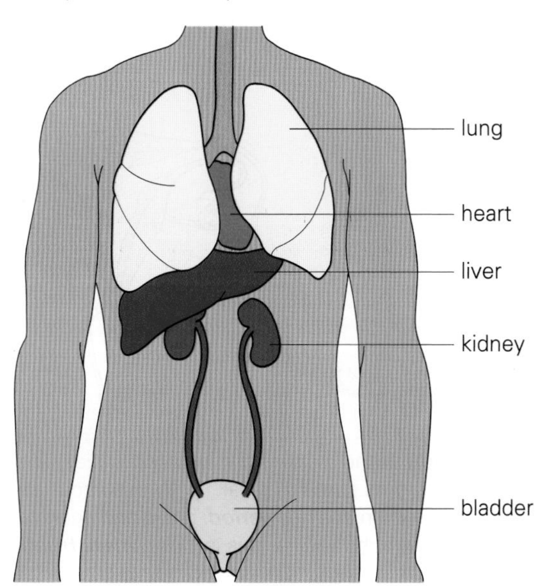

1. In the lungs:
   (a) what substance is taken into the body?
   (b) what substance is removed from the body?
2. Why are the tiny air spaces in the lungs surrounded by blood capillaries?
3. What makes your lungs expand when you breathe in?
4. Name *three* things which the liver stores.
5. Descibe *two* other jobs done by the liver.
6. What job is done by the kidneys?
7. What happens to waste substances removed by the kidneys?
8. Why are the kidneys called organs of excretion?
9. Give another example of an organ of excretion.

2.10 Making human life

By the end of this spread, you should be able to:
- *explain how a sperm and an ovum combine*
- *explain the meaning of ovulation, menstruation, menstrual cycle, and fertilization.*

A baby grows from a tiny cell in its mother. This cell is formed when a ***sperm*** from the father combines with an ***ovum*** (egg) inside the mother. A sperm combining with an ovum is called ***fertilization***.

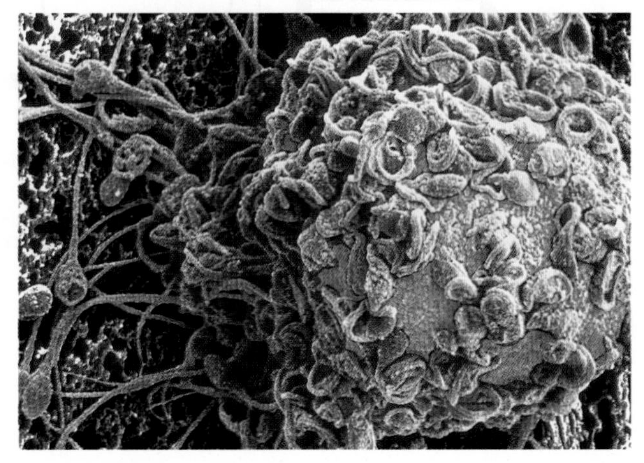

Sperms cluster around an ovum: magnification ×250

Puberty is the start of the time when a girl can become a mother and a boy can become a father. It often happens around the age of 12 to 14, but it is quite normal for it to be later than this. Girls usually reach puberty before boys. At puberty, girls start their periods. They also develop breasts which, later, if they have a baby, will produce milk. Boys produce sperms for the first time. Their voices go deeper and they grow more facial and body hair.

The female sex system

Ovulation About every 28 days, a woman releases an ovum from one of her ***ovaries***. This is called ***ovulation***. The tiny ovum moves down the ***oviduct*** (egg tube) and into the ***uterus*** (womb).

Lining growth Near the time of ovulation, the lining of the uterus thickens, and a network of blood capillaries grows in it. The uterus is now ready to receive and nourish a fertilized ovum.

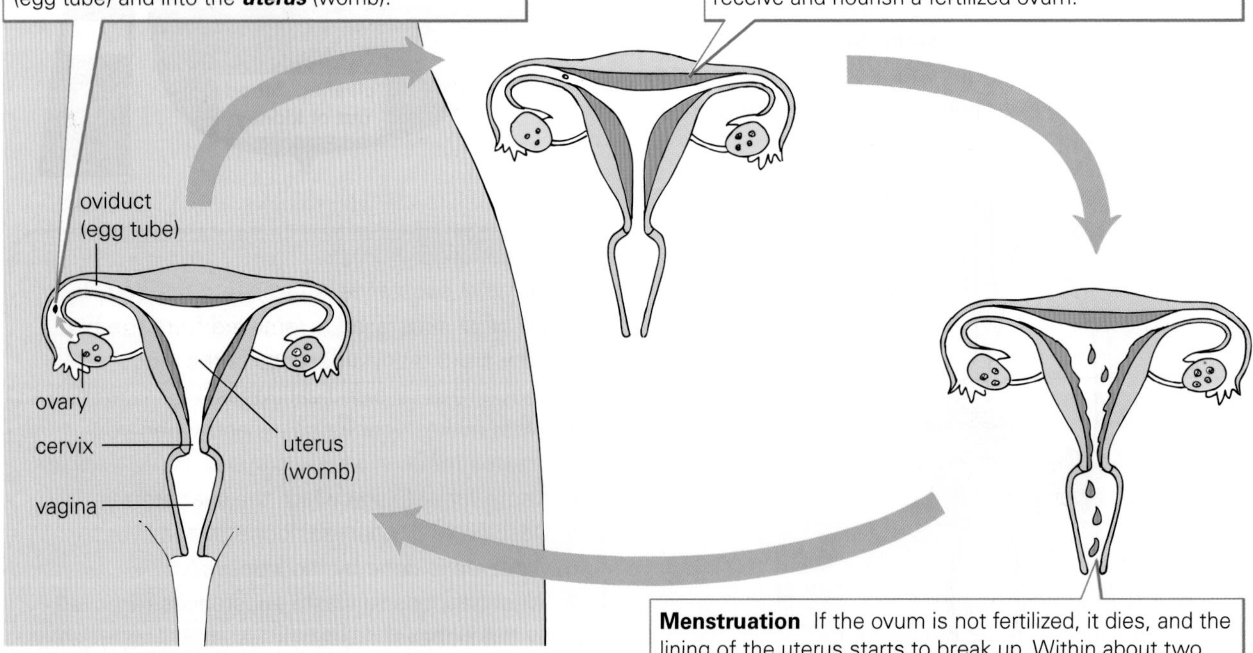

Menstruation If the ovum is not fertilized, it dies, and the lining of the uterus starts to break up. Within about two weeks, the woman has her ***period***: blood and dead cells pass out through the vagina. This is called ***menstruation***.

The 28-day cycle of ovulation, lining growth, and menstruation is called the ***menstrual cycle***.

The male sex system

Sperms are made in a man's ***testicles***. Before they leave his body, they are mixed with a liquid produced by glands. The mixture, called ***semen***, leaves the penis through the same tube as urine.

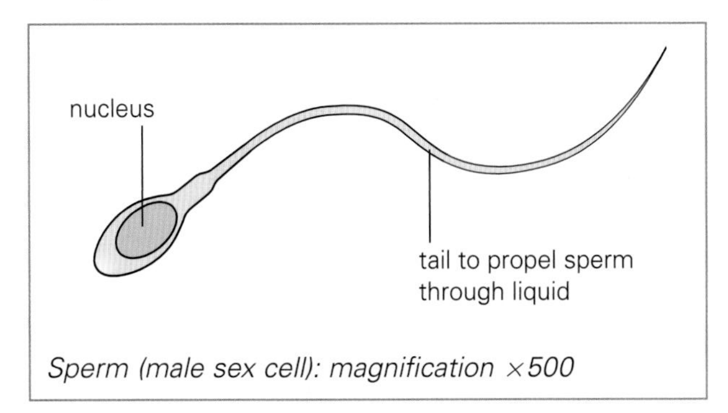

Sperm (male sex cell): magnification ×500

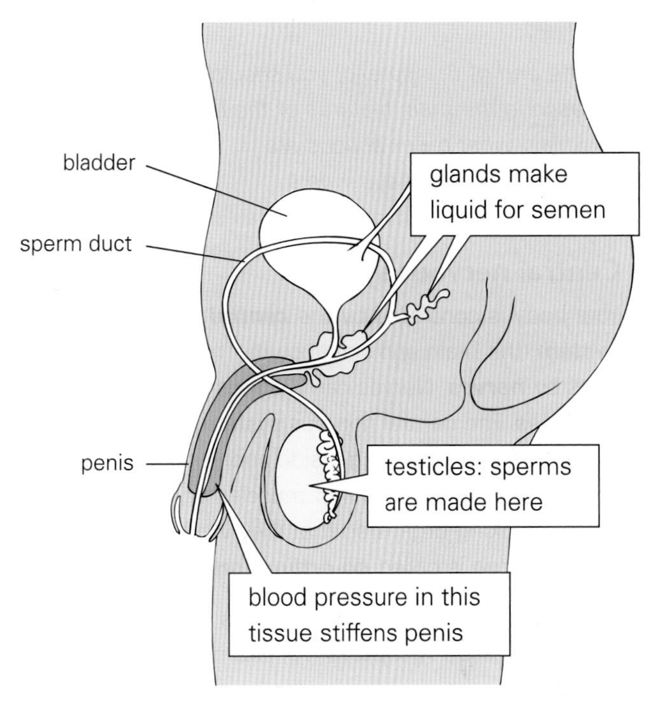

How fertilization happens

When a man and woman have sex, the man's penis goes stiff and is placed in the woman's vagina. When the man ***ejaculates***, semen is pumped from his penis. The semen contains millions of sperms. Some pass into the uterus. And some reach the oviducts, where they may meet an ovum. Only one sperm can fertilize the ovum. After fertilization, an extra 'skin' forms round the ovum to keep out other sperms.

Birth control

If parents want a small family, they may decide to use ***contraception*** (birth control). Here are some of the methods available to them:

The condom is a thin rubber cover which fits over the man's penis to trap sperms. It is more reliable if used with a ***spermicide*** (a cream with chemicals to kill sperms).

The diaphragm (cap) is a rubber-covered ring which is put over the woman's cervix. It too stops sperms reaching the uterus. Like a condom, it is best used with a spermicide.

The pill must be swallowed daily by the woman. It contains chemicals which stop the ovaries releasing ova (eggs). The method is very reliable, but it may increase the risk of heart, liver, or breast disease in some people.

The natural method The woman does tests to find out when ovulation is close and does not have sex near that time. This method can be used by people who think that other kinds of birth control are wrong.

1 Explain what each of the following means:
ovum *ovulation* *menstrual cycle*

2 About how often is an ovum released?

3 If it is not fertilized after it has been released, what happens to it?

4 What must happen to an ovum for it to be fertilized?

5 After an ovum has been fertilized, how does it stop other sperms entering?

6 Where are sperms produced?

7 Some methods of contraception are called *barrier* methods because they stop sperms reaching the uterus. Which of the methods described on this page are barrier methods?

2.11 Nerves in action

By the end of this spread, you should be able to:
- *describe the main features of the nervous system*
- *explain what reflex actions are*
- *describe how the eye works.*

Central nervous system

Your body is controlled by the ***central nervous system*** (the brain and spinal cord). This is linked to all parts by ***nerves***. Nerves carry signals called ***nerve impulses***. The central nervous system use them to sense what is happening inside and outside the body, and to control the actions of muscles and other organs. Linking organs so that they work together at the right time is called ***co-ordination***.

Impulses sent to the central nervous system come from cells called ***receptors***. For example:

Receptors in...	respond to...
eyes	light
ears	sound
nose	chemicals in air
tongue	chemicals in food
skin	touch, pressure, heat, pain

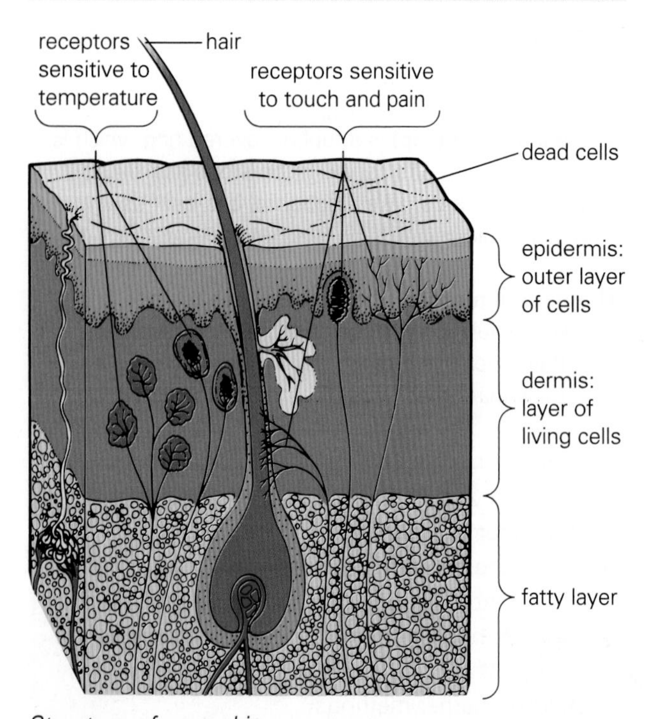

Structure of your skin

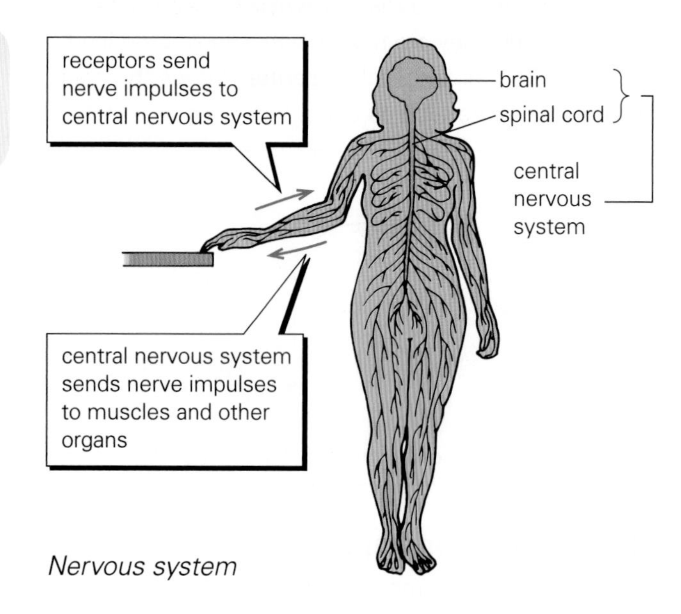

Nervous system

Nerve cells

Nerves are made up of bundles of nerve cells. The impulses they carry are tiny pulses of electricity. These travel through the cell in one direction only.

Nerve cells are also called ***neurones*** (or ***neurons***):

Sensory neurones carry information from receptors to the central nervous system.

Motor neurones carry instructions from the central nervous system to muscles or other organs. The organs they control are called ***effectors***.

Neurones (nerve cells) are very thin and can be over a metre long. A junction where one neurone meets another is called a ***synapse***. Signals are transmitted across a synapse by very rapid chemical changes.

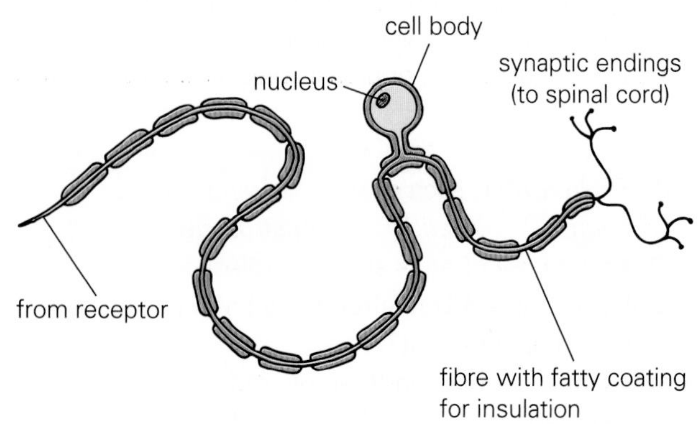

Sensory neurone (nerve cell)

Reflex actions

You do some things without having to think about them. For example, if something flashes in front of your eyes, you blink. The response is quick and automatic. Its job is to protect the body from pain or harm. Actions like this are called ***reflex actions***. Other examples include coughing and sneezing.

The brain is not involved in all reflex actions. The diagram below shows the path of the impulses if you touch something hot. The path is called a ***reflex arc***.

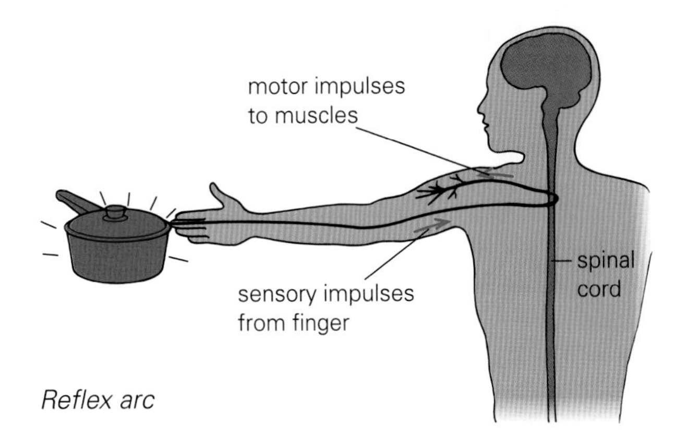

Reflex arc

The eye

For you to see something, receptors in your eyes must send nerve impulses to the brain.

Like a camera, the human eye uses a lens system to form an image at the back. The image is upside-down. However, the brain gets so used to this that it thinks the image is the right way up!

The ***retina*** is the 'film' at the back of the eye, where the image is formed. It contains millions of light-sensitive cells. These send impulses to the brain along the ***optic nerve***.

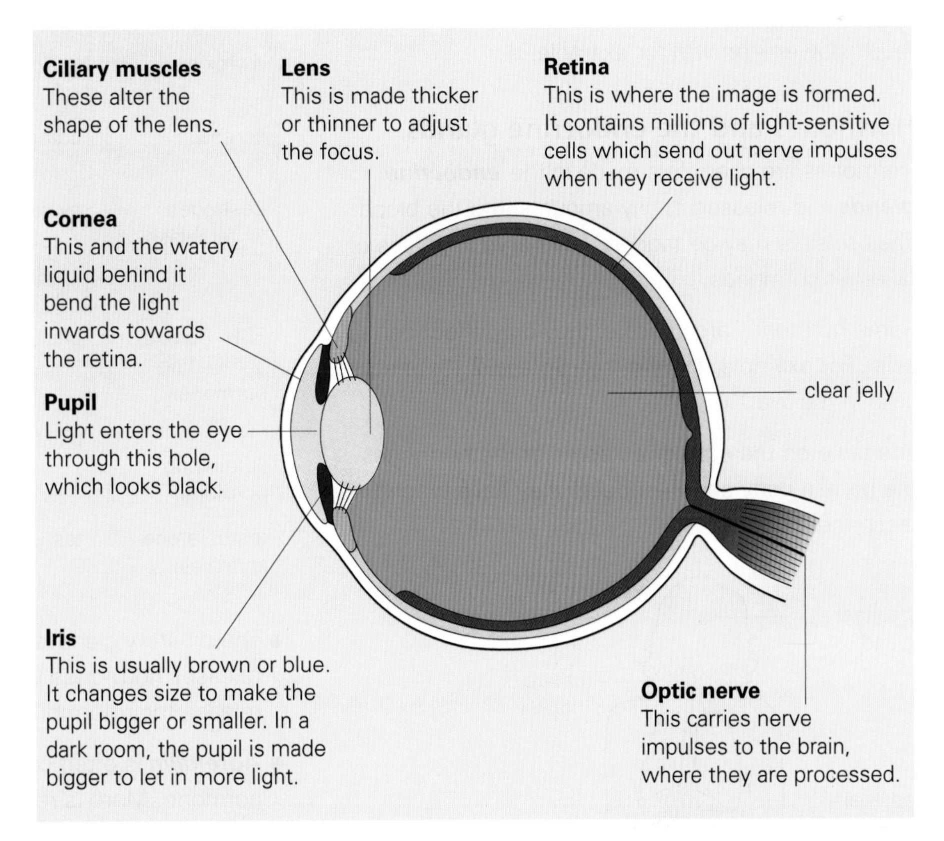

Tiny muscles alter the shape of the lens so that it can focus on things at different distances. However, many people can still not get a clearly focused image, so they have to wear spectacles or contact lenses.

1 What do receptors do?

2 What job is done by
(a) sensory neurones **(b)** motor neurones?

3 Give *three* examples of reflex actions.

4 You hear a loud noise and it makes you jump. Use your ideas about the nervous system to explain how a loud noise makes you jump.

5 *lens* *optic nerve* *cornea* *retina* *iris*
Which of the above parts of the eye
(a) contains millions of light-sensitive receptors?
(b) is used to make focusing adjustments?
(c) controls the amount of light entering the eye?
(d) carries signals to the brain?

2.12 Hormones in action

By the end of this spread, you should be able to:
- *say what hormones are and where they are made*
- *describe some of the effects of hormones*
- *explain how some hormones are used as drugs*

The body has two methods of sending messages between organs: using nerves (see Spread 2.11), and using chemical messengers called ***hormones***. The central nervous system has second-by-second control of the body. Hormones cause changes which are often much slower – growth for example.

Hormones and the endocrine glands

Hormones are chemicals made in the ***endocrine glands*** and released in tiny amounts into the blood. Their release may be triggered by nerve impulses or by other hormones.

Some hormones are specific: they only affect certain cells. For example, ***insulin*** (see Spread 2.13) affects cells in the liver.

The table on the right gives some of the hormones in the human body and the effects they have when they reach their target.

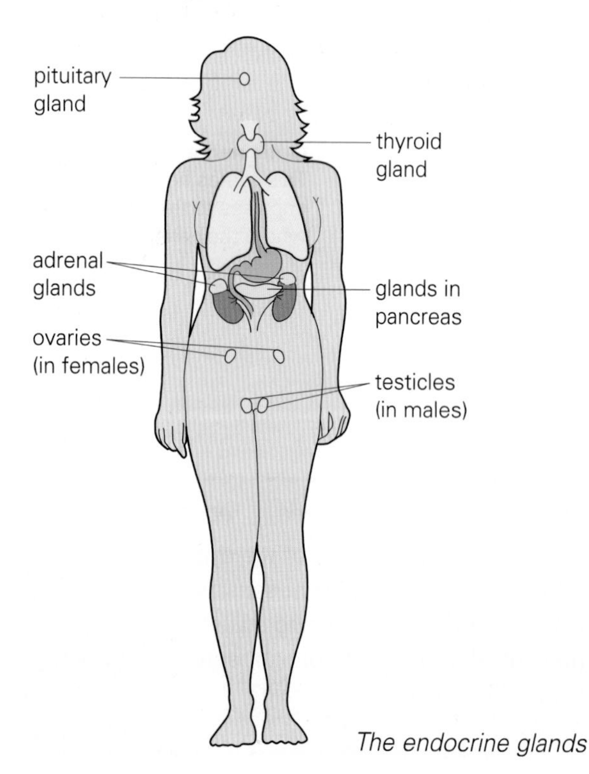

The endocrine glands

Hormone	Where made	Main effects
HGH (human growth hormone)	pituitary gland	stimulates growth
ADH (anti-diuretic hormone)	pituitary gland	makes nephrons in kidneys reabsorb more water
insulin	glands in pancreas	makes liver turn glucose into glycogen for storage
thyroxin	thyroid gland	speeds up heat production
adrenalin	adrenal glands	speeds up heart and breathing
oestrogen (in females)	ovaries	female sexual development, triggers changes during menstrual cycle
FSH (follicle stimulating hormone)	pituitary gland	triggers changes during menstrual cycle
LH (luteinizing hormone)	pituitary gland	triggers changes during menstrual cycle
testosterone (in males)	testicles	male sexual development

- The pituitary gland is the ***manager gland***. It releases hormones which stimulate other glands to release their hormones.
- ***Adrenalin*** is often called the 'fight or flight' hormone. More is released if something scares you. It prepares the body for a rapid output of energy so that you can face the danger or escape.
- For more on ADH and insulin, see Spread 2.13.
- At puberty, ***testosterone*** makes a boy develop hair on his face and body, and have a deeper voice.
- At puberty, ***oestrogen*** makes a girl develop breasts and wider hips, and start her periods. After that, oestrogen has an important part to play in the menstrual cycle.

For more about the menstrual cycle, see Spread 2.10 before reading the next page.

Hormones and the menstrual cycle

A woman's menstrual cycle is controlled by hormones. Each month, they cause an egg (ovum) to mature and be released at ***ovulation***, and the lining of the uterus to thicken, ready for a fertilized egg. If the egg is not fertilized, the hormone level drops and menstruation starts – the lining of the uterus breaks up and the woman has her period. But if the woman becomes pregnant, hormones stop menstruation and the release of more eggs.

The diagram below shows how a hormone released by the pituitary gland leads to the maturing and release of an egg. Three hormones are involved: FSH, oestrogen, and LH. To see what happens, follow the arrows, starting at the pituitary gland...

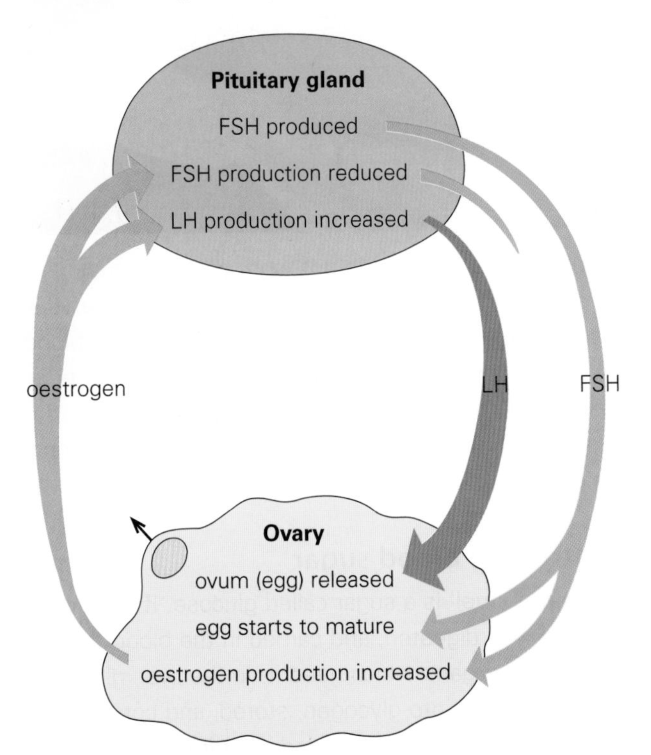

Hormones as drugs

Growth Some hormones stimulate body growth. ***Anabolic steroids*** are drugs which act like these hormones. Sensible body-builders do not use these drugs. But those who do risk heart, liver, and kidney damage, and uncontrolled behaviour.

The contraceptive pill (see Spread 2.10) This contains a chemical which acts like oestrogen. It reduces the production of the hormone LH so that eggs do not mature. The pill may increase the risk of heart and liver disease and breast cancer.

Fertility drugs If a woman cannot have a baby, it may be because her ovaries are not releasing eggs. Fertility drugs contain the hormone FSH. This triggers the process which makes eggs mature. However, it can lead to the release of several eggs at once, so there may be multiple births – such as twins or triplets.

1 How do hormones reach the organs that they control?

2 Name *two* hormones made in the pituitary gland and describe the effects they have.

3 **(a)** Where is adrenalin made?
(b) Why is adrenalin sometimes called the 'fight or flight' hormone?

4 Look at the diagram above.
(a) Where is the hormone FSH produced?
(b) What effects does FSH have when it reaches an ovary?
(c) What effects does oestrogen have when it reaches the pituitary gland?
(d) What effect does LH have when it reaches an ovary?

5 Hormones are sometimes taken as drugs. Give *two* examples of this this. Describe what problems can arise in each case.

2.13 Bodies in balance

By the end of this spread, you should be able to:
- *explain how water balance, blood sugar level, and temperature are maintained in the body.*

In your body, temperature, amount of water, and many other factors must be kept steady. Your ***internal environment*** must be in ***balance***. Maintaining this balance is called ***homeostasis***. The systems which control it work automatically.

Water balance

You lose about 2 litres of water per day in urine and sweat. This is mainly replaced by water in your food and drink, though some water is made during respiration (see Spread 2.02).

Your kidneys filter your blood and produce urine (see Spread 2.09). They contain millions of tiny ***nephrons*** like the one below. Liquid (mainly water) is absorbed from the blood, then most of this water (plus other useful substances) is reabsorbed by the blood.

The amount of water in the blood is controlled by adjusting the amount of water reabsorbed. This is done using a hormone called ***ADH***, made in the pituitary gland. ADH causes more water to be reabsorbed. If you sweat a lot or drink very little, the amount of water in your blood drops. The pituitary gland senses this and responds by making more ADH.

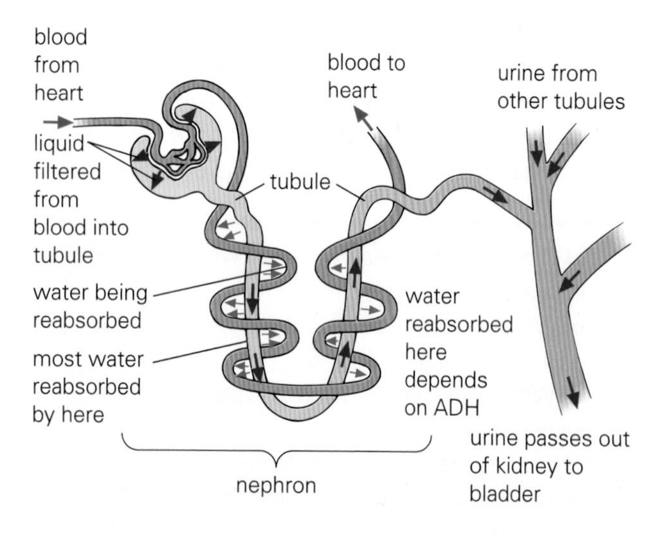

The kidneys contain millions of nephrons like this.

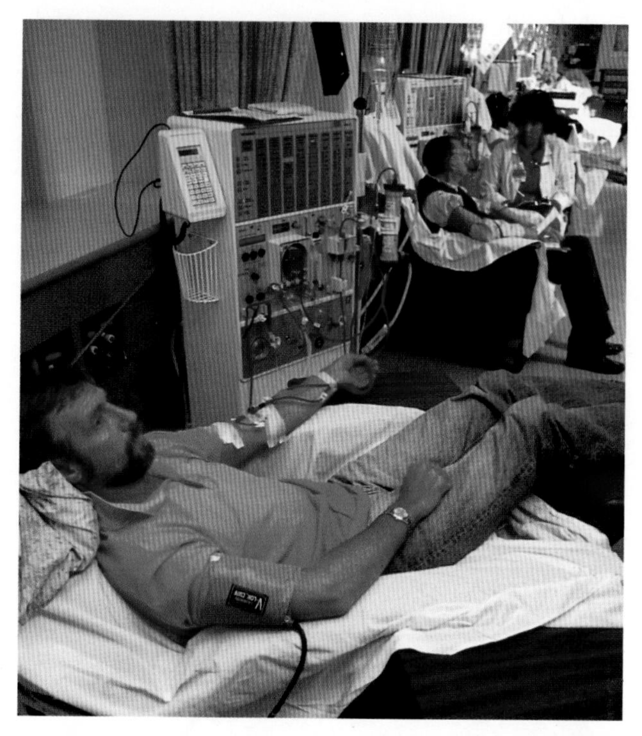

People whose kidneys do not work properly must have their blood 'cleaned' regularly by a ***renal dialysis machine*** like the one above. In the machine, the blood is pumped through a membrane tube. Excess water is pushed out through the membrane and unwanted dissolved substances diffuse across it into a liquid on the other side.

Controlling blood sugar

The body's 'fuel' is a sugar called glucose. It is made when food is digested, and carried in the blood to the tissues and organs which need it. In the liver, glucose can be changed into glycogen, stored, and converted back into glucose when needed. ***Insulin*** is the hormone which makes the liver change glucose into glycogen. It comes from glands in the pancreas.

Too much glucose is harmful. The diagram on the right shows how the blood sugar (glucose) level is automatically adjusted if someone has just eaten a big meal and their blood sugar level is too high.

People with ***diabetes*** do not produce enough insulin. They must control their own glucose level by carefully monitoring their food intake and injecting themselves with calculated doses of insulin.

Temperature control

The internal temperature of your head, chest, and abdomen is called your ***core temperature***. It must be kept close to 37 °C for the tissues and organs of the body to work properly.

Respiration releases heat. The heat is carried round the body by the blood and mainly lost through the skin. Sensors in the brain are constantly checking blood temperature to make sure that the body is not gaining or losing too much heat. On the right, you can see some of the factors affecting heat gains and losses, and how adjustments are made.

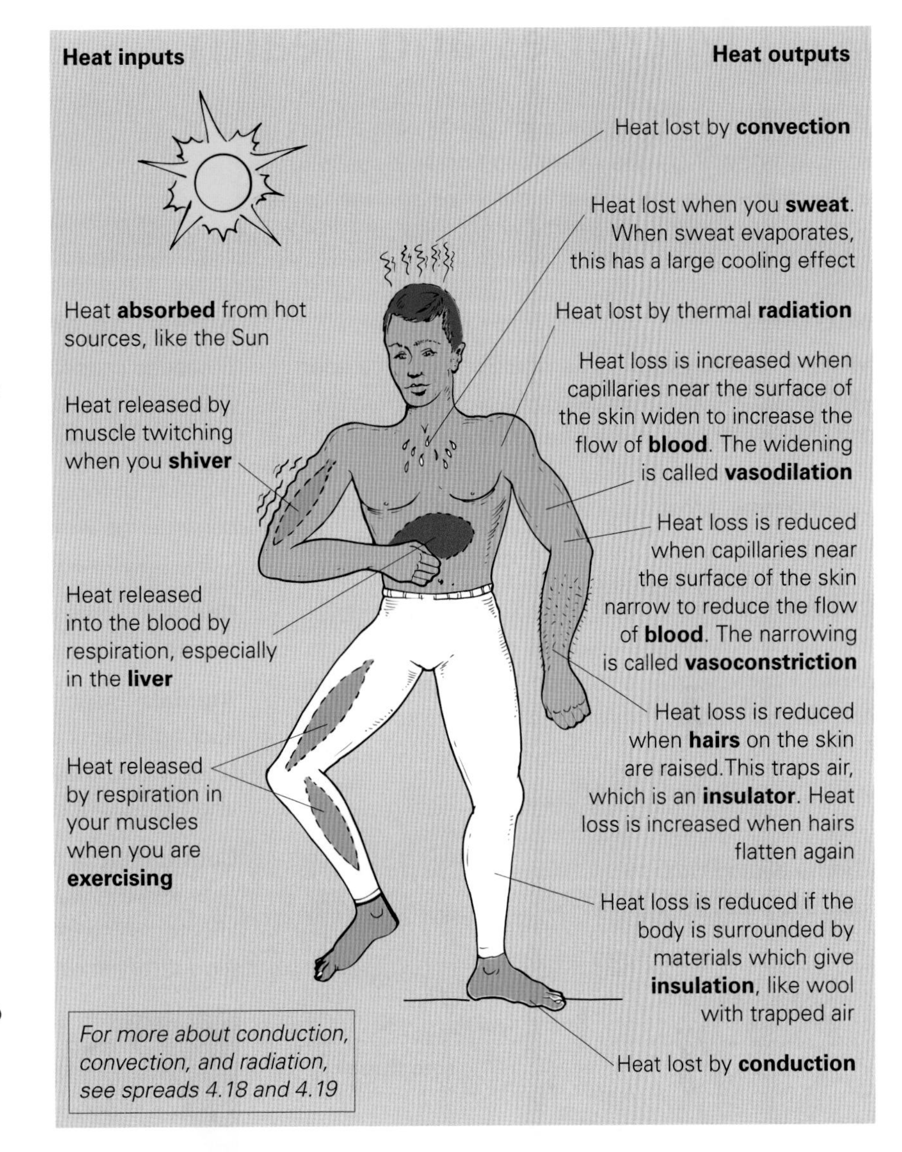

Feedback

In the system below, information on blood sugar (glucose) level is fed back to the liver, so that its glucose output can be adjusted. This is an example of ***feedback***. The body's water balance and temperature control systems also use of feedback. Each system is controlled by feeding back information about their output.

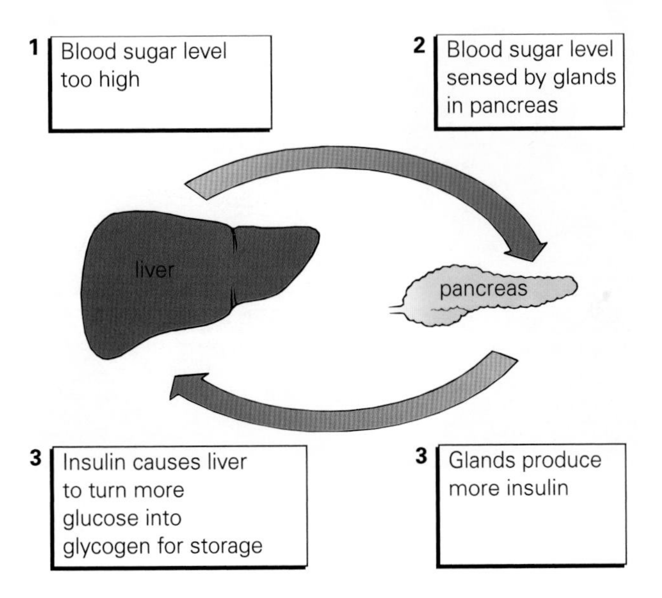

1. In what ways does the body **(a)** gain water **(b)** lose water?
2. What job is done by nephrons?
3. What is the effect of the hormone ADH?
4. Why do some people need to use a renal dialysis machine? What does the machine do?
5. What is the effect of the hormone insulin?
6. Explain how the body's blood sugar level is decreased if it starts to rise.
7. How does the blood flow near the surface of your skin change if your body is **(a)** too hot? **(b)** too cold?
8. Apart from a change in blood flow, how else can the body lose more heat when its temperature starts to rise?

2.14 Microbes and health

By the end of this spread, you should be able to:
- *explain how microbes cause and spread disease*
- *describe some of the body's defences against disease*
- *Describe the effect of AIDS and how the virus (HIV) can be spread.*

Microbes

Microbes are tiny organisms which can only be seen with a microscope. There are billions of them in air, soil, and water, and in our bodies. Some do useful jobs, but some are harmful. Harmful microbes are called ***germs***. Most diseases are caused by germs.

There are three main types of microbe:

Bacteria are living cells. If the conditions are right, they can multiply very rapidly by cell division. If harmful bacteria invade the body, they attack tissues or release poisons. They are the cause of sore throats as well as more serious diseases such as whooping cough, cholera, and typhoid.

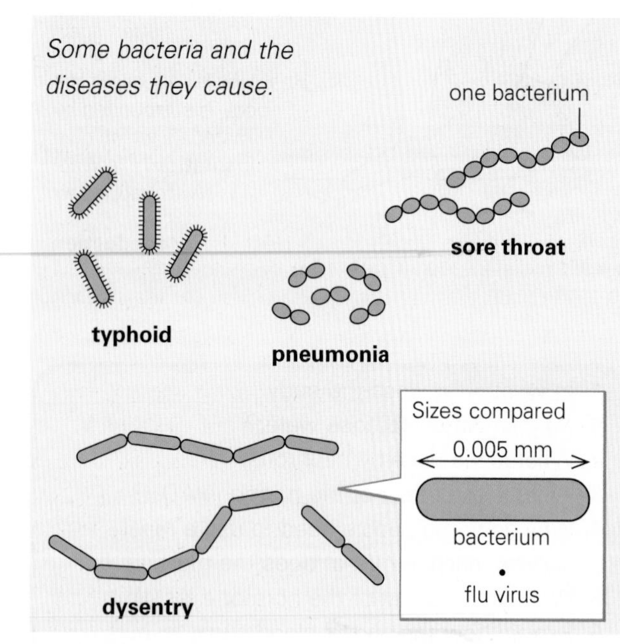

Some bacteria and the diseases they cause.

Viruses are much smaller than bacteria. Harmful viruses can invade living cells and upset the way they work. They are responsible for diseases such as flu, chicken-pox, and colds.

Fungi include moulds such as those which grow on old bread. Some skin diseases are caused by fungi, for example: athelete's foot and ringworm.

Spreading germs

Diseases caused by germs are called ***infections***. Here are some of the ways in which they can spread:

Droplets in the air When you cough or sneeze, droplets of moisture are sprayed into the air. They carry germs which are breathed in by other people. Colds and flu are spread in this way.

Contact Some diseases can be picked up by touching an infected person. Measles is one example.

Animals Insects can leave germs on food. Blood-sucking insects such as mosquitoes put germs in the blood when they bite. Malaria is spread in this way, by one type of mosquito.

Contaminated food Sewage is full of germs. If it gets into the water supply, food and drink may be affected. Also, people may contaminate food if they have dirty hands which are covered with germs.

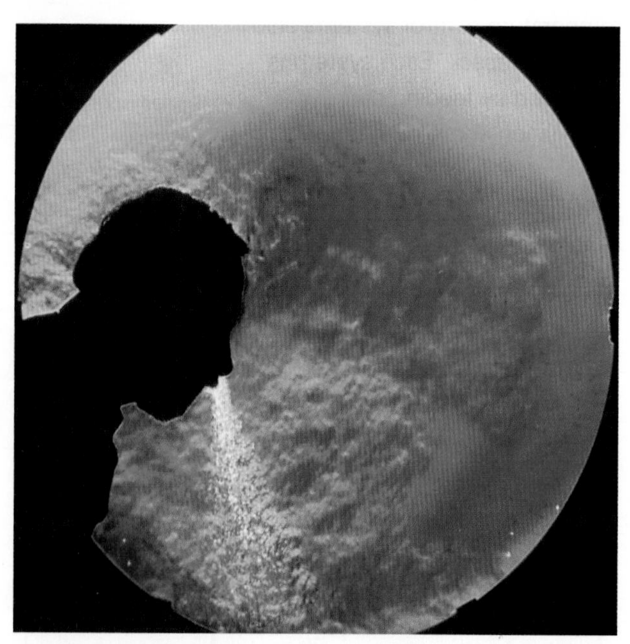

A violent sneeze. To take this photograph, a special technique was used so that air disturbances are seen as different shades and colours.

Fighting disease

Your skin stops some germs from entering the body. However, your body has an ***immune system*** for fighting invaders which do get in. Its 'soldiers' are your white blood cells. Some of these digest germs. Others make chemicals called ***antibodies*** which kill them.

Different antibodies are needed for different germs. But fortunately, your immune system has a memory. Once it has made antibodies of one type, it can make more of the same very quickly if there is another invasion. Once you have had, say, chicken-pox, you are unlikely to get it again. You have become ***immune*** to the disease. Unfortunately it is almost impossible to become immune to flu and colds. The germs keep changing, and there are many different types.

The body can be given extra help to fight disease:

Antibiotics are medicines which kill bacteria. However, they have no effect on viruses.

Vaccines contain dead or harmless germs which are similar to harmful ones. They are often given by injection. They make the immune system produce antibodies, so that the body's defences are ready if the proper disease ever attacks.

AIDS

AIDS stands for ***Acquired Immune Deficiency Syndrome***. It is a disease caused by a virus known as ***HIV***. People with the virus are ***HIV positive***. However, it may be many years before the full disease develops. There are treatments to slow its progress, but no known cure.

The HIV virus attacks white blood cells, so the immune system stops working. AIDS sufferers lose their defence against even mild diseases. Minor illnesses can kill them.

There are only three ways in which AIDS can be passed from one person to another:

- by sexual contact
- by blood-to-blood contact
- from an infected mother to her unborn child.

If a man wears a condom while having sex, this reduces the chances of the HIV virus passing between him and his partner.

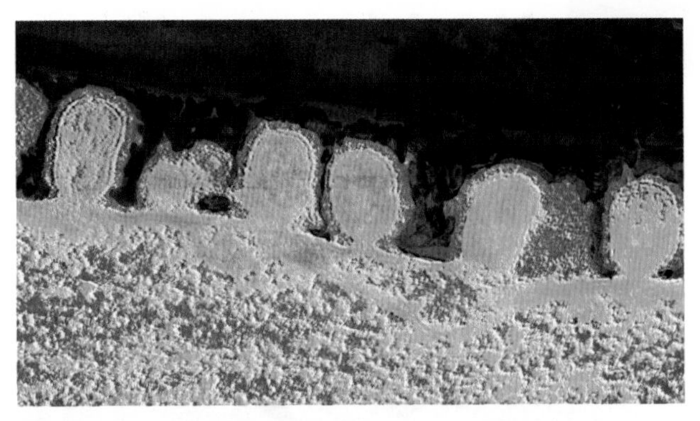

False-colour photograph of flu viruses leaving an infected cell: magnification ×27 000

More about bacteria and viruses

Animal cells have a nucleus, cytoplasm, and membrane (see Spread 2.01). Plant cells also have a cell wall. Bacterial cells are simpler than either of these.

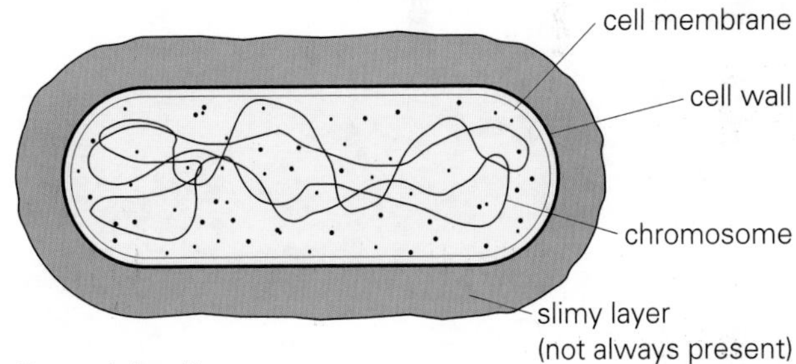

Bacterial cell

A typical bacterial cell is shown above. There is no nucleus. Instead, a single, coiled-up chromosome contains the chemical instructions (called ***genes***) which the cell needs to live and reproduce.

Viruses are little lumps of protein with a few genes inside. When they invade a living cell, they make it produce more and more copies of themselves.

1. Give *three* ways in which germs can be passed from one person to another.
2. Why is it important to wash your hands after going to the toilet?
3. How does your body deal with invading germs?
4. What are vaccines and what do they do?
5. How does AIDS affect the immune system?
6. There are only three ways in which the HIV virus can be passed on. What are they?
7. Give one difference between a bacterial cell and an animal or plant cell.

2.15 Health matters

By the end of this spread, you should be able to:
- describe the body's defences against disease
- describe how white blood cells fight germs
- explain why smoking, alcohol, drugs, and solvents can damage your health.

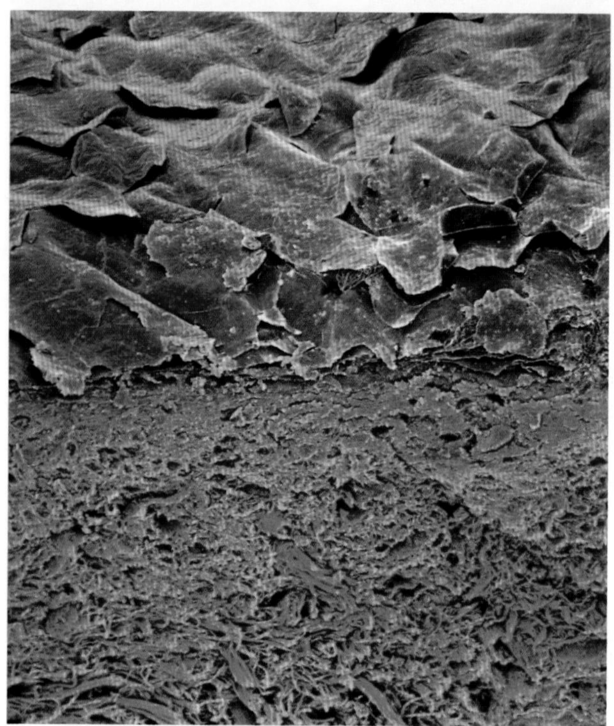

Two layers of human skin: magnification ×200

Natural barriers

As part of its defences, your body has natural barriers to keep out dirt and germs. For example:

Your skin has an outer layer of dead cells. Oils in this layer keep it flexible and waterproof, and also help kill germs. The dead cells gradually wear away, but they are replaced by new cells growing just beneath.

Dust and germs in the air you breathe in are trapped by a sticky ***mucus*** on membranes lining the nose, trachea (windpipe), and air passages in the lungs. Tiny hairs called ***cilia*** make movements which carry the mucus, germs, and dust to the back of the mouth (see Spread 2.01). There, they are swallowed. They pass out of the body with other waste matter.

If germs do get past the natural barriers and enter your body, the immune system is there to fight them.

The immune system in action

Below, you can see how your immune system uses white blood cells called ***phagocytes*** and ***lymphocytes*** to deal with invading germs (harmful microbes):

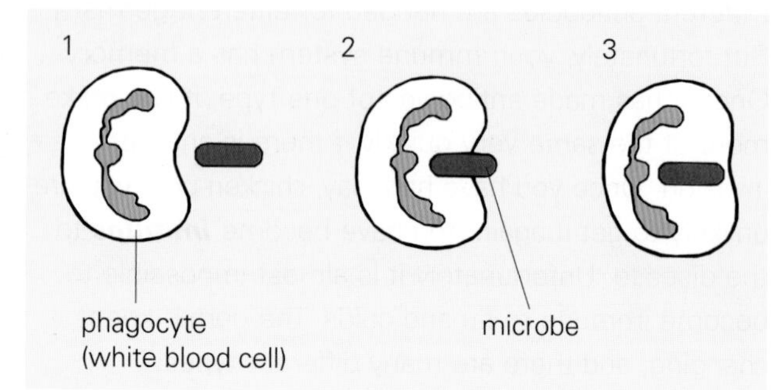

Phagocytes engulf and digest microbes, such as the germs which get into your blood when you cut yourself.

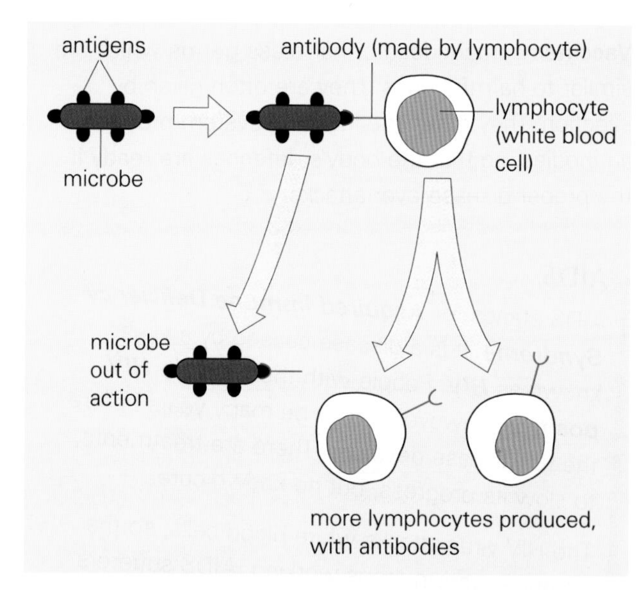

Lymphocytes detect chemicals called ***antigens*** on microbes. The antigens stimulate the lymphocytes into making and releasing antibodies. These chemicals stick to the antigens. They put the microbes out of action or make them easier for phagocytes to digest. When antigens are detected, your body makes lots of lymphocytes, all with the same antibodies. Some are stored, in case the same disease attacks again.

Some antibodies are ***antitoxins***. They are chemicals which the destroy the toxins (poisons) produced by microbes.

Health risks

To help you stay healthy, you need to eat sensibly, take plenty of exercise, and avoid health risks like these:

Cigarette smoke This contains tar, nicotine, and poisonous gases which irritate and damage the lungs (see below). Nicotine is a poison which damages the heart, blood system, and nerves. Smokers become ***addicted*** to it. They find it very difficult to give up.

Alcoholic drinks The alcohol in these is a chemical called ***ethanol***. Small amounts slow people's reactions. Too much makes them drunk. Years of heavy drinking can cause liver, brain, and heart disease.

Solvents These are used in glues, paints, and cleaning fluids. Some people sniff the fumes, but this can damage the heart and cause heart attacks.

Drugs Some drugs are medicines. But others are taken to make people feel happier, or more lively or aggressive. Some, such as ***heroin***, are very addictive. Addicts want regular injections. But dirty needles cause infections, and shared needles can transmit diseases such as AIDS.

More about smoking

If you smoke, you are at risk of the following:

Cancer Most lung cancer victims are smokers.

Heart disease Smokers are much more likely to suffer from heart disease than non-smokers.

Emphysema Lung tissue is destroyed, so breathing becomes more and more difficult.

Bronchitis Air passages inside the lungs become inflamed, and mucus collects in them. Smoking stops the cilia working properly to clear the mucus.

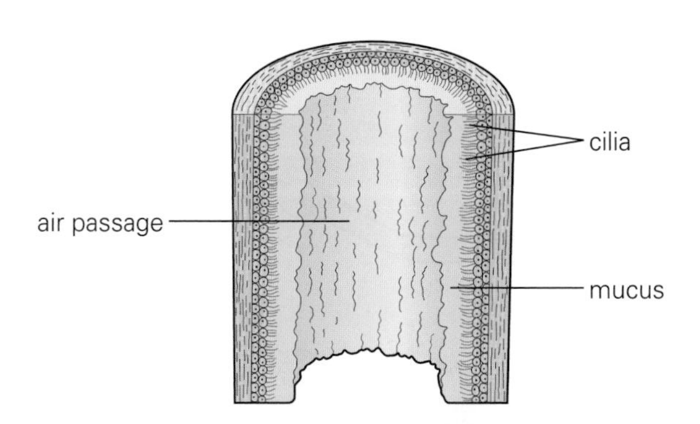

Part of an air passage in the lungs

1 How does your skin help to protect you against disease?
2 Your nose and other air passages are lined with mucus.
(a) How does mucus help protect you against disease?
(b) How does your body get rid of mucus when its job is done?
3 What do antibodies do?
4 What are antitoxins?
5 Give *two* ways in which white blood cells deal with invading germs.
6 Some substances are addictive.
(a) What does addictive mean?
(b) Give *two* examples of addictive substances.
7 Give *five* harmful effects of smoking.

2.16 Pass it on

By the end of this spread, you should be able to:
- *describe how living things show variation*
- *describe how characteristics depend on genes and on the environment*
- *explain some uses of selective breeding.*

Varying features

Your different features are called your ***characteristics***. Some, like eye colour, are easy to see. Others, like your blood group, are not so obvious. Many of your characteristics are passed on to you by your parents. They are ***inherited***.

No two people are exactly alike. Characteristics like height, weight, and eye and skin colour show ***variation***. Identical twins are more alike than most. But even they are not *exactly* alike (see photo).

Continuous variation Humans can be short, or tall, or any height in between. Height shows continuous variation.

Discontinuous variation Some people can roll their tongues, others can't. There is nothing in between. This is an example of discontinuous variation.

All animals and plants show variation, not just humans.

Genes

A complicated set of chemical instructions is needed to build a human body. Nearly every cell in your body has these instructions. They are stored in the nucleus, in 23 pairs of thread-like ***chromosomes***. Small sections within these chromosomes are called ***genes***. The study of genes is called ***genetics***. You have about 30 000 genes altogether. Each gene carries the chemical instructions for a different characteristic.

Genes normally work in pairs. One gene in each pair is inherited from each parent. For example, you may have inherited a gene for black hair from your mother and a gene for blond hair from your father. Only one of these genes can control your hair colour. A black hair gene is ***dominant*** over a blond hair gene, so you end up with black hair.

Human cell

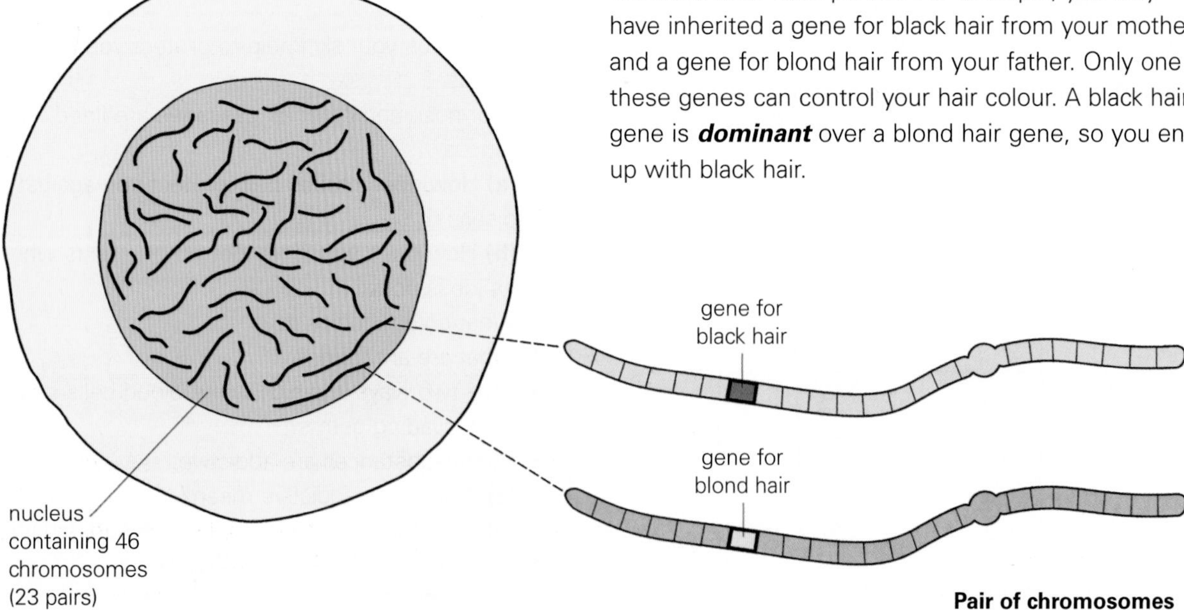

Pair of chromosomes

Each of your parents also inherited two genes for each characteristic. However, they only passed on one of the two to you. Which one was a matter of chance. Unlike other cells, sperms and ova (eggs) only carry one gene from each pair. But when a sperm combines with an ovum, they make a new cell with a full set of genes. When an ovum is fertilized, millions of gene combinations are possible. That is partly why people can vary so much, even in the same family.

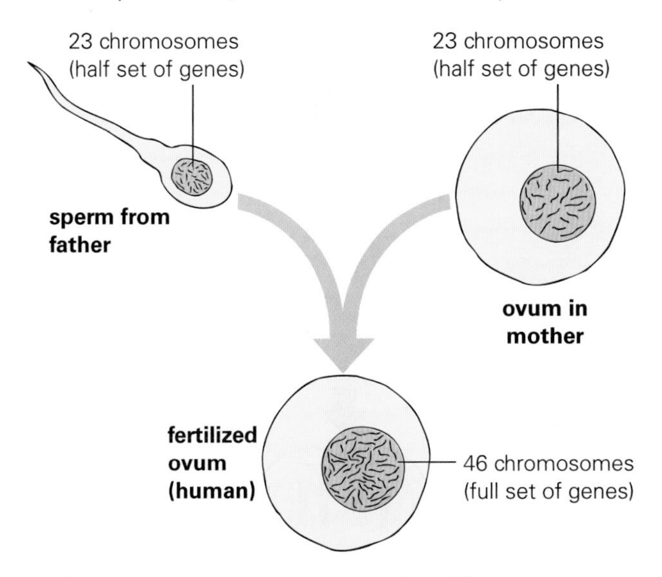

In humans, each sperm or ovum has 23 chromosomes: this is called the ***haploid number***. Other cells have 46 chromosomes, called the ***diploid number***.

Genes and the environment

Identical twins inherit exactly the same set of genes. Yet one may be heavier than the other because he or she eats more. So, your characteristics depend partly on your genes and partly on your ***environment*** (your conditions and surroundings). This is true for other living things as well. For example, two plants may have the same genes. But one may grow better than the other because it has more light, or a better supply of water and minerals from the soil.

Selective breeding

People often try to breed animals with special characteristics: for example, sheep with plenty of wool, or horses that can run fast. To do this, they select the animals which will be mated. This is called ***selective breeding***. The idea is that the offspring ('babies') may inherit the best features of both parents. But chance still affects the result. If two champion racehorses mate, their offspring will not necessarily be a champion.

Selective breeding is also used with plants. For example, one variety of wheat may grow faster than another, or be more resistant to disease. By controlling how the wheat is pollinated, scientists can breed varieties with the characteristics they want. This can also be done using ***genetic modification*** (see 1.04 and 2.19), but the technique is controversial.

Farmers try to preserve rare breeds of farm animal in order to save their genes. If disease strikes a common breed, all the animals might die. However, if a rare breed has genes which make it resistant to the disease, it can be used to improve the first breed.

Genes worth saving: Dartmoor Greyface Sheep from the Cotswold Farm Park, Gloucestershire.

1 Give *two* examples of continuous variation.
2 Give *two* examples of discontinuous variation.
3 What are genes? Where are they in your body?
4 Genes are in pairs. Where do the genes in each pair come from?
5 **(a)** What makes 'identical' twins look alike?
(b) Why are 'identical' twins not exactly alike?
6 Give an example which shows that your characteristics can depend on your environment as well as on your genes.
7 Give *two* examples of the use of selective breeding.
8 Why is it important to preserve rare breeds of animal or rare varieties of plants?

2.17 Genes in action

By the end of this spread, you should be able to:
- *explain how genes are inherited, and the effects which different combinations can produce*
- *describe how sex cells are produced.*

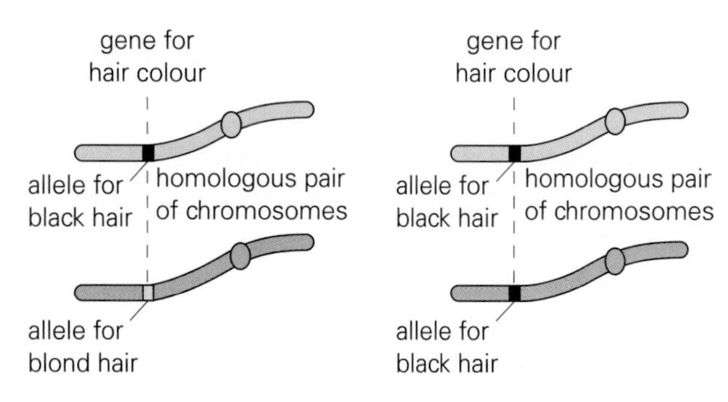

The pairs of chromosomes above are from different people:

Chromosomes A are identical in size and shape. They are a ***homologous pair***. Genes for the same characteristic (for example, hair colour) are in matching positions along them. However, the gene for hair colour is present in two versions – black and blond. Different versions of a gene are called ***alleles***.

Person A will have black hair because the allele for black hair is ***dominant*** over that for blond hair. The allele for blond hair is ***recessive***.

With some characteristics, neither allele is dominant. Each has an effect. In some animals, for example, a brown hair allele and a white hair allele will produce light brown hair.

Chromosomes B are also homologous. Here, however, the alleles for hair colour are the same. Both are for black hair, so person B will have black hair.

Phenotypes and genotypes

A characteristic which is seen, like black hair or big muscles, is called a ***phenotype***. The combination of alleles which produces it is called a ***genotype***. Genotypes are inherited. But phenotypes can depend on conditions and behaviour – for example, how much exercise you take. Identical twins have the same genotypes but different phenotypes.

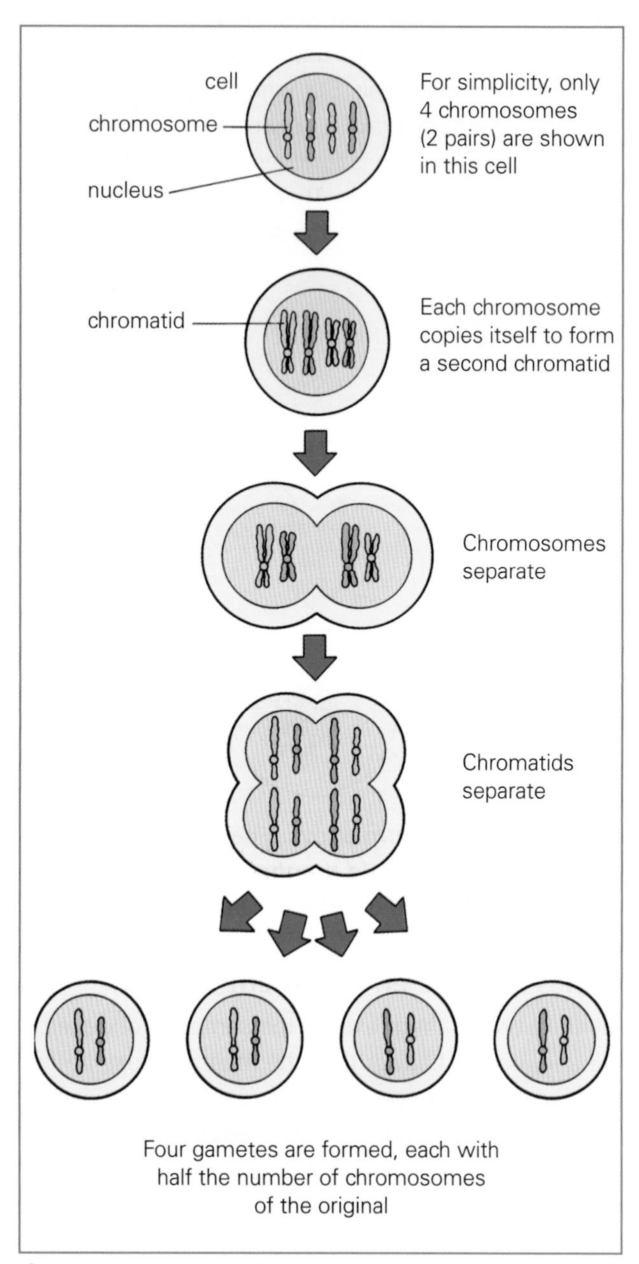

Cell division by meiosis

Cell division – meiosis

When a sperm combines with an ovum, a single cell called a ***zygote*** is formed. A baby develops from this. Sperms and ova are called ***gametes***. Above, you can see how these sex cells are made by cell division. The process is known as meiosis. Each gamete receives one allele from each pair in the original cell. But which one is a matter of chance.

Inheriting genes

The charts on the right give examples of how genes are inherited. Capital letters stand for dominant alleles, small letters for recessive ones.

In chart 1, the mother has two alleles (HH) for black hair. The father has two alleles (hh) for blond hair. When a sperm and ovum join, there are four possible combinations. All give a zygote with Hh alleles. H is dominant, so the child will have black hair.

In chart 2, both parents have Hh alleles and black hair. Again, there are four possible combinations. Three give black hair. The fourth gives blond hair. So there is a 1 in 4 chance that the child will have blond hair, even though both its parents have black hair.

These are simplified examples. Characteristics are usually controlled by more than one pair of alleles.

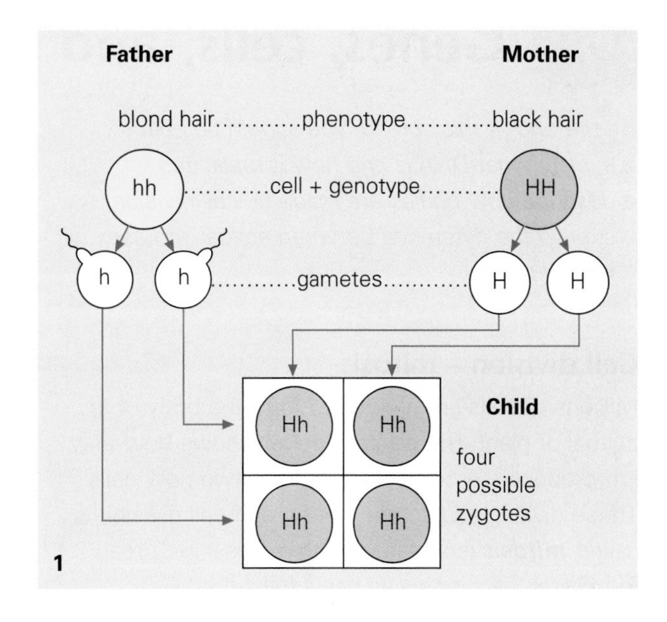

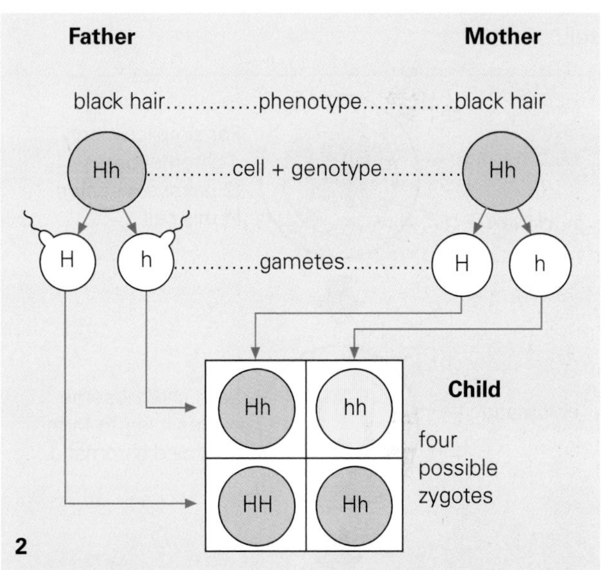

Male or female?

In human cells, 22 pairs of chromosomes are homologous. But the 23rd pair do not always match. These are the ***sex chromosomes***. They include genes which control whether someone is male or female. Females have two **X** chromosomes per cell. Males have an X chromosome and a shorter **Y** chromosome.

In males, meiosis produces about equal numbers of X and Y sperms. In females, it produces ova with one X chromosome each. The chart below shows the four possibilities when a sperm combines with an ovum. Two give males and two give females. So there is the same chance of the child being a boy as a girl.

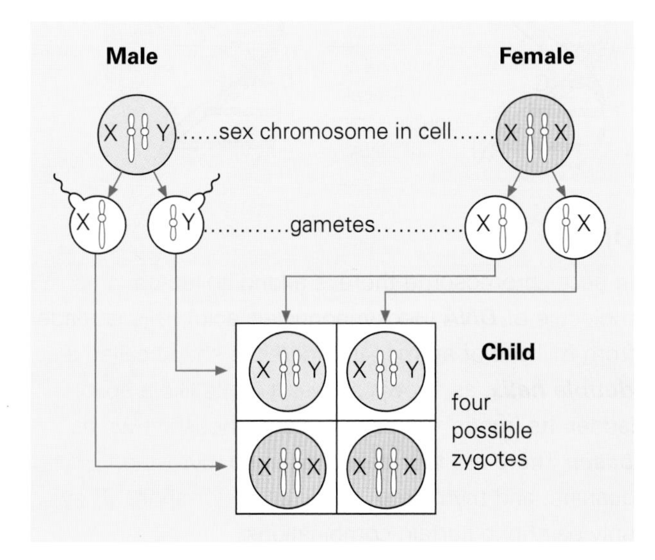

1 What is the difference between a phenotype and a genotype? Give an example of each.

2 *alleles meiosis zygote homologous pair*
Which of the above means **(a)** cell division in which gametes are formed? **(b)** different versions of a gene? **(c)** cell formed when male and female gametes combine?

3 How many chromosomes are there in **(a)** a human gamete? **(b)** a human zygote?

4 What is the sex of someone whose sex chromosomes are **(a)** XX? **(b)** XY?

5 A mother has blond hair (genotype hh). A father has black hair (genotype Hh). What are the chances of their child having blond hair?

2.18 Genes, cells, and DNA

By the end of this spread, you should be able to:
- *explain what DNA is and how it replicates*
- *describe how bodies are made by cell division*
- *explain the difference between sexual and asexual reproduction.*

Cell division – mitosis

Millions of cells are needed to form the body of an animal or plant. The diagram below shows how they are produced. A cell divides to form two new cells. These divide...and so on. This type of cell division is called ***mitosis*** (compare it with meiosis in Spread 2.17.) In mitosis, each new cell has a full set of chromosomes which are copies of those in the original cell.

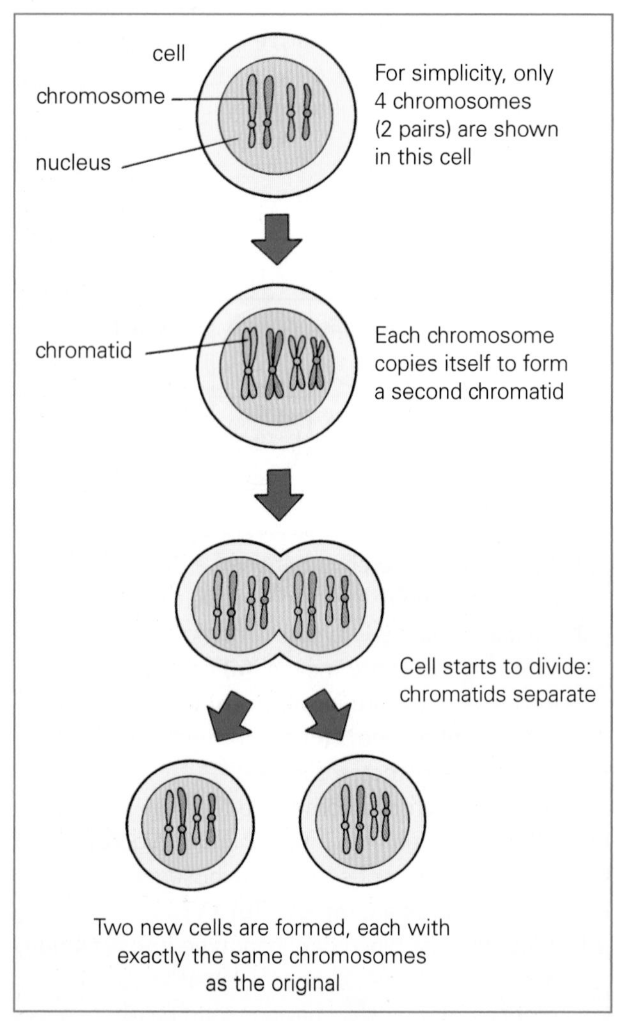

Cell division by mitosis

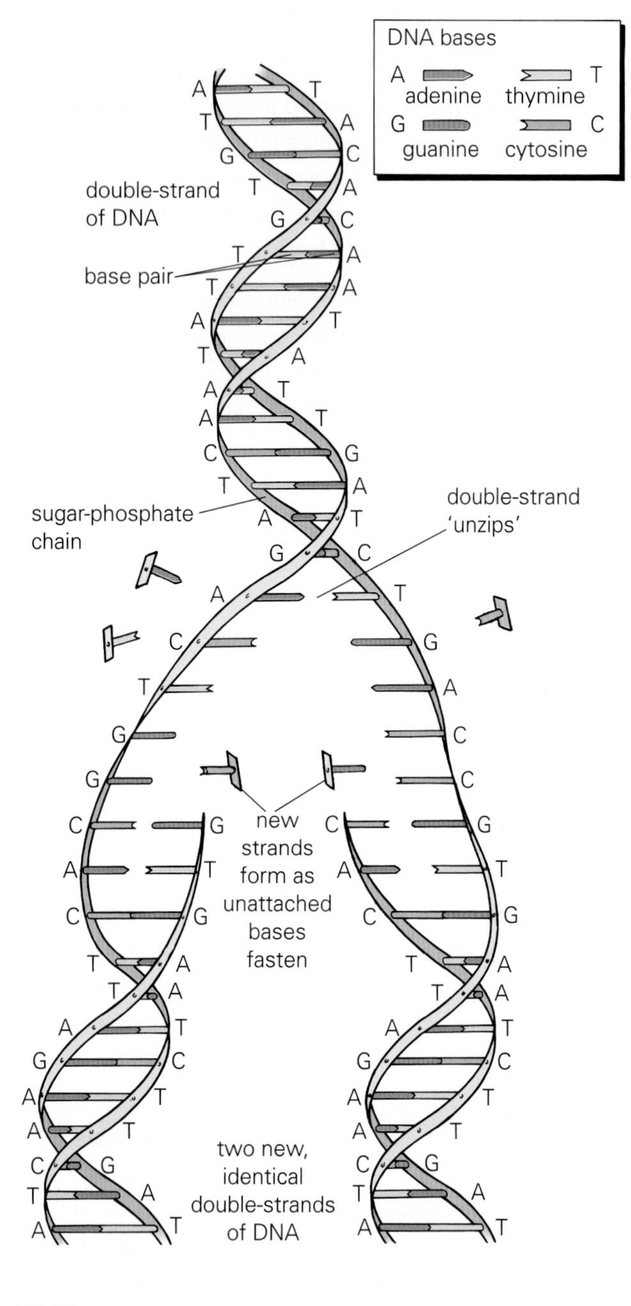

DNA

In each chromosome there is a long coiled-up molecule of ***DNA*** (deoxyribonucleic acid). This is made from millions of atoms, arranged in a shape called a ***double helix***, as shown above. It looks like a spiral ladder. Its 'rungs' are pairs of substances known as ***bases***. There are four bases: adenine, cytosine, guanine, and thymine (A, C, G, and T for short). They only pair up in certain combinations.

Look at either half of the DNA ladder and you will see that the order of the bases varies. The sequence forms a set of coded instructions for building the complete ***organism*** (animal or plant). Genes are different sections of this sequence. The full set of genes in all the chromosomes is called the ***genome***.

Translating the code

Cells are largely built from ***proteins***. These are made from twenty ***amino acids***, linked in different combinations. DNA stores information about which amino acids must be produced in each type of cell. The code works like this:

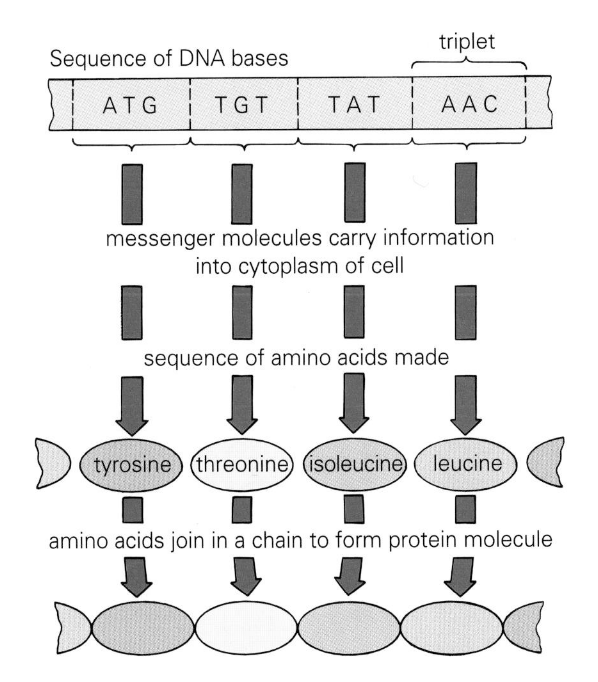

Along a strand of DNA, each sequence of three bases is called a ***triplet***. Different amino acids are represented by different triplets. 'Messenger' chemicals carry information to the cytoplasm (the 'factory' part of the cell). There, the proteins are synthesized (made) from incoming materials.

Copying the code

During cell divison, copies of chromosomes are made. This is possible because a DNA molecule can replicate (copy) itself as shown on the opposite page. First, the DNA molecule starts to 'unzip'. Then new 'pieces' of chemicals become attached to the two halves so that two identical molecules of DNA are formed.

Sexual and asexual reproduction

For humans to reproduce, a sperm must combine with an ovum (egg). Reproduction like this, involving male and female gametes, is called ***sexual reproduction***. Each baby inherits half its DNA (and therefore half its genes) from each parent. This produces a wide variety of individuals who are different from their parents. At every new generation, the genes are reshuffled.

Flowering plants reproduce sexually (see Spread 2.04). However, some also reproduce in another way: they develop parts (such as bulbs and runners) which separate and grow as new plants. This is called ***asexual reproduction***. Each new plant is genetically identical to its parent because it is produced only by mitosis. Gametes are not involved.

Asexual reproduction

1 Cells can divide by *mitosis* or by *meiosis* (see Spread 2.17). Which type of cell division
(a) produces cells with the same number of chromosomes as the original cell?
(b) is used to produce sex cells (gametes)?

2 **(a)** Where is DNA found?
(b) What is DNA used for?

3 What do amino acids form when they join in a chain?

4 How does a DNA molecule replicate itself?

5 Compared with its parent(s), what would you expect a plant to be like if it had been produced by **(a)** sexual **(b)** asexual reproduction?

2.19 More genes in action

By the end of this spread, you should be able to:
- *explain the results of cross-breeding plants*
- *describe what genetic diseases are*
- *explain what cloning and genetic engineering are.*

Breeding generations

Remember: P stands for a dominant allele, p for a recessive allele (see Spread 2.17). An animal or plant is an ***organism***.

Growers often cross-pollinate plants (see Spread 2.04) to produce different characteristics. On the right, a grower crosses a green pod (PP) plant with a yellow pod plant (pp) to produce a new generation of plants with Pp alleles. Then the grower crosses these Pp plants with each other to produce another generation. On average, 3/4 of the new plants have green pods (PP or Pp) and 1/4 have yellow pods (pp).

If alleles are the same (PP for example), then scientists say that the organism is ***homozygous*** or ***pure bred*** for that characteristic. If alleles are different (Pp for example) then the organism is ***heterozygous*** or ***hybrid***. On the right, the grower crosses two pure bred varieties to produce a generation of hybrids. This is the ***F_1 generation***. Crossing these plants produces the ***F_2 generation***.

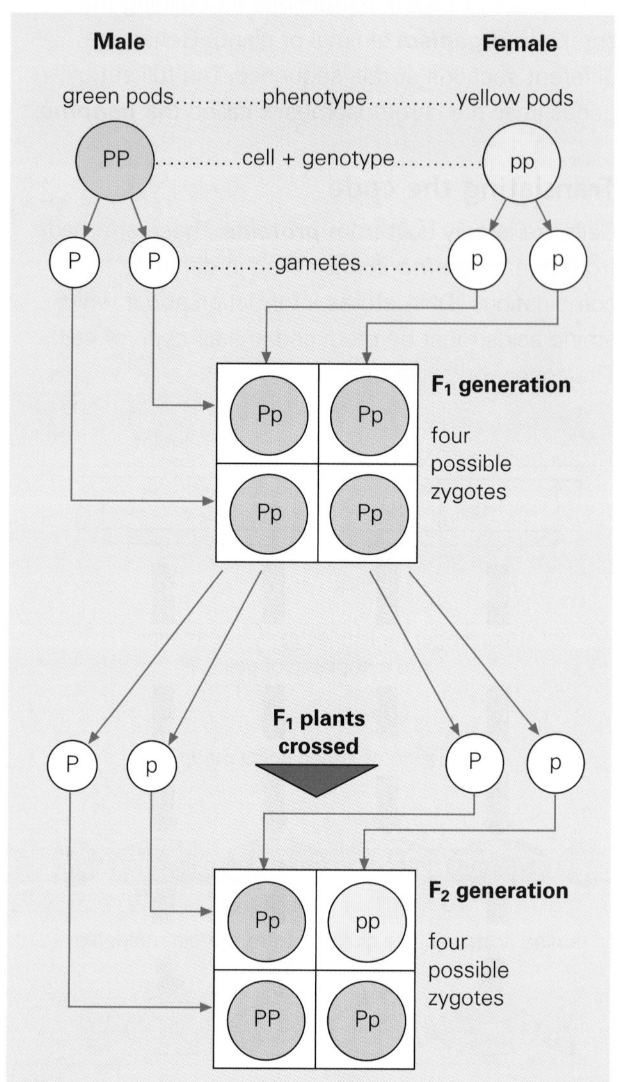

Cloning

When gardeners take cuttings and grow them, the new plants are genetically identical to the original. This is useful if, say, cross-breeding produces a pure bred disease-resistant plant and the grower then wants lots more plants with the same characteristic.

Making genetically identical copies is called ***cloning***. Here are three ways of cloning:

- Taking cuttings from plants (as on the left)
- Taking a group of cells and growing these into a larger organism. This is called ***tissue culture***.
- Splitting the first small group of cells (for example, when an animal embryo starts to develop).

Genetic diseases

Some diseases are inherited, and caused by faulty genes. ***Genetic diseases*** in humans include cystic fibrosis, sickle-cell anaemia, and haemophilia.

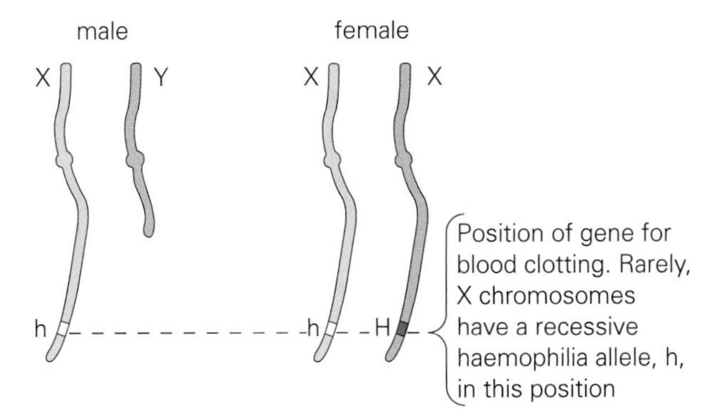

If someone has haemophilia, their blood does not clot properly. The condition is caused by a rare, recessive allele on the X chromosome. Most X chromosomes have a dominant, non-harmful allele in that position. A female has two X chromosomes, so the chances of both carrying the harmful allele are slight. However, a male has only one X chromosome, so if this carries the haemophilia allele, the person will have the disease. This means that virtually all haemophilia sufferers are male. However, a mother can be a carrier. She can pass the allele on to her children though not being haemophilic herself.

Scientists say that haemophilia is ***sex linked*** because it affects one sex more than the other. Colour blindness is another sex linked condition which affects mainly males.

1 If someone is homozygous for hair colour, what does this tell you about their alleles?
2 What is cloning? Give an example.
3 Why are males more likely to have haemophilia than females? How can a female be a carrier?
4 How can bacteria be genetically engineered to produce substances normally made in the human body?
5 A smooth-seed (SS) pea plant is crossed with a wrinkled-seed (ss) pea plant. What are the possible zygotes for the F_1 and F_2 generations? What proportion of each generation would you expect to have wrinkled seeds?

Genetic engineering

Scientists have discovered how to transfer genes from one organism to another. Using enzymes, they can 'cut' pieces from one strand of DNA and insert them into the DNA of a different organism. Techniques like this are called ***genetic engineering***.

Some drugs and hormones can be produced by genetic engineering. For example, the diagram below shows how genetically engineered bacteria can be used to make human insulin. (Diabetics need a supply of insulin because their own bodies do not make enough. Where bacteria are not used for insulin production, the insulin is obtained from pigs or cows.)

With genetic engineering, it may be possible to replace the harmful genes that cause genetic diseases. However, genetic engineering is controversial. For example, foods made from ***genetically modified (GM)*** crops have been welcomed by some but opposed by others. For more on GM foods, see Spread 1.04.

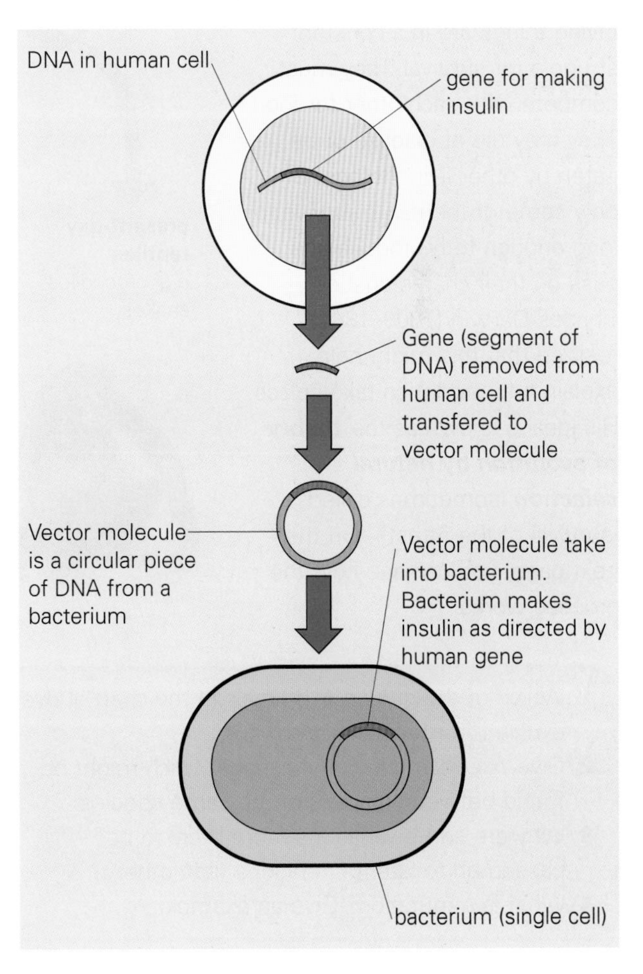

2.20 Evolution

By the end of this spread, you should be able to:
- *describe how natural selection works*
- *explain what mutations are and how they occur.*

There are over a million different types of living thing on Earth. Each type is called a ***species***. In time, as generation follows generation, a species can gradually change. This process is called ***evolution***.

If members of a species begin to live and breed in separate groups, evolution can take a different course with each group and new species may develop. For example, the chart on the right shows some of the stages in the evolution of today's mammals.

Scientists think that all species existing today (or in the past) evolved from simple life-forms which first developed on Earth more than 3000 million years ago.

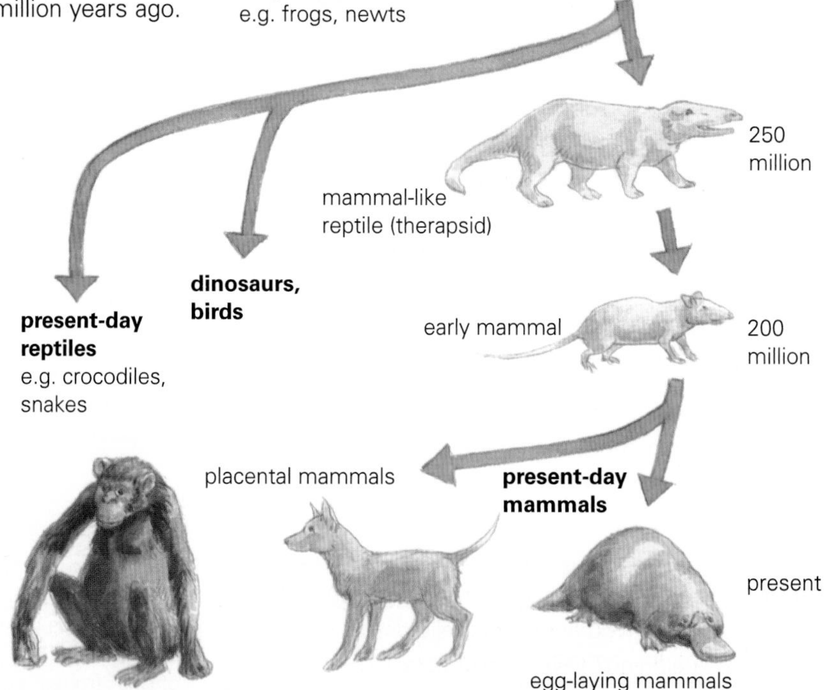

Natural selection

Living things are in a constant struggle for survival. They must compete with each other for food. They may die of disease or be eaten by other living things. So only some of them will survive for long enough to reproduce and pass on their characteristics. Charles Darwin (1809–1882) realized that this might help explain how evolution takes place. His idea is known as the ***theory of evolution by natural selection*** (sometimes called 'survival of the fittest'). On the next page, you can see how the process works.

1 Which of the groups of animals in the chart above evolved from early reptiles?
2 Give *two* examples of variations which might be found between animals of the same species.
3 Why are some variations more likely to be passed on to later generations than others?
4 What is a mutation? Give an example.
5 What *two* things make mutation more likely?
6 In the photographs opposite, which moth is better adapted to its environment? Why?
7 Why, in areas affected by factory soot, are there more dark peppered moths than light ones?
8 Why is it possible for several species to evolve from a single species?

Variation In any species, individuals vary in colour, weight, height, shape, and other features. (Sexual reproduction produces more variations than asexual reproduction.)

Selection Some variations help in the struggle for survival and some do not. So some individuals survive to reproduce and some do not. In this way, nature 'selects' living things for survival.

Adaptation The survivors pass on their characteristics to later generations. So the species evolves to cope with its living conditions. It becomes adapted to its environment.

Natural selection in action

Peppered moths live in woodlands and are eaten by birds. There are light and dark varieties, as on the right. In the 1850s, moths with the allele for dark colouring were rare. But then soot from factories started to blacken tree trunks. By 1895, 98% of the peppered moths in Manchester were dark. They were difficult for birds to see, and survived to reproduce. So the allele for dark colouring was passed on to more and more moths in later generations.

Mutations

Sometimes, genes can change at random. These changes are called ***mutations***. They happen when DNA is copied inaccurately during cell division.

Chemicals or radiation (see Spread 4.30) can greatly increase the chances of mutations.

Most mutations are harmful, but some may be helpful. The peppered moth's allele for dark colouring was a mutation which turned out to be helpful when the moth's environment changed.

The ***gene pool*** is the complete range of genes and their alleles in a population. Mutations add to the gene pool, so they increase variation.

2.21 Evolving ideas

By the end of this spread, you should be able to:
- *describe some of the evidence for evolution*
- *explain how genetics has improved our understanding of evolution.*

Before developing his theory of evolution, Darwin spent many years studying fossils, as well as living plants and animals. With the idea of natural selection, he could explain why evolution takes place. However, variation is needed for natural selection to operate, and Darwin was unable to explain how variation occurs. That problem was only solved when the basic principles of genetics had been established.

Fossil evidence

Fossils are the remains or traces of ancient animals or plants (or microbes) found in rocks. They provide evidence of the changes that have taken place over many millions of years as species have evolved.

Here are some of the ways in which fossils can form:

- Hard parts of animals, such as shells and bones, remain when softer body tissues have rotted away.
- Parts of the body, trapped in the ground, are gradually replaced by hard minerals. In this way, a 'stone copy' of the animal or plant is formed.
- Footprints or other tracks are left in sediment which later hardens into rock.
- Rarely, parts of an animal or plant (or even a whole body) may be preserved because the conditions for decay are not present. For example, the remains of mammoths have been found frozen in ice.

Adaptation or extinction

From his fossil studies, Darwin realized that a huge number of species had died out in the past. They had become ***extinct***.

Although, through natural selection, species tend to become adapted to their environment, that environment can sometimes change too rapidly for them. For example, the climate may change or new predators may emerge. However, while the new conditions may lead to the extinction of some species, they may provide new opportunities for others to flourish.

Fossil of an ammonite, found in rocks in Germany. The original creature lived about 150 million years ago.

Dinosaurs died out 65 million years ago. Their extinction may have been caused by climate changes, produced when a huge meteorite hit the Earth.

Mendel and the birth of genetics

In Darwin's day, most scientists assumed that offspring ('babies') inherited a blend of their parents' features. This caused problems for Darwin's theory. If offspring were an 'average' of their parents, there would never be enough variation for natural selection to work.

The solution to this problem came from the work of Gregor Mendel, an Austrian monk. In the 1860s, Mendel discovered the basic principles of inheritance, by carrying out pollination experiments on pea plants. Unfortunately, his ideas were ignored at the time, and Darwin never found out about them.

An experiment similar to one of Mendel's is described at the start of Spread 2.19. Two stages of pollination produce a 3:1 ratio of green to yellow pods. From ratios like this, Mendel concluded that the plants carried two 'factors' of inheritance for each characteristic. One was dominant and caused the feature seen. These factors would later be called genes. They could produce much more variation than blending.

If inheritance worked by blending features, crossing red flowers with white flowers would produce pink flowers. But that doesn't happen.

1 Giraffes' long necks help them reach their food. How is the development of longer necks explained by **(a)** Lamarck's theory? **(b)** Darwin's theory?
2 Give one reason why Lamarck's theory is not generally accepted today.
3 Give one reason why Darwin's theory was so strongly opposed when it first appeared.
4 Read the section in Spread 1.04 on antibiotics. Use your ideas about natural selection to explain why the over-use of antibiotics has led to the appearance of antibiotic-resistant 'superbugs'.

Milestones

1750 Most people believe that different species are unrelated and unchanging. However, as more and more fossils are found, some scientists begin to think that evolution might be possible.

1809 Lamarck puts forward a theory of evolution to explain how a species can change. Animals might acquire certain characteristics because of the way they live. For example, giraffes could develop longer necks by stretching to reach their food. Such characteristics might then be passed on to their offspring ('babies'). Lamarck's theory seems reasonable, but no evidence emerges to support it. Within a century, it will be known that acquired characteristics are not passed on.

1859 Darwin puts forward his theory of evolution. His views are opposed by some in the Church who believe that humans are God's special creation. They cannot accept that humans might share the same ancestors as apes or other animals. In time, the Church drops its opposition to Darwin, who, on his death in 1882, is buried in Westminster Abbey.

1866 Mendel puts forward his theory of inheritance.

1909 Johannsen develops Mendel's ideas and uses the term 'gene' for the first time.

1933 Brachet shows that animal cells and plant cells contain DNA.

1937 Dobzhansky applies the principles of genetics to Darwin's theory of evolution. His work helps explain how variation occurs and how natural selection produces changes.

1944 Avery, McCarty, and MacLeod find evidence that DNA is the chemical that carries genetic information.

1953 Crick and Watson discover the structure of DNA.

1985 DNA analysis begins to show how closely different species are related.

2000 The Human Genome Project produces the first 'map' of a full set of human genes.

2.22 Living together

By the end of this spread, you should be able to:

- *describe some of the factors which affect living things and their environment*
- *explain how living things are adapted to their environment.*

The place where an animal or plant lives is called its ***habitat***. It is usually shared with other animals and plants. All the organisms (living things) in one habitat are called a ***community***. Their ***environment*** is everything around them which affects their way of life. Together, a community and its environment are known as an ***ecosystem***.

Here are some of the factors which affect an organism's environment:

1 Explain what these words mean:
habitat *population* *community* *predator*

2 Give an example of how an animal's environment can change from one part of the day to another.

3 Give an example of how a plant's environment can change from season to another.

4 How do plants compete with each other?

5 If an animal is *adapted to its environment*, what does this mean? Give an example.

6 In the graph on the opposite page
(a) why does the slug population fall?
(b) why does it then start to rise again?

7 What would happen to the number of toads if a chemical killed off half of the plants? Why?

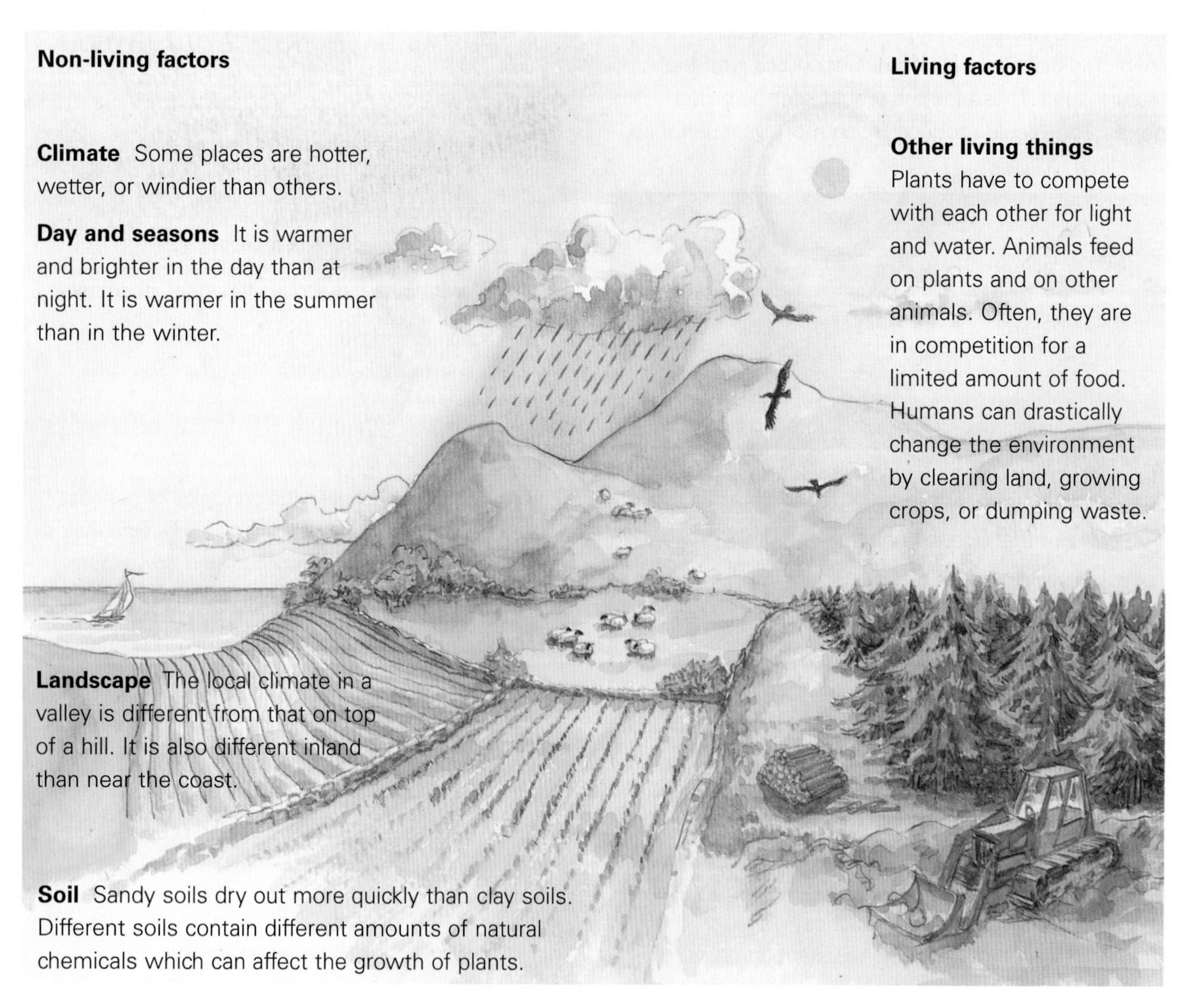

Adapted for living

Over many millions of years, animals and plants have developed special features to help them cope with their way of life. They have become ***adapted*** to their environment. Here are some examples:

- Dormice and many other small mammals hibernate in the winter months so that they can survive when food is scarce. They go to sleep with their life processes slowed right down.
- Hawks have claws and a beak which are specially shaped for gripping small animals and tearing them apart.
- Many trees lose their leaves in the autumn. This means that they do not need to take up so much water during the months when the ground might be frozen.

Adapted for hunting: the chameleon has a long tongue which it can flick out to catch insects

Changing populations

A group of animals of the same kind is called a ***population***. Animals depend on plants or other animals for their food. So a change in one population may affect several other populations.

Animals which feed on other animals are called ***predators***. The animals they eat are their ***prey***.

Imagine a garden with a stream running through it. This is a habitat for toads, slugs, and plants. The toads feed on the slugs, and the slugs feed on the plants. Normally, the numbers of toads, slugs, and plants will reach a balance. The graphs below show what happens if too many toads start to develop. Over a period of time, the balance is restored.

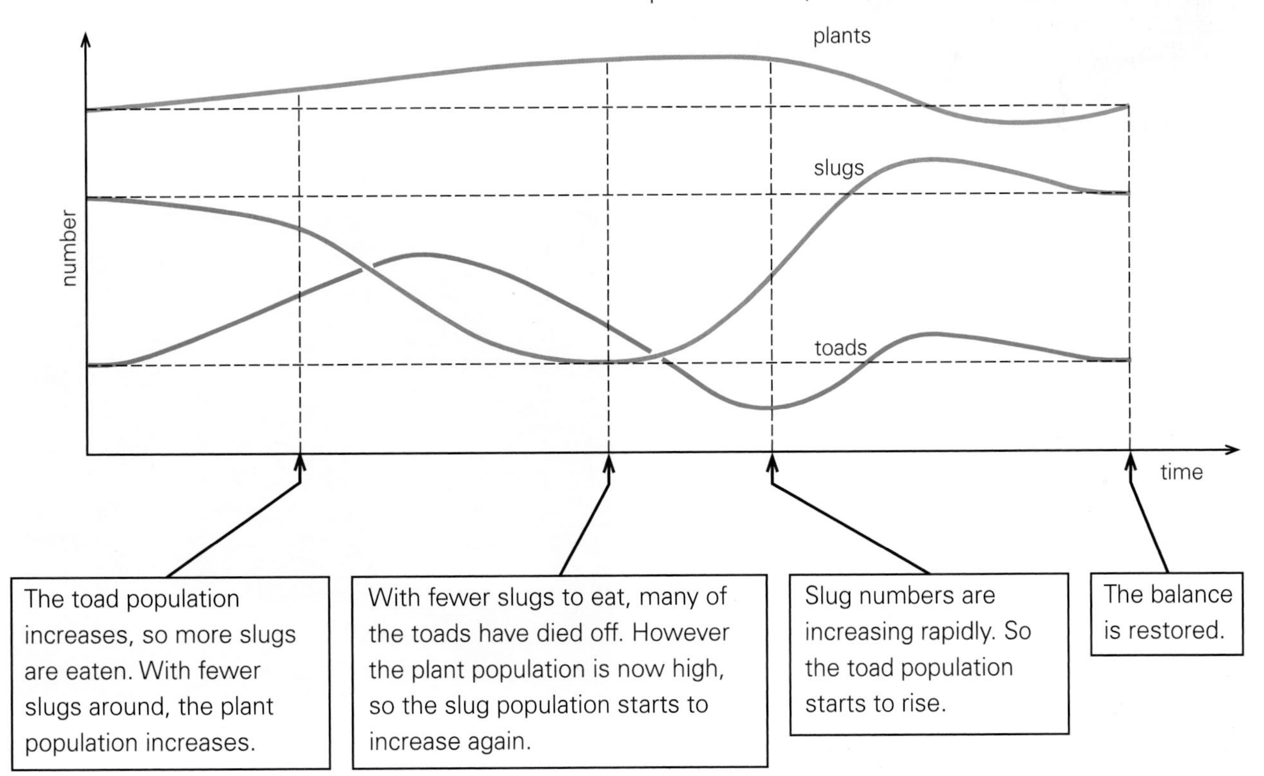

2.23 Populations and problems

By the end of this spread, you should be able to:
- *explain what limits the size of a population*
- *describe how human activity can affect other populations and the environment.*

Animals eat plants or other animals, and compete for food and shelter. So the size of one population affects others (see Spread 2.22). But humans and their activities have the biggest effects of all.

Population limit

Harvest mouse

There is a limit to the size of a population. This example shows why:

Harvest mice feed on grain. If a few mice enter a new cornfield, they have plenty of food, so they breed quickly and the population grows. But as more mice need food and shelter, survival becomes more difficult. Eventually, mice die at the same rate as new ones are born and the population stops growing. Here are some of the factors which limit it:

Predators Mice may be caught and eaten by predators such as foxes and hawks.

Food supply and shelter These may be limited.

Competitors Other animals may be competing for the same food or nesting places.

For more information about some of the topics in the chart on the right, see the next spread, 2.24.

The need for protection

The world's human population is growing all the time. As it does so, it needs more crops, meat, wood, fuels, and minerals. The chart on the next page shows why many populations need protecting from the harmful effects of human activity.

Sustainable development

To develop their homes and cities, humans process natural materials like rocks (for building materials, metals, and minerals), oil (for fuels and making plastics), and wood (for buildings, furniture, and papermaking).

Some of these resources – oil and rocks, for example – took millions of years to form and cannot be replaced. Others are replaceable. Softwoods (such as pine) grow quickly, and with careful planting and management, new timber can be produced at the same rate as old timber is used up. In this case, the development of the resource is ***sustainable***.

Timber for buildings, furniture, and papermaking should come from sustainable forests.

Harmful effects of human activity

Human activity on Earth is causing problems for many wildlife populations, and for humans as well. Anything produced by humans which causes a harmful change to the environment is called ***pollution***.

Air pollution

Carbon dioxide is the main gas given off when engines and power stations burn fuel. It isn't directly harmful to living things. But it traps the Sun's heat, and the extra amount in the atmosphere may be causing global warming.

Other waste gases Engines and power stations also produce gases which can cause acid rain, smog, and lung disease. Catalytic converters reduce the problem.

Smoke particles from factories and diesel engines damage health.

Destroying populations and habitats

Hunting and fishing Humans hunt and fish for some of their food. But killing too many animals means that not enough are left to breed. So whole populations can die out.

Cutting down forests This is done for timber or to make space for agriculture or industry. But trees supply the world with some of its oxygen and provide shelter for wildlife. Also, when trees are removed, the soil is easily eroded (worn away).

Digging up land Mining and quarrying destroy wildlife habitats and produce huge heaps of waste materials. Some contain toxic metals which can harm plants.

Building When people build houses, factories, and roads, they destroy wildlife habitats.

Growing crops When hedges are cut down to create huge fields for crops, wildlife habitats are destroyed.

Water pollution

Chemical waste is sometimes dumped into rivers or the sea.

Fertilizers and pesticides are sprayed onto fields and crops. But they can get into streams and rivers and harm wildlife.

Oil can spill from tankers. It kills sea-birds and other marine life.

Sewage is often dumped at sea. It can be a health hazard.

1 Twenty rabbits are released into a grassy area. At first, the rabbit population rises rapidly. Give two reasons why the population will eventually stop rising.

2 A Quarrying for stone
B Cutting down hedges to make fields larger
Give one reason for doing each of the above. Then give one problem caused by each.

3 Give one reason why dumping rubbish might be harmful to **(a)** animals **(b)** plants.

4 Softwood, oil, and limestone are all resources used by humans.
(a) If the use of a resource is sustainable, what does this mean?
(b) Are any of the three resources mentioned sustainable? If so, which? Explain your answer.

5 Give an example of how our demand for more materials might threaten a population of animals.

6 Give *two* ways in which a river might become polluted.

2.24 Problems in the biosphere

By the end of this spread, you should be able to:
- *explain how human activity is causing damage in the biosphere.*

On Earth, life exists on the land, in the sea, and in the lower atmosphere. Together, the regions in which life exists are known as the ***biosphere***.

The Earth's human population is more than 5000 million, and growing all the time. Humans need food, fuel, and materials for their industries. But their activities are causing damage in the biosphere:

Global warming

Plants take in carbon dioxide gas from the atmosphere. Animals give out carbon dioxide, and so do burning fuels. But fuel-burning is upsetting the balance. More carbon dioxide is being added to the atmosphere than is being removed.

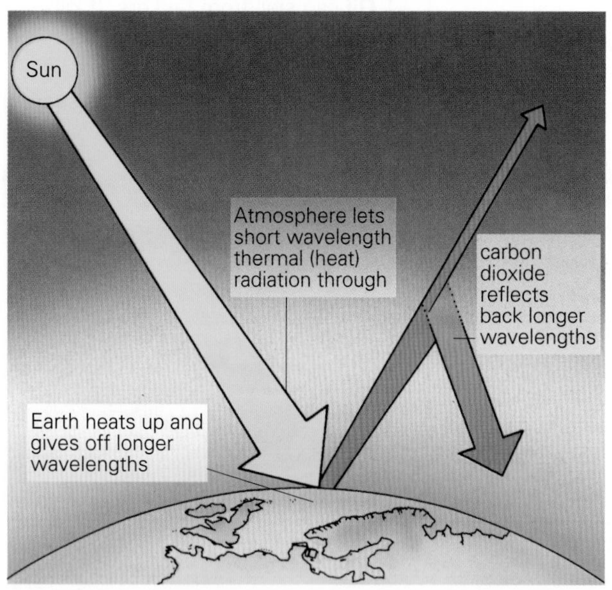

Carbon dioxide acts like the glass in a greenhouse and trap the Sun's heat. This is known as the ***greenhouse effect***, and it keeps the Earth much warmer than it would otherwise be. However, the extra carbon dioxide going into the atmosphere may be causing ***global warming*** – the Earth may be slowly warming up. Over the next hundred years, average temperatures may rise by only a few degrees, but this could shrink the polar ice caps and drastically affect world climates.

Methane (natural gas) is another greenhouse gas. It comes from swamps, animal waste, paddy fields, and oil and gas rigs. So keeping cattle, growing food, and drilling for oil and gas can all add to global warming.

Destruction of the ozone layer

High in the atmosphere, there is a band of gases called the ***ozone layer***. It screens us from some of the Sun's ultraviolet rays, which cause skin cancer. Chemical processes in the atmosphere are constantly making and destroying ozone. But some of the gases we produce on Earth are destroying too much. These include ***CFCs*** (chlorofluorocarbons). They have been used in aerosols, fridges, and freezers, and in making foam packaging. However, many manufacturers are now using more 'ozone friendly' gases instead.

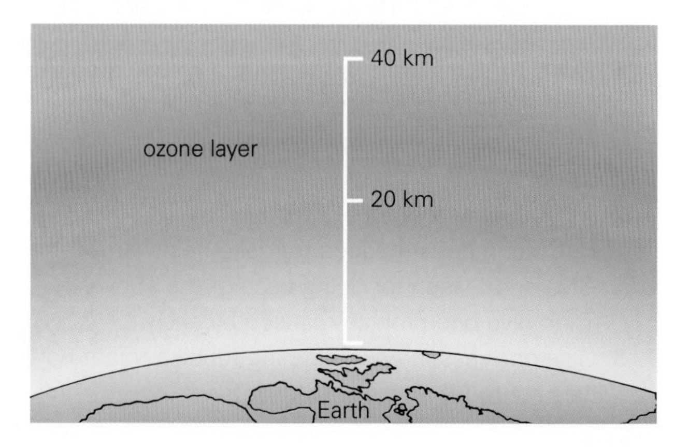

Acid rain

Rain is naturally slightly acid. But its acidity is increased by pollutants such as sulphur dioxide and nitrogen oxides from factories, power stations, and car exhausts. When these gases dissolve in rainclouds, they form sulphuric and nitric acids. The acid rain attacks stone and steelwork. Catalytic converters on exhausts reduce emissions of nitrogen oxides. Desulphurization units can be fitted to coal-burning power stations to reduce sulphur dioxide emissions.

The effects of acid rain

Lake and river pollution

Human activity can upset the ecosystem of a lake or river, and harm animal and plant life:

- Chemical pollution kills water life. The chemicals may also end up in people's drinking water supply.
- If sewage gets into a lake or river, the nutrients in it encourage the growth of algae and water plants. The effect is called ***eutrophication***. Microbes feeding on dead algae use up all the oxygen, so fish and other forms of water life die.
- Fertilizers can be washed out of the soil by rainwater. Like sewage, fertilizers can put unwanted nutrients into a lake or river.
- Factories and powers stations may discharge warm water into a river. Warm water has less oxygen dissolved in it, so fish are affected. The warmth also encourages the growth of weeds.

Forest destruction

Rainforests covers 7% of the Earth's land surface. But an area the size of Wales is being destroyed every year. Some is cut down for timber. Some is burnt to clear land for agriculture and industry.

- Destroying forests reduces the number of plants which can take in carbon dioxide and make oxygen.
- Burning timber adds to global warming by putting more carbon dioxide into the atmosphere.
- Removing trees removes nutrients which would otherwise be returned to the soil.
- Removing trees exposes soil which is then lost by erosion. The lack of vegetation also affects the amount of water vapour in the air. In time, large areas of land may be reduced to desert.

1 Why are levels of carbon dioxide in the atmosphere rising?
2 Why is the burning of fuels likely to cause global warming?
3 How does the ozone layer protect life on Earth?
4 What causes acid rain? What are its effects?
5 Describe *three* ways in which pollution can harm animal and plant life in streams and rivers.
6 Describe *two* ways in which forest-burning can add to the greenhouse effect.
7 In what other ways can forest destruction damage the environment?

2.25 Chains, webs, and pyramids

By the end of this spread, you should be able to:
- *explain what food chains and food webs are*
- *draw pyramids of numbers and biomass*
- *describe what decomposers are and what they do.*

Food chains

All living things need food. It supplies them with their energy and the materials they need for building their bodies.

Plants are ***producers***. They produce their own food. But animals are ***consumers***. They have to get their food by consuming (eating) other living things.

A ***food chain*** shows how living things feed on other living things. For example, if a blackbird feeds on snails, and these feed on leaves, then the food chain looks like the one on the right.

Pyramid of numbers

In a food chain, only a fraction of the energy taken in by one organism reaches the next. So fewer and fewer organisms can be fed at each stage. For example, it might take 30 000 leaves to feed 300 snails, and 300 snails to feed one blackbird. This can be shown using a ***pyramid of numbers*** like the one below.

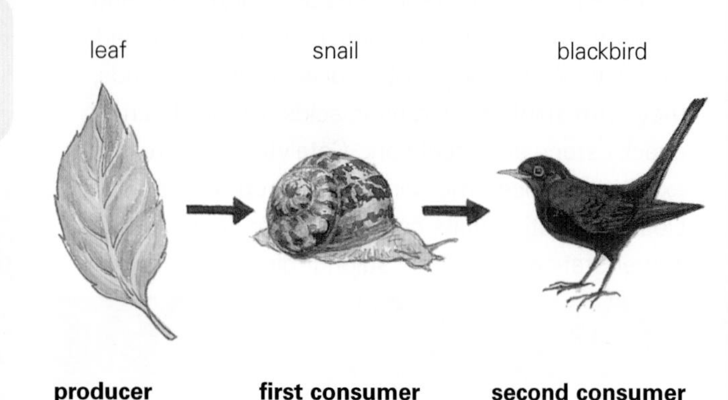

A simple food chain

Pyramid of biomass

A leaf is lighter than a snail. So the numbers of leaves and snails do not really give an accurate picture of how much food is being eaten at each stage of the chain. For this, scientists use the idea of ***biomass***. In a food chain:

The biomass is the total mass of each type of organism.

The biomasses of the leaves, snails, and bird have been worked out in the table below. Using this information, a ***pyramid of biomass*** can be drawn.

	A Number	***B*** Mass of each in g	***A × B*** Biomass in g
blackbirds	1	250	250
snails	300	50	15 000
leaves	30 000	20	600 000

(g = gram)

Pyramid of numbers

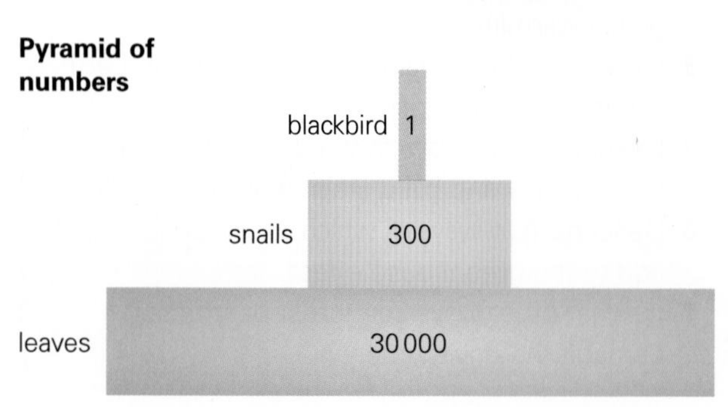

Pyramid of biomass

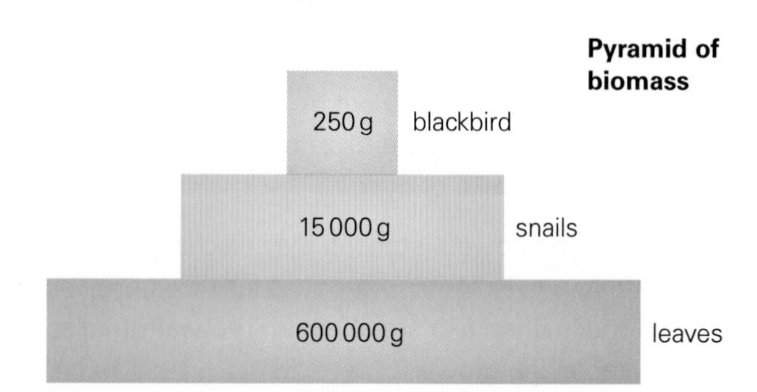

Food webs

Many animals eat more than one type of food. So organisms can be part of several food chains. The result is a network of linked food chains called a ***food web***. Here is an example:

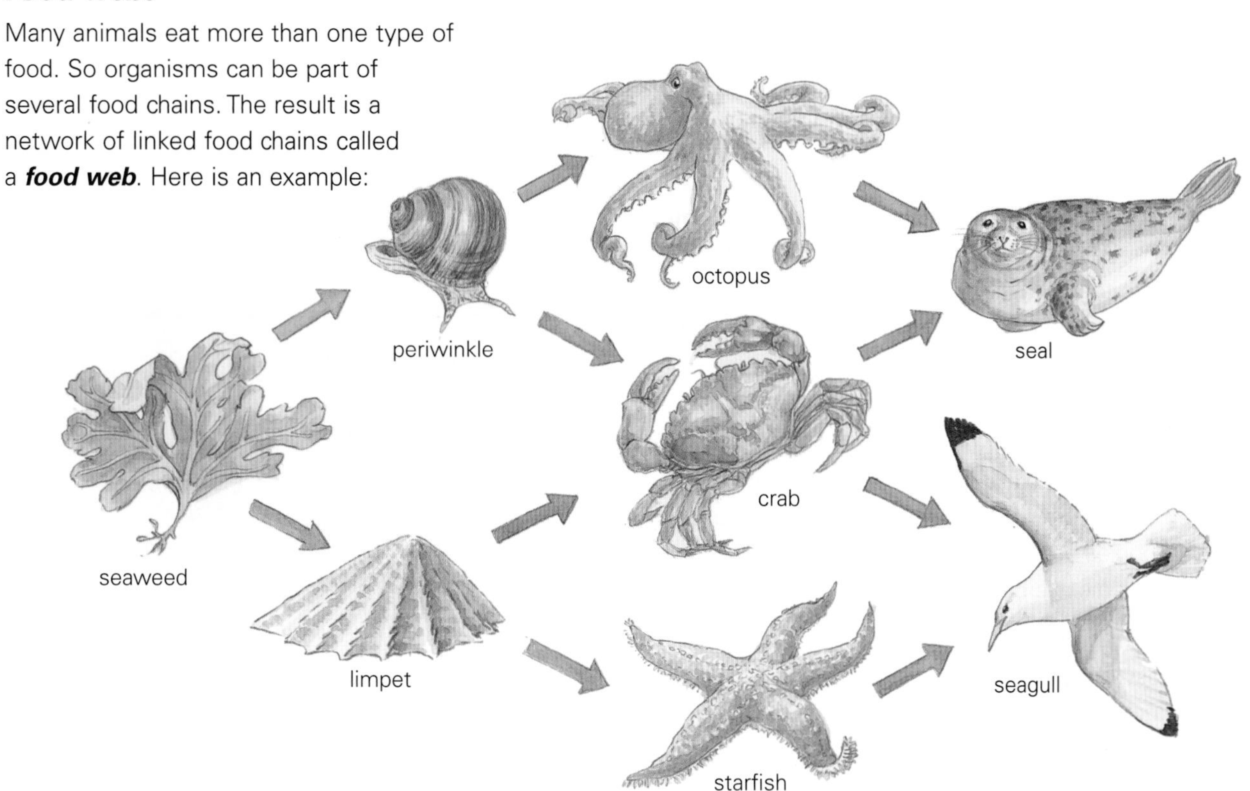

Decomposers at work

Many microbes (bacteria and fungi) feed on the remains of dead plants and animals. They produce enzymes which make the dead things decompose (rot) into a liquid. Then they feed on the liquid.

Microbes which make things rot are known as ***decomposers***.

Decomposers are important because:

- they get rid of dead plants and animals
- they put useful chemicals back into the soil.

Materials which rot are called ***biodegradable*** materials. As well as dead plants and animals, they include things made from plant or animal matter, such as paper, wool, and cotton.

1 What is the difference between a *producer* and a *consumer*? Give an example of each.
2 What are *biodegradable* materials? Give *two* examples?
3 In the food web on this page, what other organisms would be affected if periwinkles were poisoned by chemical waste? Explain what might happen to these organisms.
4 A frog feeds on 250 worms. These feed on 25 000 leaves. Draw the pyramid of numbers.
5 The frog in the last question has a mass of 200 g, a worm 40 g, and a leaf 20 g. Draw the pyramid of biomass. (Hint: start by making a table like the one on the opposite page.)

2.26 Recycling atoms

By the end of this spread, you should be able to:
- *explain how atoms of carbon and nitrogen are recycled by living things.*

Like everything else, living things are made of ***atoms***. These join together in different ways to form different materials in the body. As plants and animals grow and die, most of their atoms are used over and over again.

The carbon cycle

There is a small amount of carbon dioxide in the atmosphere. Plants take in some for photosynthesis, so their bodies are partly carbon. Animals eat plants, so their bodies are partly carbon as well. Respiration puts carbon dioxide back into the atmosphere. So does burning.

As photosynthesis, respiration, and burning take place, carbon atoms are used over and over again. This process is called the ***carbon cycle***.

Atoms in living things

Mainly .... oxygen, carbon, hydrogen

with some .... nitrogen

and smaller amounts of calcium, phosphorus, others

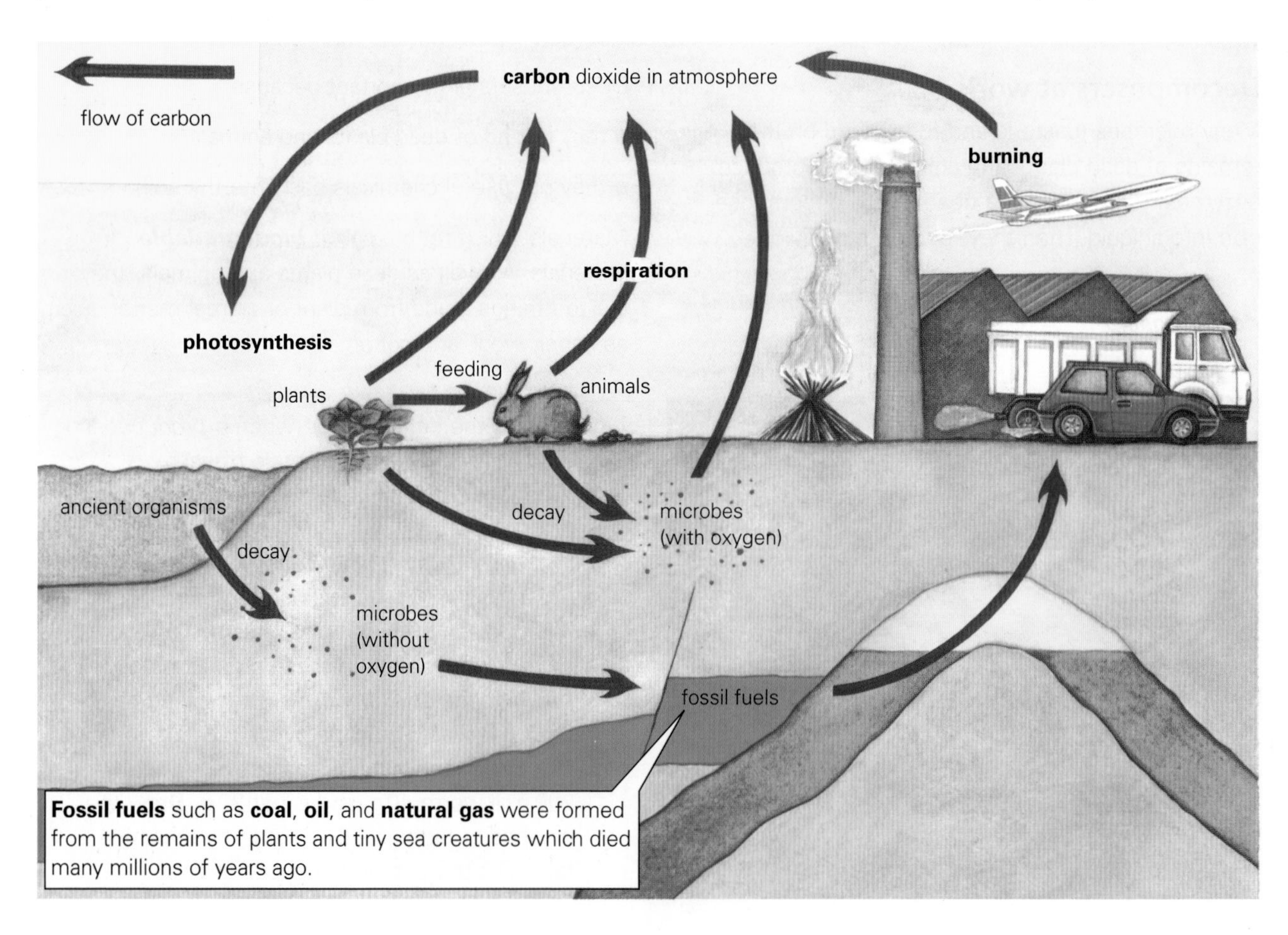

Fossil fuels such as **coal**, **oil**, and **natural gas** were formed from the remains of plants and tiny sea creatures which died many millions of years ago.

The nitrogen cycle

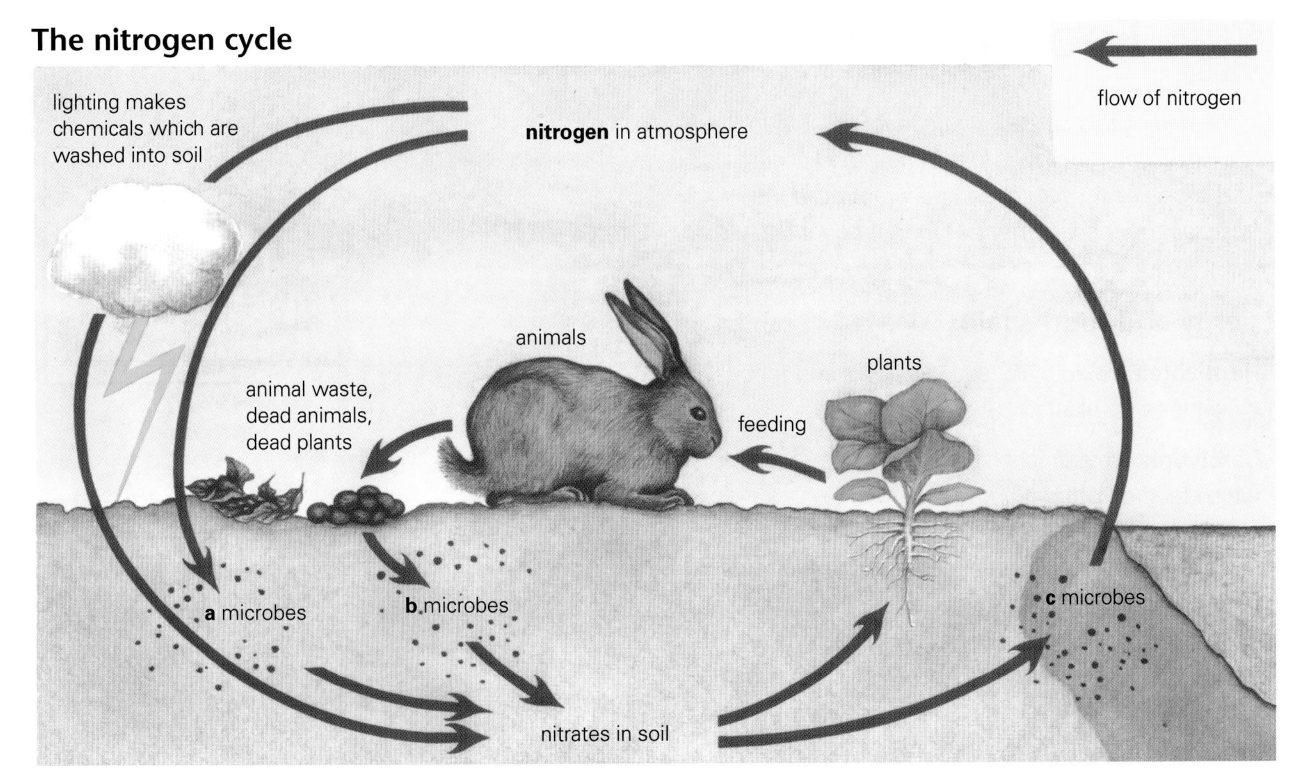

Living things need nitrogen to help make the proteins in their body tissues. There is plenty of nitrogen in the atmosphere, but it is of no direct use to plants or animals because it does not easily take part in chemical changes.

Plants get their nitrogen by taking in chemicals called ***nitrates*** from the soil. These are partly made from nitrogen. They dissolve easily and are absorbed through roots. Animals get their nitrogen by eating plants (or by eating other animals which have fed on plants).

Nitrogen is used over and over again by living things. The process is called the ***nitrogen cycle***. Microbes in the soil have an important part to play:

a Some microbes use nitrogen from the air to make nitrates. They release these into the soil.
b Some microbes make nitrates from animal waste, dead animals, and dead plants.
c Some microbes in wet soil remove nitrogen from nitrates and release it into the atmosphere.

The nodules on the roots of this pea plant contain microbes which take in nitrogen and make nitrates.

1. *Respiration photosynthesis burning*
   (a) Which of the above processes takes carbon dioxide from the atmosphere?
   (b) Which of the above processes put carbon dioxide into the atmosphere?
2. Describe how the carbon atoms in a lump of coal can end up as part of the body of an animal.
3. Why do living things need nitrogen?
4. How do plants get their nitrogen?
5. How do animals get their nitrogen?
6. If plants take nitrates from the soil, how are these nitrates replaced?

2.27 Food and waste

By the end of this spread, you should be able to:
- *describe how energy is lost from a food chain*
- *describe decay factors, and pollutant build up*
- *discuss the management of food production.*

Energy and food chains

Herbivores are animals (such as cows and rabbits) which feed on plants.

Carnivores are animals (such as foxes and cats) which feed on other animals.

Each stage of a food chain is called a ***trophic level***. At the first trophic level, plants get energy from the Sun. Only about 5% of the energy at one trophic level reaches the next. A food chain rarely has more than four or five trophic levels because there is not enough energy left to support any more.

Energy is eventually lost from a food chain, but the materials needed for growth can be used over and over again (see 2.25 and 2.26). Decomposer microbes feed on dead animal and plants, and on animal waste. They return vital elements such as nitrogen and phosphorus to the soil. However, when crops are grown and picked, this natural recycling does not happen. That is why farmers add fertilizers to the soil.

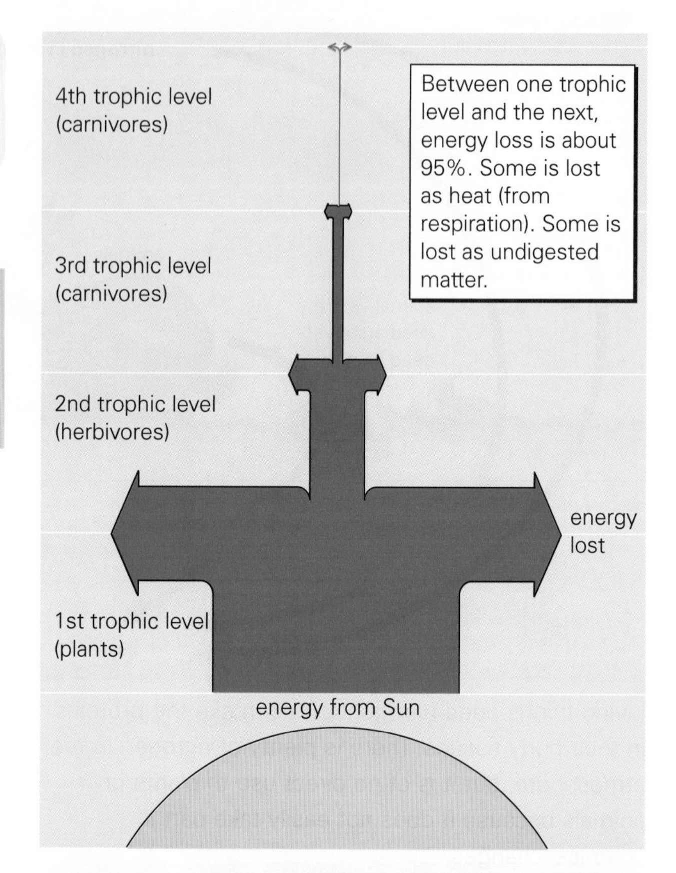

Decay factors

The rate at which plant or animal matter ***decays*** (decomposes) depends on several factors:

Temperature Most decomposer microbes thrive best in warm conditions (about 35 °C). At lower temperatures, the decay process is slower.

Water Decomposer microbes need water to live, grow, and multiply. So they require moist conditions.

Air Most decomposer microbes need oxygen. Their respiration is ***aerobic*** (oxygen-using) and they give off carbon dioxide gas.

Some decomposer microbes do not need oxygen. Their respiration is ***anaerobic*** (without oxygen) and they give off methane gas. This happens deep in a swamp or rubbish tip, where there is no air.

Humans put decomposer microbes to work in several ways. Here are two examples:

- Gardeners often put their plant clippings on a ***compost heap*** like the one above. The clippings decay and makes compost for the soil.
- In a sewage works, decomposer microbes break down waste from humans and turn it into fertilizer.

Pollution in a food chain

If toxic (poisonous) chemicals are released into the air, water, or soil, they can enter food chains and harm or kill many things.

Some chemicals get trapped in the body tissues of living things, so they accumulate (build up). Examples include some pesticides, and heavy metals such as mercury and lead. As they pass along a food chain, they become more and more concentrated. The diagram on the right shows an example of this.

In the diagram, the water in a pond has been polluted by pesticide. As materials pass along the food chain, chemicals absorbed by thousands of microscopic plants eventually build up in the body of each water fowl.

pesticide
microscopic water plants
small water creatures
fish
water fowl

Food from the land...

Humans grow crops to supply themselves with food. Some crops are fed to animals, which humans then eat as meat. But this is a very inefficient way of feeding people. The land needed to produce beef for one person could produce enough wheat for twenty.

Here are some of the methods used to increase food production. Some cause environmental damage. Some are not good for animal welfare, and some may not be good for human health.

- Adding fertilizers to the soil.
- Killing off weeds, pests, and diseases with herbicides, pesticides, and fungicides.
- Using high-yield crop varieties produced by selective breeding or genetic modification (see Spreads 1.04 and 2.19).
- Using hormones to control the ripening of fruit so that there is less wastage (see Spread 2.04).
- Injecting animals with hormones so that have more young, grow fatter, or produce more milk.
- Keeping animals warm and restricting their movement so that they waste less energy and fatten more quickly.

...and sea

Fishing must be carefully managed to preserve stocks. If too many young fish are taken from the sea, there are not enough left to breed, so the fish population falls. To prevent this happening, trawlers must limit the size of their catches and use nets which let smaller, younger fish pass through.

1 What is the difference between a *herbivore* and a *carnivore*? Which would you expect to find at the lower trophic level in a food chain? Why?

2 Why do food chains not normally have more than four or five trophic levels?

3 Why do poisons become more and more concentrated as they pass along a food chain?

4 Describe *three* conditions which would encourage decomposer microbes to thrive.

5 Why are decomposer microbes useful?

6 To feed more people, why is it better to grow wheat than to raise cattle?

7 What problems would be caused if trawlers used nets with a very small mesh?

3.01 Elements, compounds, and mixtures

By the end of this spread you should be able to:
- *describe what elements are made of*
- *explain the differences between compounds and mixtures*

Elements

Elements are substances that cannot be broken down into anything simpler by chemical action. There are about 100 different elements. The smallest bit of an element is an ***atom***. Atoms are so small that the head of a pin, for instance, contains more than 100 billion billion atoms.

Atoms of elements are often shown as symbols. The symbol is often the first letter of the name (sometimes the Latin name), written as a capital. A second, small letter, is used when more than one element has a name starting with the same letter: for example, Fe for iron (*Ferrum* in Latin). Other examples are shown in the table below – which also divides the elements into their two main types: metals and non-metals.

Metals		**Nonmetals**	
Element	*Symbol*	*Element*	*Symbol*
aluminium	Al	bromine	Br
calcium	Ca	carbon	C
copper	Cu	chlorine	Cl
gold	Au	fluorine	F
iron	Fe	helium	He
lead	Pb	hydrogen	H
magnesium	Mg	iodine	I
potassium	K	nitrogen	N
silver	Ag	oxygen	O
sodium	Na	phosphorus	P
tin	Sn	silicon	Si
zinc	Zn	sulphur	S

Metals are usually hard, strong solids which are shiny when polished. They can be bent and shaped and are good conductors of heat and electricity.

Non-metals are usually gases or brittle solids which melt easily.Most are insulators.

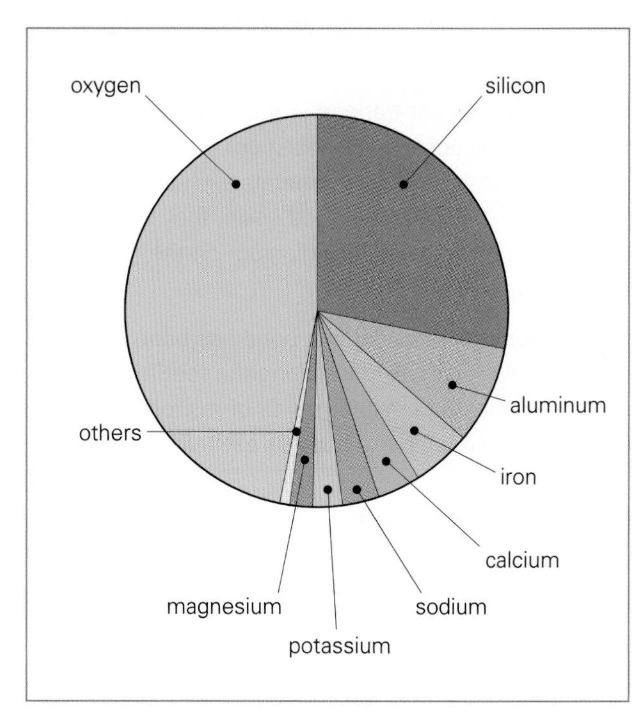

The eight most common elements in the Earth's crust

Molecules of elements

Very few elements exist as separate atoms. The smallest units of most non-metals are ***molecules*** – particles made up of two or more atoms stuck together. In elements, the atoms are identical.

Air consists mainly of the elements nitrogen and oxygen: both gases have molecules made up of two atoms strongly bound together. The yellow, powdery element sulphur consists of molecules having eight atoms each.

Element	Molecule	Formula
nitrogen	N N	N_2
oxygen	O O	O_2
sulphur	S S S S S S S S	S_8

Compounds

Molecules are also formed when the atoms of two or more *different elements* join together. These are the smallest bits of new substances called ***compounds***. Carbon dioxide is a compound and its molecules are made up of two oxygen atoms stuck to one atom of carbon. The ***formula***, in this case CO_2, shows the numbers of each kind of atom present in the molecule. Note that the number 1 is never shown.

A pure compound always has exactly the same composition by mass. Water (H_2O), for instance, always contains 11.1% of hydrogen linked with 88.9% of oxygen.

It is usually difficult to break down a compound into the elements or substances from which it was made.

Some compounds are made up of different atoms which have reacted with each other to become electrically charged ions (see Spread 3.07).

Mixtures

Most natural materials are not pure elements or compounds. Instead, they are ***mixtures*** of two or more separate substances. Because the substances are not chemically linked, they are easy to separate and the amounts present can vary.

Air is a mixture of mainly nitrogen, oxygen, and carbon dioxide. It contains less oxygen and more carbon dioxide after it has been breathed in and out.

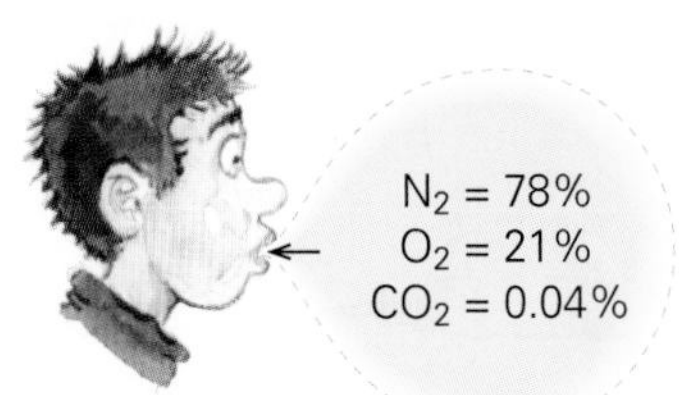

Air breathed in

Air breathed out

N_2 = 78%
O_2 = 16%
CO_2 = 4%

Oil and vinegar shaken together will separate into two layers – they do not mix. Shaken with an ***emulsifier***, such as egg yolk, the oil breaks up into tiny droplets. These remain suspended in the watery vinegar as an ***emulsion*** – a thick liquid mixture (salad cream).

Compound	Molecule	Formula
water		H_2O
carbon dioxide		CO_2
ammonia		NH_3
methane		CH_4
sulphuric acid		H_2SO_4

Rocks

Most rocks are mixtures of natural compounds called ***minerals***. Quartz, feldspar, and mica are the main minerals in ***granite***, which can have many forms depending upon the amounts of minerals present.

Different minerals in granite

1. What are **(a)** elements **(b)** the two main types?
2. Which of the eight most common elements in the Earth's crust are metals?
3. Why is Fe the symbol for iron?
4. Which element exists as molecules containing eight atoms?
5. What do molecules of compounds contain?
6. The molecule of phosphoric acid has one phosphorus atom linked with four oxygen atoms and three hydrogen atoms. Write the formula.
7. Give *two* differences between compounds and mixture.

3.02 Solids, liquids, and gases

By the end of this spread, you should be able to:
- *describe the arrangement of particles in solids, liquids, and gases*
- *explain the links between particles, temperature, change of state, diffusion, and gas pressure.*

Particle theory of matter

All substances consist of tiny particles – atoms, molecules, or ions – which are constantly on the move.

In a ***solid*** the particles are closely packed in regular patterns. They are held together by strong forces of attraction and can only vibrate in their fixed positions.

The particles of a ***liquid*** are more spaced out and irregular. They still attract each other but vibrate more and have enough energy to move around. So, a liquid can flow and take up the shape of its container.

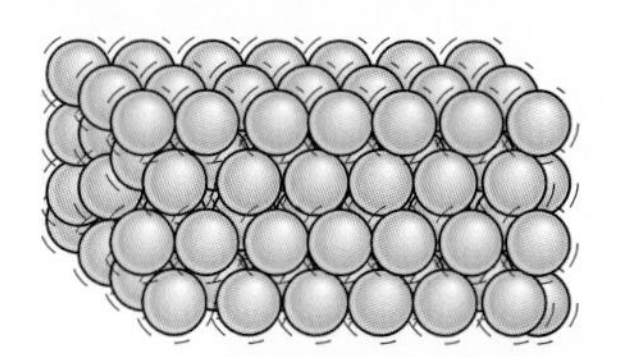

Solid Particles vibrate about fixed positions

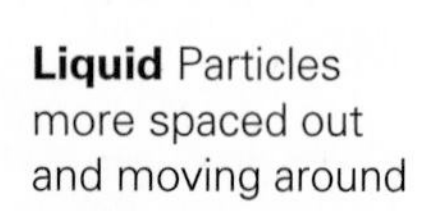

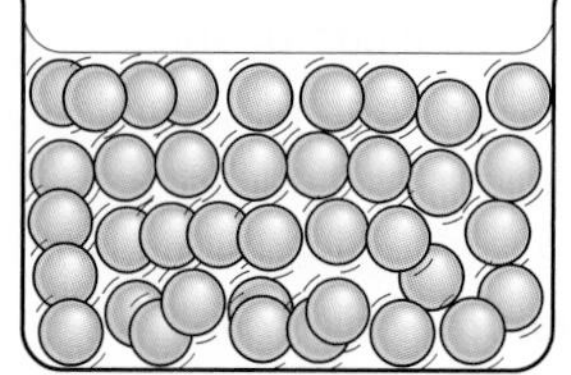

Liquid Particles more spaced out and moving around

In a ***gas***, the particles are widely spaced and free to move at high speeds. They always spread out to fill any space available. The bouncing of the particles off the walls of the container gives the ***pressure*** of the gas.

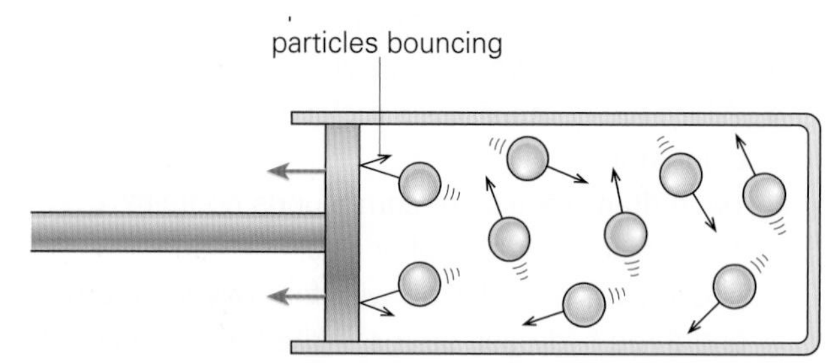

Gas Particles widely spaced and free to move

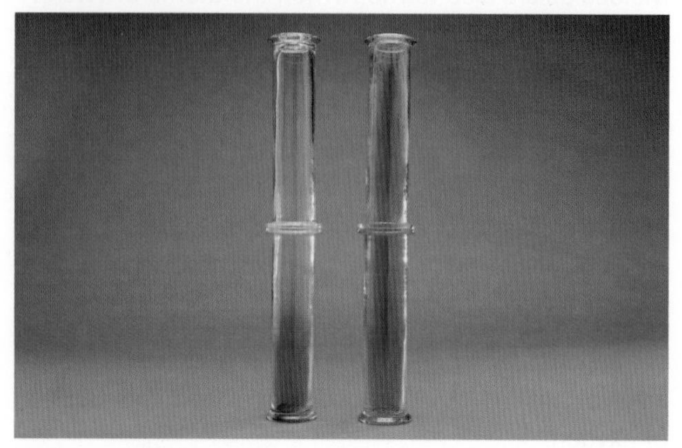

Before and after: bromine diffusing in air

Diffusion

If a capsule of bromine is broken inside a jar of air, the brown bromine molecules quickly spread out until the colour is the same throughout. This is ***diffusion***. Gases with light particles diffuse more quickly than those with heavy ones.

Solutions (see Spread 3.03) will also diffuse into each other but more slowly because the particles move slowly.

Temperature

When something gets hotter, its particles move faster and its temperature rises. Everyday temperatures are measured on the ***Celsius scale*** (sometimes called the ***centigrade*** scale). Its unit is the ***degree Celsius (°C)***.

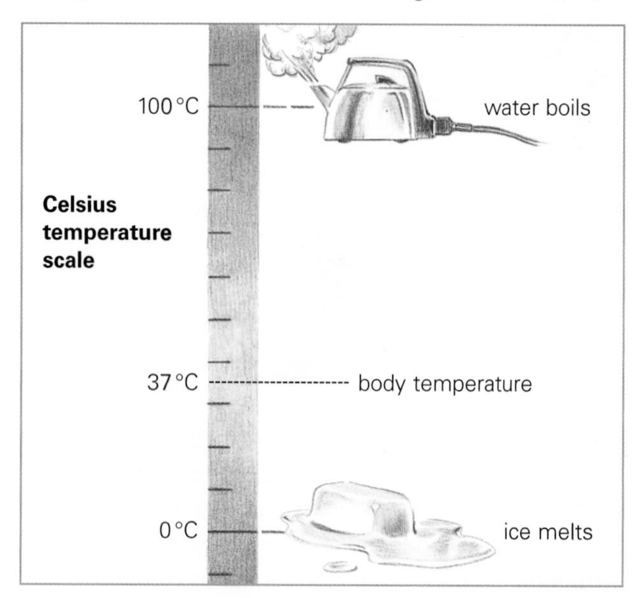

Changing state

When a very cold block of ice is heated, its molecules vibrate more and more until some of them begin to move around as liquid. This is ***melting*** and it happens at a different temperature for every solid. The ***melting point*** of pure ice is 0 °C.

As water is heated, its molecules move even faster until some have gained enough energy to break free of the attractions holding them together. They escape as steam as the water ***evaporates***. When water ***boils***, bubbles appear everywhere in the liquid as it quickly changes to gas. Every liquid has its own ***boiling point***. For pure water it is normally 100 °C.

These changes of state can be very useful. Marathon runners often douse themselves so that they cool down as the evaporating water draws heat from them.

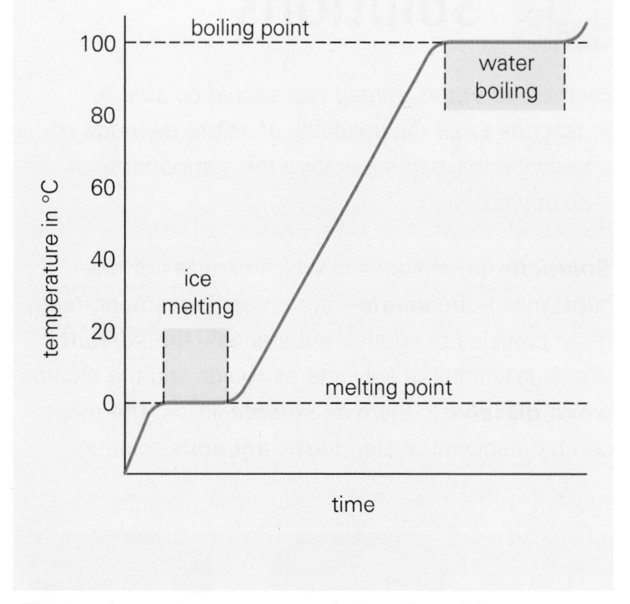

Change in temperature: gentle heating of ice

Density

The amount of matter in something is called its ***mass***. It is measured in ***kilograms (kg)***.

The amount of space something takes up is its ***volume***. This is measured in ***cubic metres (m^3)***.

There is more mass in a cubic metre of graphite (a form of carbon) than there is in the same volume of water. The ***density*** of the graphite is greater because its particles are more closely packed together. Air is much less dense than water mainly because its particles are so spread out.

You can calculate density with this equation:

$$\text{density} = \frac{\text{mass}}{\text{volume}}$$

mass in kg
volume in m^3
density in kg/m^3

For example, if a block of steel has a mass of 15 600 kg and a volume of $2\,m^3$, its density is:

$$\text{density} = \frac{\text{mass}}{\text{volume}} = \frac{15\,600}{2} = 7800\,kg/m^3$$

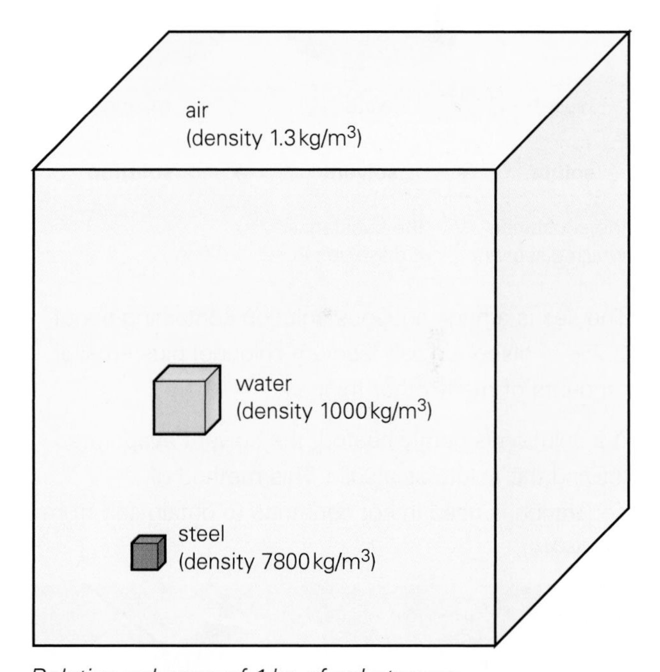

Relative volumes of 1 kg of substances

1. What differences are there in the behaviour of the particles in solids, liquids, and gases?
2. What is gas pressure due to?
3. What is meant by *diffusion*?
4. What happens to the particles in a substance if its temperature rises?
5. Given a thermometer and a Bunsen burner, how could you tell if the water in a beaker was pure?
6. Why does evaporation have a cooling effect?
7. If $3\,m^3$ of aluminium has a mass of 8100 kg, what is its density?
8. What volume would 2 kg of water occupy?

3.03 Solutions

By the end of this spread you should be able to:
- *describe what the solubility of solute depends on*
- *describe ways of separating the components of solutions.*

Solutions are mixtures in which particles of one substance – the ***solute*** – are spread out among many more particles of another substance – the ***solvent***. We usually think of solvents as liquids and the solutes which ***dissolve*** in them as ***soluble*** solids. The most common solvent, water, forms ***aqueous*** solutions.

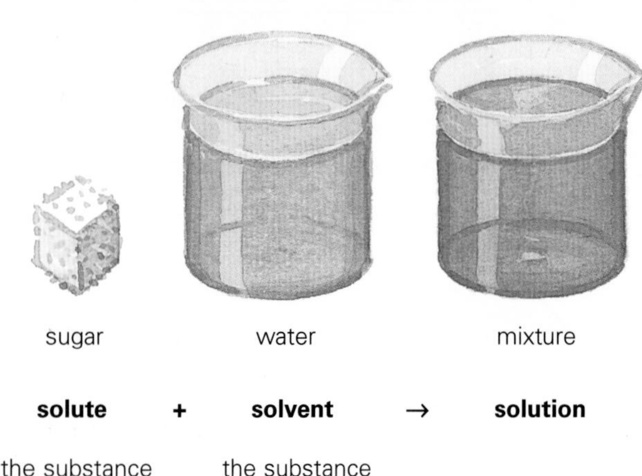

The sea is a huge aqueous solution containing about 2.7% of dissolved salt (sodium chloride) plus smaller amounts of many other things.

If a solution is gently heated, the solvent evaporates leaving the solute as a solid. This method of separation is used in hot countries to obtain salt from sea water.

The sea is an aqueous solution

Solubility

When small portions of salt are stirred into water, there comes a point when no more will dissolve. The solution is ***saturated***. The ***solubility*** of the salt is the mass which will dissolve in 100 g of water to produce a saturated solution. It is important to state the temperature because the solubilities of most solid solutes increase when the water is hot. Every solid has its own ***solubility curve*** – some are shown below.

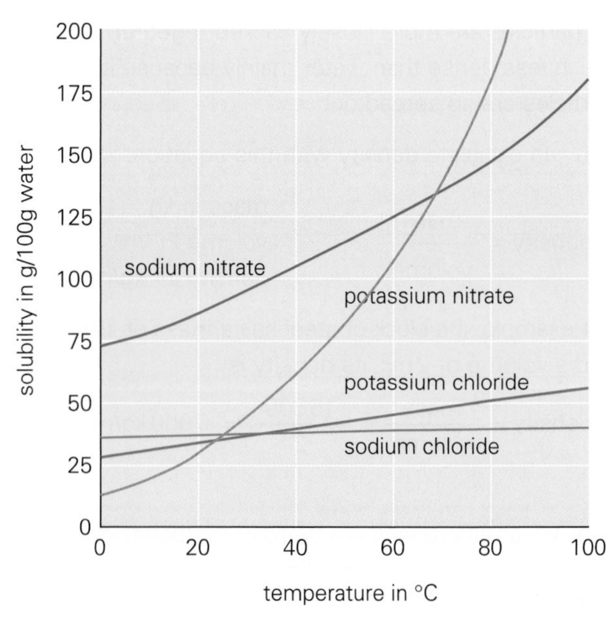

Most gases dissolve in water but their solubilities are generally very much lower than those of solids. Unlike solids, gases are less soluble in hot water than in cold.

Other common sovents

Some everyday substances do not dissolve much in water. Biro ink and nail varnish stains can be removed, however, using appropriate solvents. Ethanol dissolves Biro ink and propanone (acetone) is nail varnish solvent. These solvents, on the other hand, do not dissolve salts:

Solvent	Solute
ethanol	Biro ink
propanone	nail varnish
trichloroethane	grease
white spirit	gloss paint
methyl benzene	sulphur

Separating components of solutions

Crystallization If a solution of copper(II) sulphate is left for several days, the water slowly evaporates and the solution becomes more concentrated. Eventually it reaches saturation and blue crystals of hydrated copper(II) sulphate begin to form. The slower the rate of evaporation, the larger are the crystals that form.

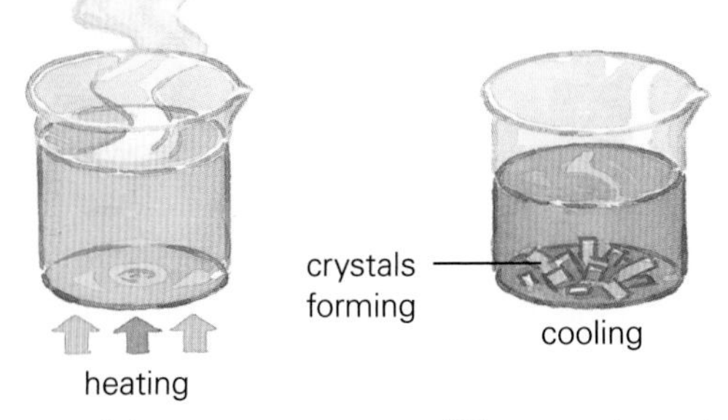

It is possible to separate two different solutes in the same solution by crystallizing at different temperatures.

1. Give the meaning of **(a)** *solute* **(b)** *solvent* **(c)** *saturated solution* **(d)** *aqueous solution*.
2. What is meant by the *solubility* of a substance? How does temperature affect it?
3. Use the graphs on the opposite page to find the solubilities of sodium chloride and potassium nitrate at **(a)** 20 °C **(b)** 80 °C.
4. Name *two* common solvents other than water.
5. How could you make powdered copper(II) sulphate into large crystals?
6. What sorts of mixtures can be separated by **(a)** chromatography **(b)** fractional distillation?
7. How can sea water be made fit to drink?

Chromatography is a way of identifying small amounts of different solutes in the same solution. Dyes in a spot of black ink on a filter paper can be separated by dipping the paper into a beaker of water. As the water soaks into the spot and up the paper, it carries the dyes with it. The filter paper slows down the dyes by different amounts, and they separate into bands of colour.

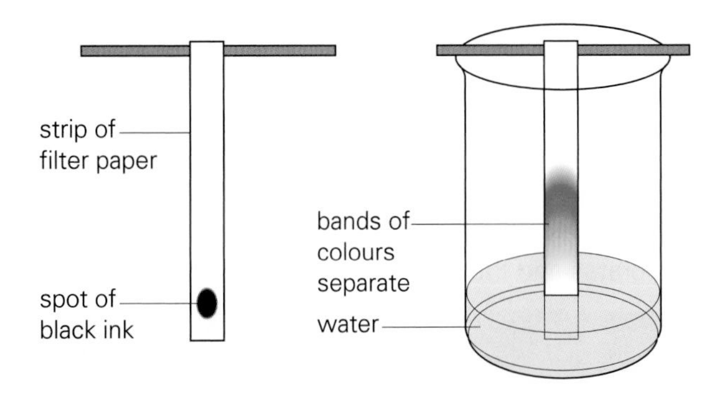

This principle is used a great deal in hospitals to separate and identify substances in body liquids. Sugar in urine, for example, is a sign of diabetes. The substances are not usually coloured but are shown up by spraying them with a locating agent.

Distillation separates and recovers the solvent from a solution. Distilled water is obtained by boiling water and condensing the vapour. All the dissolved impurities are left behind and the ***distillate*** is pure water. Sea water can be made fit for drinking by distillation.

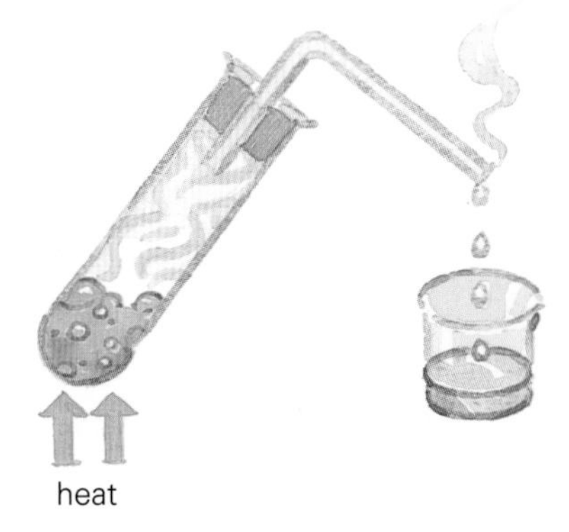

Fractional distillation separates mixtures of liquids which boil at different temperatures. Purified air, cooled sufficiently, will condense to a liquid. When slowly warmed, the liquid first releases helium and neon, then oxygen at –218 °C, nitrogen at –196 °C, and finally the other noble gases (see Spread 3.06).

3.04 Physical and chemical changes

By the end of this spread you should be able to:
- *describe the differences between a physical change and a chemical change*
- *recognize a chemical reaction.*

Physical changes

If you crush up some pure rock salt, its appearance changes from colourless crystals to a white powder. This powder disappears into solution if you shake it with water. It is still there, of course, but you cannot see it. Left to stand for a few days, the water slowly evaporates and the colourless crystals reform.

Physical changes such as those above are changes which:
- do *not* form new substances, and
- are easy to reverse.

Such changes do not cause any change in mass.

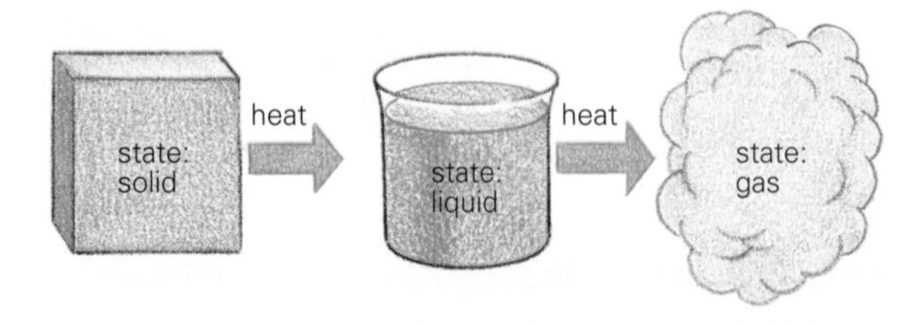

Changes of state (see Spread 3.02) are physical changes. When water is heated enough it boils and becomes steam. As soon as the steam is cooled it changes back to water.

Very rapid chemical reactions

Chemical changes

A mixture of iron filings and sulphur can be separated by passing a magnet over it. The black iron sticks to the magnet leaving behind the yellow sulphur powder.

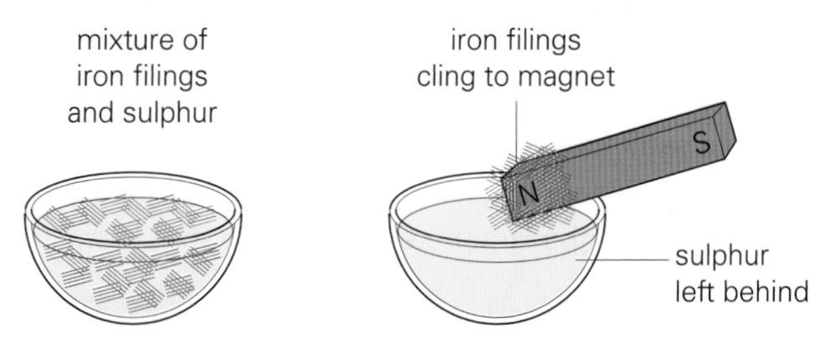

If the mixture is heated in a test tube, it glows red hot and changes into a grey solid compound called iron(II) sulphide. This completely new substance is not attracted to a magnet. The iron and sulphur have combined and a chemical change has taken place.

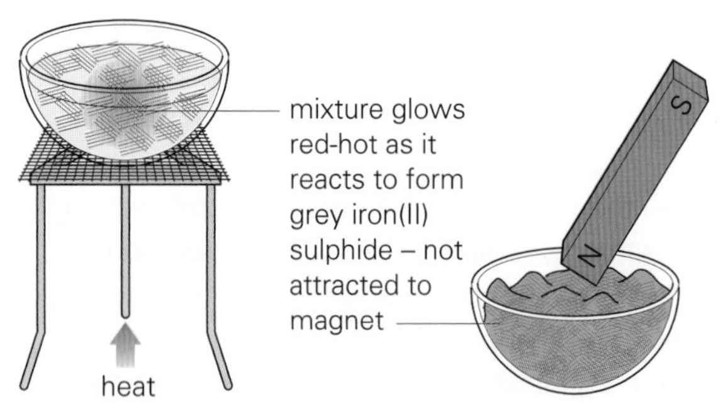

Chemical changes are changes which:
- form new substances, and
- are usually difficult to reverse.

Like physical changes, chemical changes involve no overall change in mass. There are the same number of atoms present before and after the change. They are just arranged in different ways.

Signs of chemical change

When a chemical change has taken place, scientifically speaking, there has been a ***chemical reaction***. Here are some of the signs:

Permanent change Solutions of silver nitrate and sodium chloride are both colourless. When mixed, they react to produce silver chloride. This is insoluble and appears as a white solid suspended in the liquid. Spread 3.10 shows the equation for the reaction.

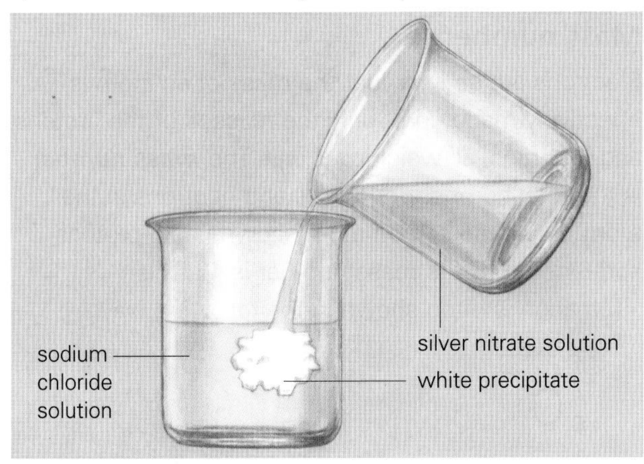

In many reactions, a gas is formed when solids and liquids react. If copper(II) carbonate is added to a tube containing dilute nitric acid, a violent fizzing occurs as carbon dioxide gas is produced and released (see also Spread 3.14). Gases can be seen as bubbles in reactions involving liquids, or sometimes smelt.

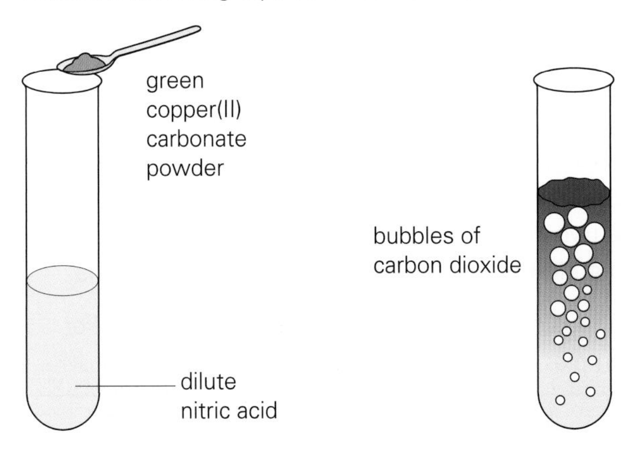

Colour change This shows that new substances are being formed even if no changes of state occur. A few drops of purple potassium manganate(VII) solution lose their colour when added to colourless sodium sulphate(IV) solution. In this reaction the manganate oxidizes (Spread 3.12) the sulphate(IV) to sulphate(VI).

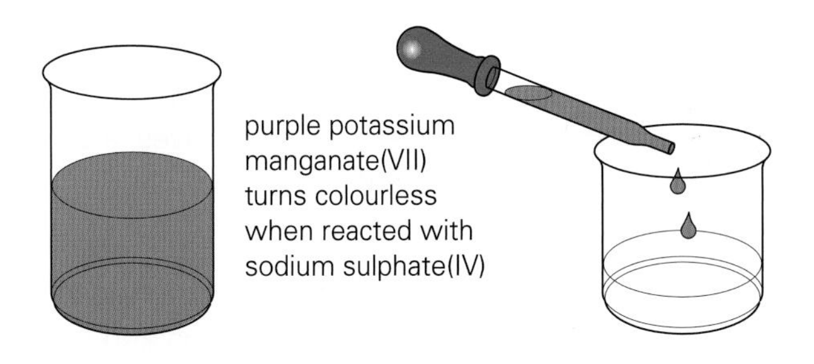

Energy change When chemical changes occur, the energy possessed by the new particles of the ***products*** is different from that of the starting particles – the ***reactants***. Often, the products have less total energy than the reactants. The difference is given out as heat, and the reaction vessel gets hot. This is an ***exothermic*** reaction. Some reactions are ***endothermic***; they take in heat, and the surroundings feel colder.

When dilute hydrochloric acid is added to sodium hydroxide solution, nothing appears to happen. However, a marked rise in temperature is a sure sign that a reaction has occurred.

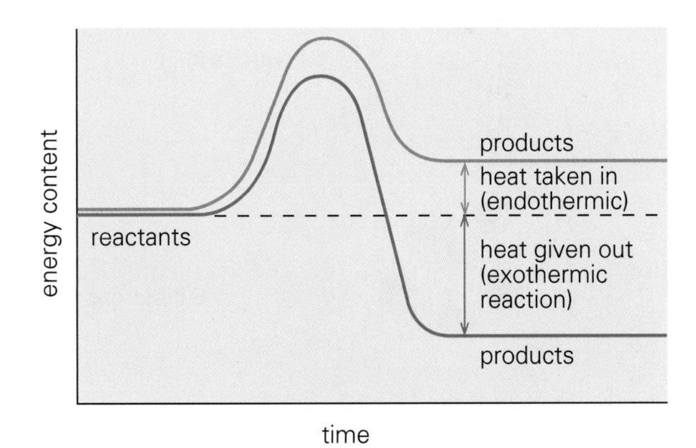

1 (a) How do physical and chemical changes differ?
(b) Give one example of each type of change.
(c) What is the same about the two types?

2 What mass of iron sulphide is formed by heating 56 g of iron filings with 32 g of sulphur?

3 Describe all you can see when copper(II) carbonate is added to dilute nitric acid.

4 Give *three* ways of telling that a chemical reaction is taking place.

5 What are *exothermic* and *endothermic* reactions?

3.05 Atoms and isotopes

By the end of this spread you should be able to:
- *describe the structure of an atom*
- *explain that elements are mixtures of isotopes.*

Atomic number

Although atoms are the smallest particles that can take part in chemical changes, they, themselves, are made up of even smaller bits. These bits are not normally found permanently on their own.

The mass of an atom is concentrated in a tiny, heavy ***nucleus*** containing ***protons*** and ***neutrons*** packed tightly together. A proton has a positive electric charge and a mass called (for simplicity) 1 unit. A neutron also has a mass of 1 unit but has no charge.

Every positive proton is balanced electrically by a negatively charged ***electron***. Electrons have virtually no mass and whizz around the nucleus in spaces called ***orbitals***. These behave like clouds of charge.

The chemical nature of an element depends upon the number of electrons there are around the nuclei of its atoms. This is the same as the number of protons in the nuclei and is called the ***atomic number*** of the element.

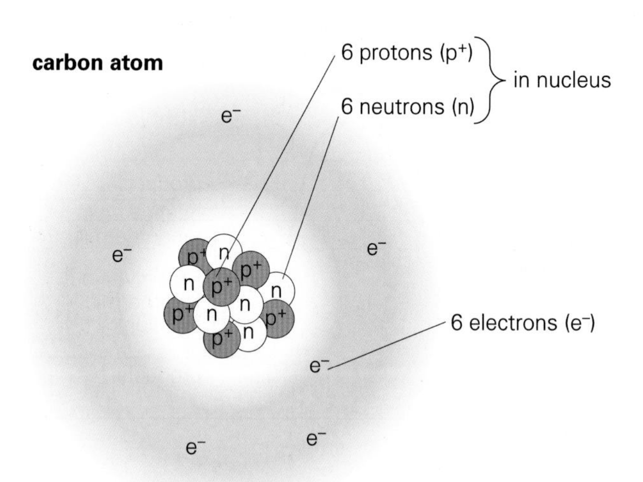

All carbon atoms, like the one above, have 6 protons in the nucleus balanced by 6 electrons in the surrounding cloud. The atomic number is 6. Nitrogen is the next element (atomic number = 7) with 7 protons (and 7 electrons) in its atoms.

Particle	proton	electron	neutron
Where found	nucleus	orbital cloud	nucleus
Mass	1 unit	negligible	1 unit
Charge	1+	1–	0

Mass number

Electrons are so light that the mass of an atom is effectively made up of only the masses of the protons and neutrons, all weighing 1 unit. The ***mass number*** is the total number of protons and neutrons in the atom. Since the carbon atom shown has 6 protons and 6 neutrons, its mass number is 12. This infomation is often shown in the following way:

$$^{12}_{6}C$$

mass number = number of protons + neutrons

chemical symbol for element

atomic number = number of protons = number of electrons

The mass of this atom is defined as 12.0000 units and is the standard to which all other atoms are compared.

Isotopes

All the atoms making up an element must have the same atomic number – the same number of protons. The atoms may not have the same mass number, however, because they can have different numbers of neutrons. For example, there is another kind of carbon atom with 8 neutrons in its nucleus and a mass of 14. The two kinds are called ***isotopes*** – atoms of the same element with different numbers of neutrons.

Isotopes are named using their mass numbers. For example: carbon-12 and carbon-14:

Carbon-12	Carbon-14
6 electrons (–)	6 electrons (–)
6 protons (+) 6 neutrons	6 protons (+) 8 neutrons
12 total mass	14 total mass

Relative atomic mass (A_r)

All chlorine atoms have 17 protons and electrons, but some have 18 neutrons in their nuclei and others have 20. The total mass of the first isotope is 17 + 18 = 35, whilst the mass of the second is 17 + 20 = 37.

Three-quarters, 75%, of the atoms in chlorine gas are chlorine-35 and one-quarter, 25%, are chlorine-37. So:

$$\text{average mass per atom} = \frac{35 \times 75}{100} + \frac{37 \times 25}{100}$$

$$= 26.25 + 9.25$$

$$= 35.5$$

This average is called the ***relative atomic mass*** (***A_r***) of chlorine and it is the figure used in calculations. Many elements occur as mixtures of isotopes. Even hydrogen has three isotopes: ${}^{1}_{1}H$, ${}^{2}_{1}H$, and ${}^{3}_{1}H$.

Therefore, relative atomic masses are seldom whole numbers, although they are usually rounded up to the nearest one. This is often the same as the mass number of the most common isotope.

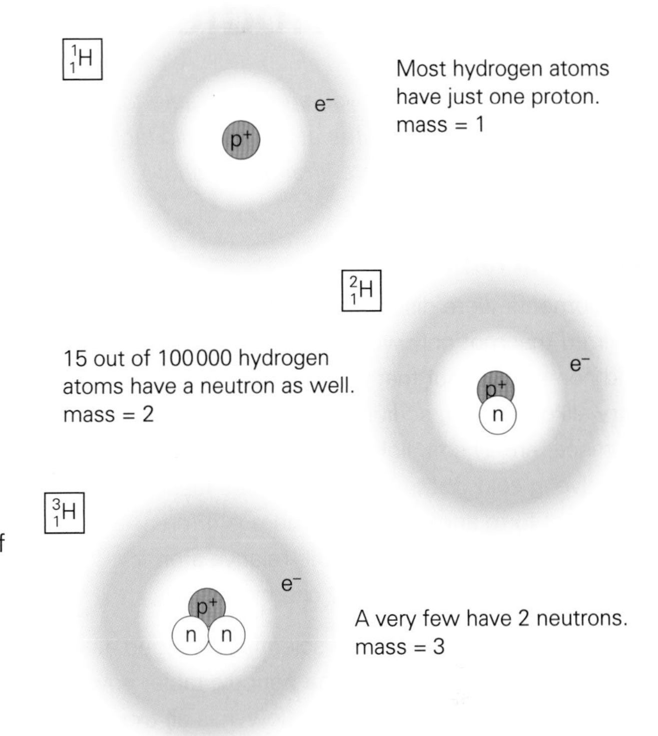

Radioactive isotopes

One of the isotopes of an element is often radioactive (see Spread 4.30) and can be very useful. Cobalt-60, for instance, is used in treating cancer and for sterilizing medical instruments. Carbon dating is a method of finding the ages of wooden and cloth articles by measuring the amount of carbon-14 left in them.

1 Give the names, charges and masses of the three particles making up all atoms.
2 Draw a diagram to show the arrangement of the 3 types of particle in carbon-12 atoms.
3 Give the meaning of the terms: *atomic number, mass number* and *isotopes.*
4 How many of each of the three particles are present in the lithium atom ${}^{7}_{3}Li$?
5 What is the term for *the average mass per atom of the isotopes in an element?*
6 Bromine consists of two isotopes: 50% is bromine-79 and 50% bromine-81. What is its A_r?
7 What is *carbon dating?* Use the ICT network to find out more about it.

Models of the atom

Atoms are impossible to draw! Pictures like those above are just simple ***models*** that are useful for describing the main features of atoms. Later pictures will use even simpler models, without the electron clouds. In advanced work, scientists use mathematical models that don't require pictures.

Remains from a peat bog: carbon dating showed that this man died around 220–240 BC.

3.06 The periodic table

By the end of this spread you should be able to:
- *describe the main features of the periodic table*
- *explain how electrons are arranged in shells.*

Chemical families

As elements were discovered in the past, people realized that, not only were there two sorts, metals and non-metals, but that small groups of them were very alike. Lithium, sodium, and potassium, for instance, are all soft, light metals which fizz and dissolve in cold water. The noble gases (described opposite) form another family. There is often a mathematical ratio between the relative atomic masses of the members of a chemical family.

Dmitri Mendeleev, a Russian scientist, arranged all the known elements in a table. The periodic table, now modified into the form shown below, has all the elements arranged in ascending order of atomic number across it in rows called ***periods***. Elements with similar properties are arranged under one another in vertical columns called ***groups***.

Metal elements are found in the groups at the left of the table and non-metals in the right-hand groups.

Originally there were just 8 columns, but the table now has Group 3 separated from Group 2 by a block of ***transition metals***.

This lighthouse uses a xenon lamp.

The noble gases – Group 0

If all the nitrogen, oxygen, carbon dioxide, and water vapour are removed from air, about 1% remains and will not react with anything. It may be separated into a family of similar gases. These ***noble gases*** are unusual elements because their atoms exist singly and do not normally join with others. At low pressure, they glow when electricity is passed through them. This, and their unreactivity, can be put to use.

Noble gas	Symbol	Use
helium	He	balloons, divers' 'air', lasers
neon	Ne	advertising signs, lasers
argon	Ar	light bulbs, arc welding
krypton	Kr	lights, strobe lamps
xenon	Xe	lighthouse lamps

Periodic table

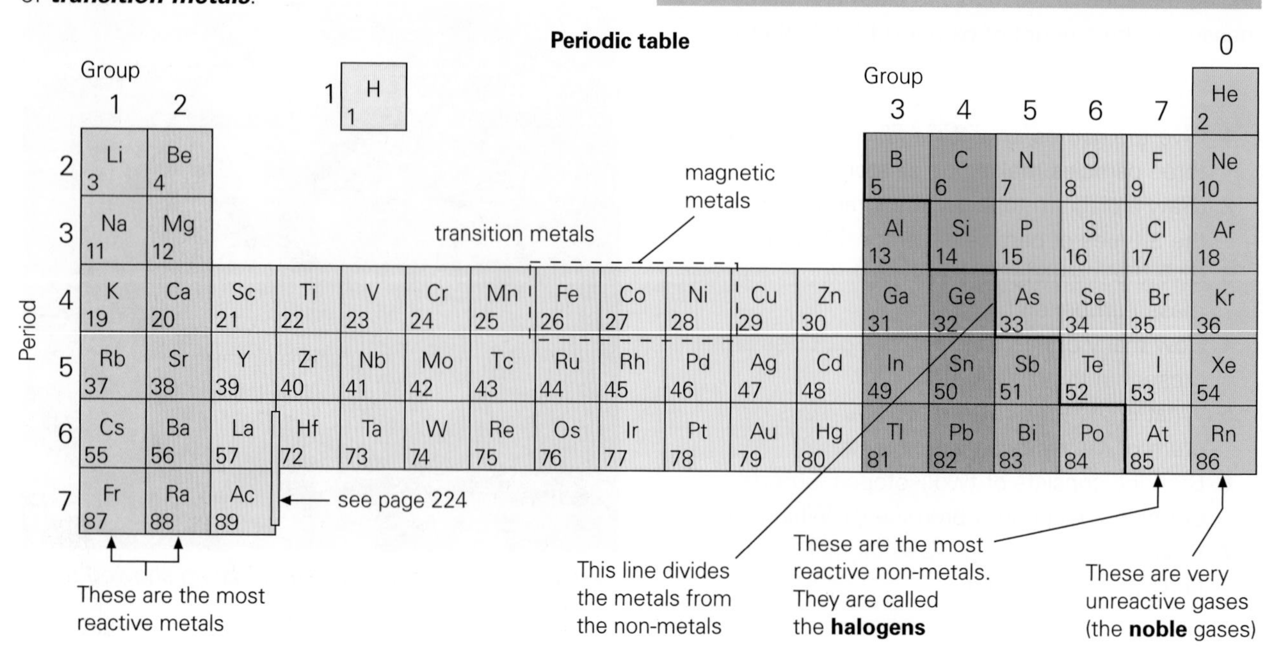

Electron distribution

Electrons in the cloud around the nucleus of an atom exist in defined regions known as ***energy levels*** or ***shells***. The first shell, closest to the nucleus, can hold only two electron paths. The second shell is further from the nucleus and holds up to eight electrons of higher energy. Subsequent shells can also hold eight electrons. The hydrogen atom has only one electron and it exists in the first shell. Helium, the next atom, has two electrons in the first shell surrounding the two protons (and the neutrons) in the helium nucleus.

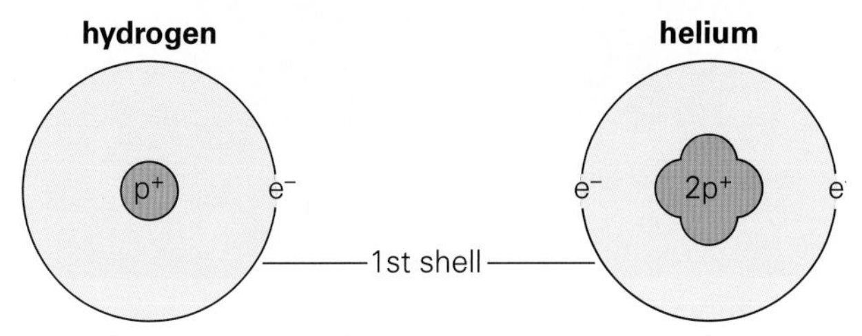

p^+ = proton; e^- = electron; neutrons are not shown

The lithium atom has three electrons. Two of these fill the first shell and the third occupies the second shell. This electron has more energy than the first two.

Sodium has 11 electrons. Two of these fill the first shell, eight fill the second shell, and the last electron starts the third shell.

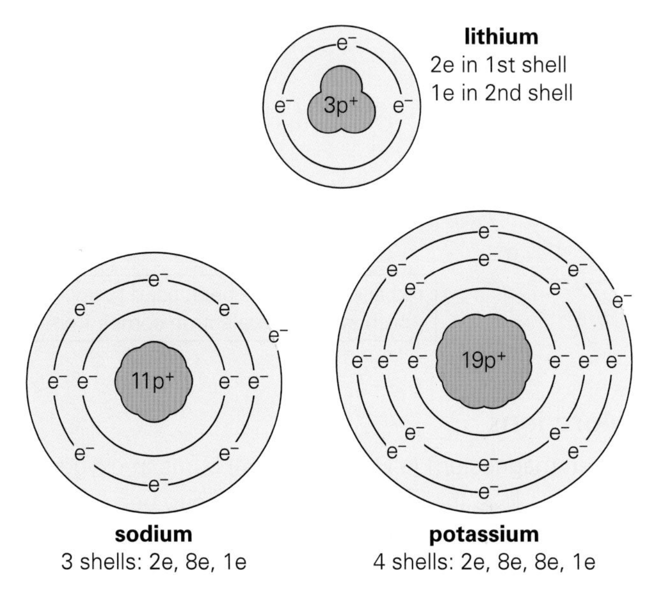

All the Group 1 metals have one electron in their outermost shells. Similarly, all the Group 2 elements have two electrons in their outermost shells. The group number tells you how many outer electrons the atom of the element has.

Up to eight

The chart below shows electron arrangements of some of the elements in Groups 7, 0, and 1.

Element	Group	Number of electrons (e) in shell				
		1st	2nd	3rd	4th	5th
helium (H)	0	2e				
lithium (Li)	1	2e	1e			
fluorine (F)	7	2e	7e			
neon (Ne)	0	2e	8e			
sodium (Na)	1	2e	8e	1e		
chlorine (Cl)	7	2e	8e	7e		
argon (Ar)	0	2e	8e	8e		
potassium (K)	1	2e	8e	8e	1e	
bromine (Br)	7	2e	8e	18e	7e	
krypton (Kr)	0	2e	8e	18e	8e	
rubidium (Rb)	1	2e	8e	18e	8e	1e
iodine (I)	7	2e	8e	18e	18e	7e
xenon (Xe)	0	2e	8e	18e	18e	8e

Note that the almost totally unreactive noble gases all have full outer shells of electrons – 2 in the first and 8 in all the others. The full outer shell is a very stable arrangement. Atoms of other elements react by transferring or sharing outer electrons in order to achieve noble gas structures.

The chart also shows that the atoms of very reactive elements, such as the Group 1 metals and the Group 7 non-metals, are only a step away from having full outer shells.

1 Give *three* features which are **(a)** common to the elements in Group O – the noble gases **(b)** common to the metals in Group 1.
2 Name *four* noble gases and give a use of each.
3 Name the elements in Period 2 of the periodic table.
4 How many electrons are there in the outer shells of **(a)** sodium (Na) **(b)** carbon (C)?
5 Explain what is meant by *electron shells*.
6 Draw a table to show how the electrons are arranged in shells for the atoms of **(a)** sodium (Na) **(b)** sulphur (S) **(c)** chlorine (Cl) **(d)** argon (Ar).

3.07 Ions

By the end of this spread you should be able to:
- *describe what ions are and how they form giant lattice structures*
- *write fomulae for ionic compounds.*

Electron transfers

Hot sodium, placed in a jar of chlorine, burns brightly and forms white solid salt, sodium chloride. In this reaction, the one outer electron of the sodiun atom is transferred to join the 7 outer electrons of the chlorine atom. The sodium is left with electron shells of 2e and 8e – the structure of neon. The chlorine has gained the argon structure – full shells of 2e, 8e, and 8e. Both resulting particles are now extremely stable.

Sodium burning in chlorine

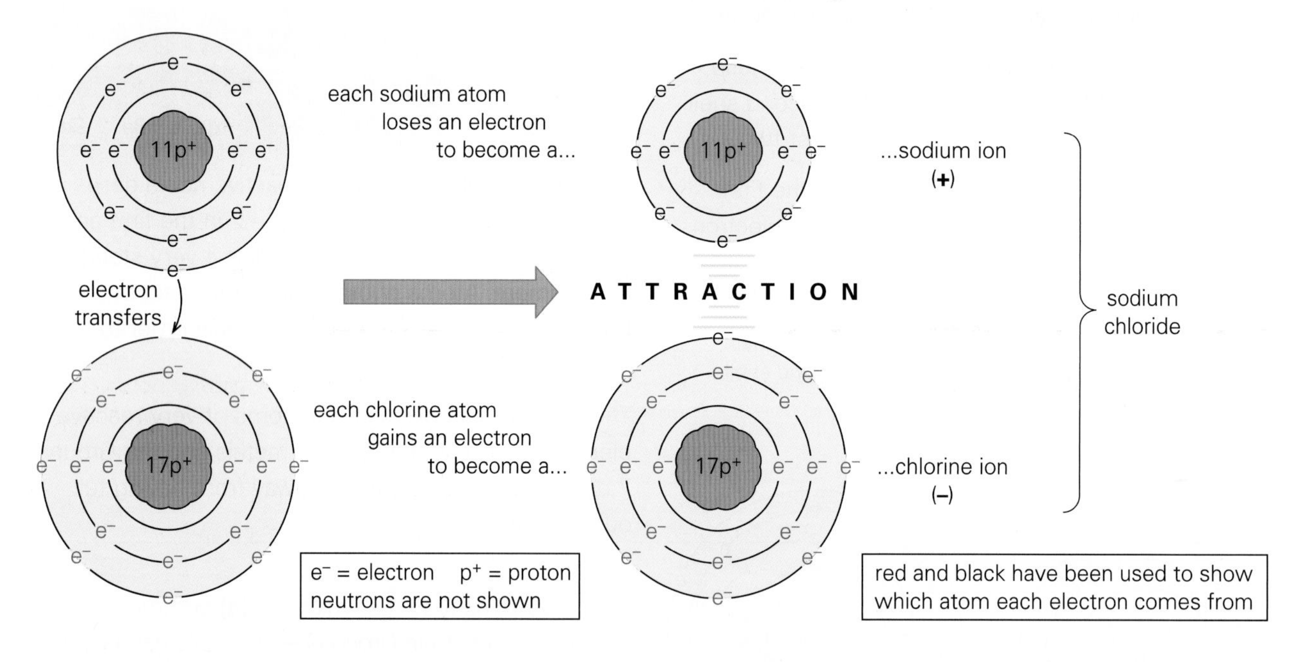

The sodium particle has a positive charge because it contains 11 positive protons and only 10 negative electrons. It is now a sodium ***ion,*** $\boldsymbol{Na^+}$. Conversely, the chlorine has gained an extra negative charge to become a chloride ion, $\boldsymbol{Cl^-}$. The solid compound is held together by the strong force of attraction between the oppositely charged ions. This is ***ionic bonding***.

Ions are atoms (or groups of atoms) having an overall positive or negative charge. Metals produce ***cations*** (positive) and non-metals give ***anions*** (negative).

More ions

When magnesium (a Group 2 metal) burns in oxygen, its atoms lose their two outer electrons to become cations with a 2+ charge. The 2 electrons mingle with the 6 outer electrons of the oxygen atoms to produce anions with full shells of 8e and an overall charge of 2. The formula for the product, magnesium oxide, is $\boldsymbol{Mg^{2+}O^{2-}}$ or simply ***MgO***.

The formula of an ionic compound is the smallest number of each ion needed to balance the charges.

Dot–cross diagrams

This is a simplified way of showing how atoms combine, drawing *only the outer shell electrons* as dots from one atom and crosses from the other. (Actually, the electrons are identical.) The formation of sodium chloride, magnesium oxide, and magnesium chloride are illustrated on the right using this method.

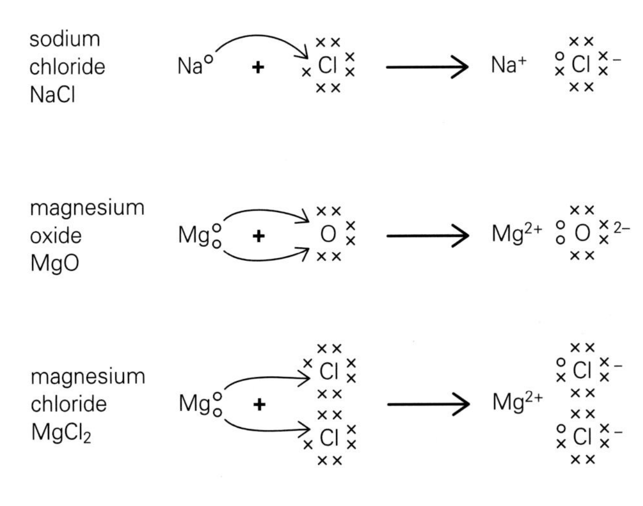

Giant ionic lattices

In sodium chloride crystals, every sodium ion attracts and is surrounded by 6 negative chloride ions. In turn, each chloride is surrounded by 6 positive sodium ions. Billions of them cluster together as closely as possible building up a giant three-dimensional lattice structure – cubic in this case.

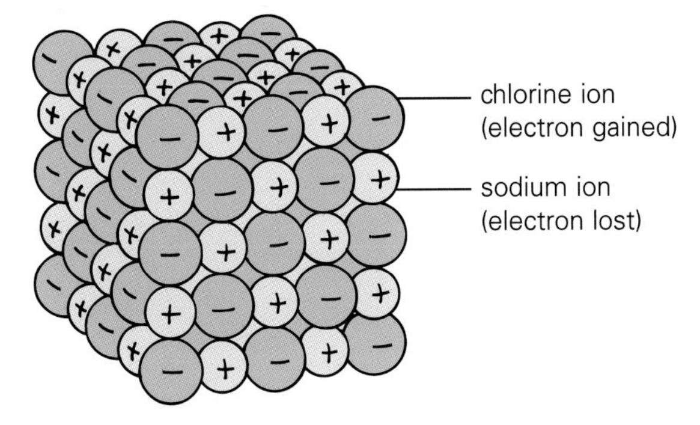

Crystal of sodium chloride

Other ionic compounds have different lattice structures which depend upon the sizes and charges of their ions.

A lot of energy (heat) is needed to break down the very strong forces in giant ionic lattices. As a result, they have high melting points. Once molten, however, the ions are free to move and will conduct an electric current.

Ions are also 'unlocked' when lattices dissolve in water (see next spread), again allowing them to conduct electricity.

Compound ions

Ions do not always come from single atoms – they can also be groups of joined atoms carrying an overall charge. Some of the most common ones are included in the table below.

Cations (+)	Anions (–)
ammonium NH_4^+	hydroxide OH^-
potassium K^+	nitrate NO_3^-
calcium Ca^{2+}	carbonate CO_3^{2-}
iron(III) Fe^{3+}	sulphate SO_4^{2-}
iron(II) Fe^{2+}	phosphate PO_4^{3-}

The formula of a substance formed from compound ions is arrived at by making sure that the positive and negative charges cancel each other. Ammonium carbonate, for example, must have two 1+ ammonium ions to balance one 2– carbonate ion.

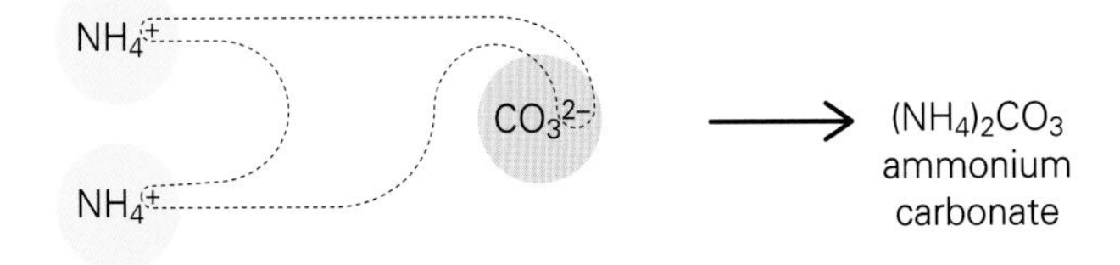

1 What happens to the electrons in the centre shells of the atoms when sodium burns in chlorine?
2 Draw the electron shells of the sodium and chloride ions.
3 What do you understand by the terms *ion*, *cation*, and *ionic bonding*?
4 Use the dot–cross method to show the ions present in magnesium chloride.
5 Why does sodium chloride form a giant ion lattice and why is it so difficult to melt?
6 Write the formulae for **(a)** sodium hydroxide **(b)** ammonium sulphate **(c)** iron(III) oxide.

3.08 Molecules

By the end of this spread you should be able to:
- describe how atoms are bonded in molecules
- explain why water is such a good solvent.

Electron sharing

When two atoms of non-metals join together as a molecule, they *both* need to gain electrons to fill their outer shells. They do this by sharing pairs of electrons so that each atom attains a noble gas structure.

Two chlorine atoms, having 7 electrons in their outer shells, will share one from each between them to make up full shells of eight. The attraction of the shared pair to both nuclei holds the atoms together and is called a ***covalent bond***.

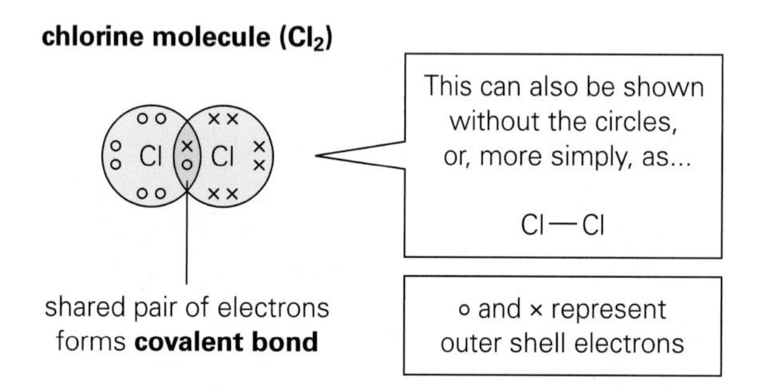

The stable unit of chlorine is therefore a ***diatomic*** (two atom) molecule Cl_2. Hydrogen and oxygen also exist as diatomic molecules.

hydrogen molecule (H_2)

H H

hydrogen needs only 2 electrons for a full shell

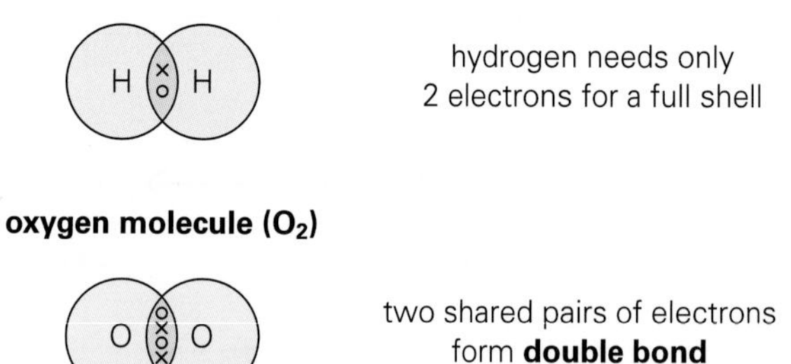

Although they vary a bit, the covalent bonds within these molecules are very strong. There is very little attraction, however, between the separate molecules, and so they are free to move about as gases.

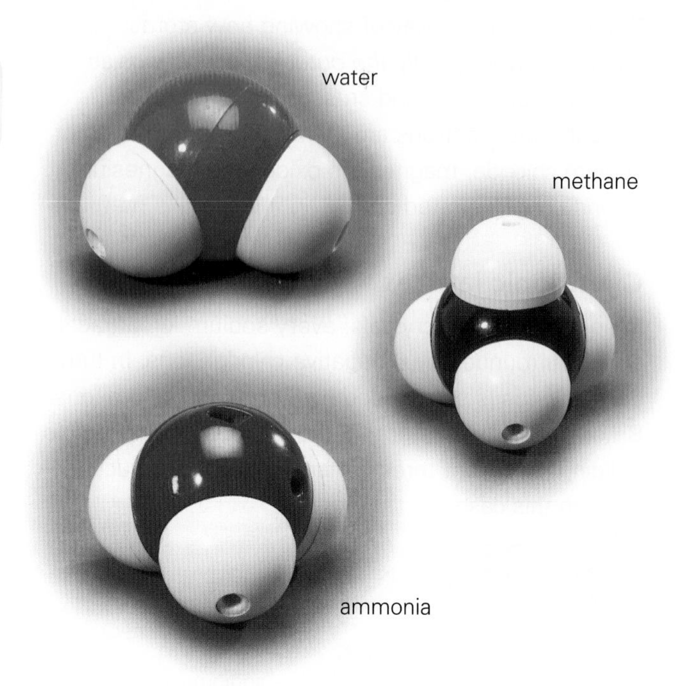

Models of molecules

Compounds

Covalent bonds between atoms of different elements are just as strong – as in the three N–H links in the ammonia molecule NH_3.

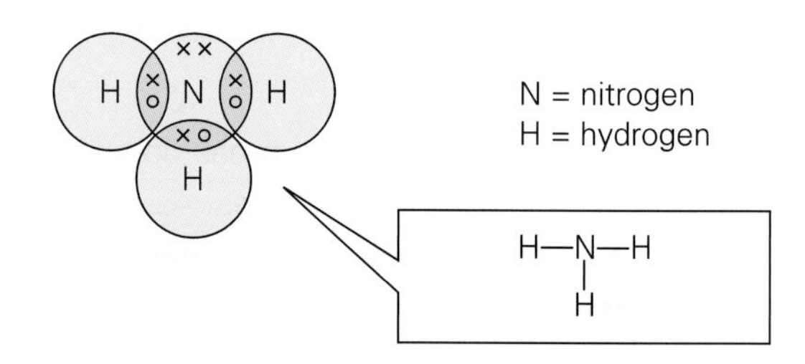

Carbon dioxide molecules have ***double bonds*** – two lots of shared electron pairs – attaching the carbon atom very firmly to each oxygen atom.

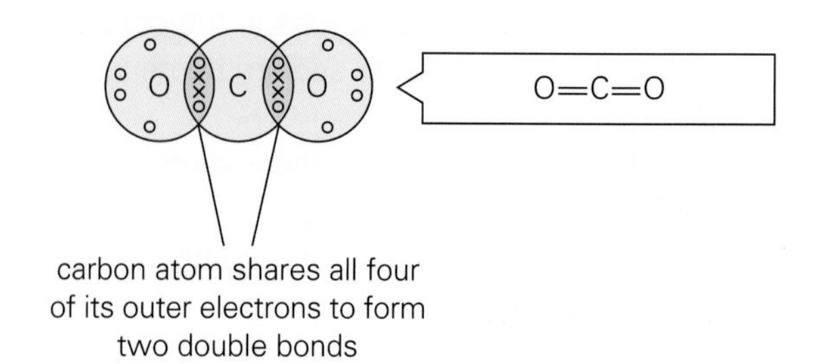

Water

Sometimes the electrons of a covalent bond are not shared equally between the two nuclei. In the water molecule, for instance, the oxygen nucleus attracts the shared electrons far more than the hydrogen. As a result, the oxygen end of the molecule becomes slightly negative and the hydrogen slightly positive. The charged ends are called ***poles***, and charges on them balance out so that the molecule as a whole is neutral.

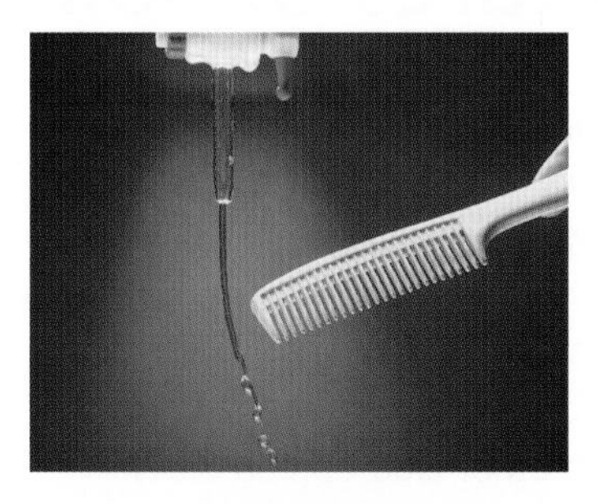

*Water is attracted to charged comb because its molecules are **polar**: they have charged ends.*

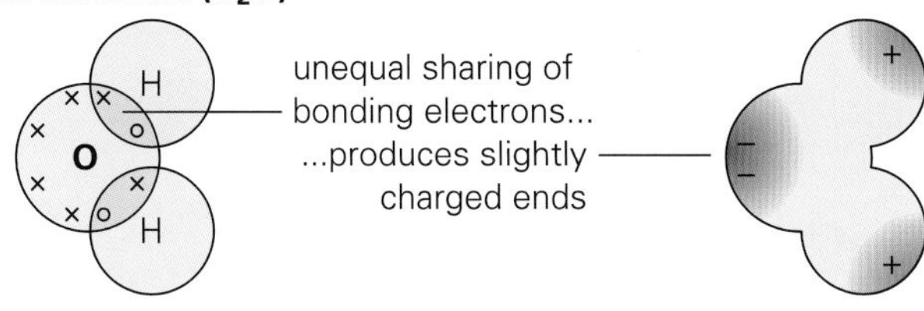

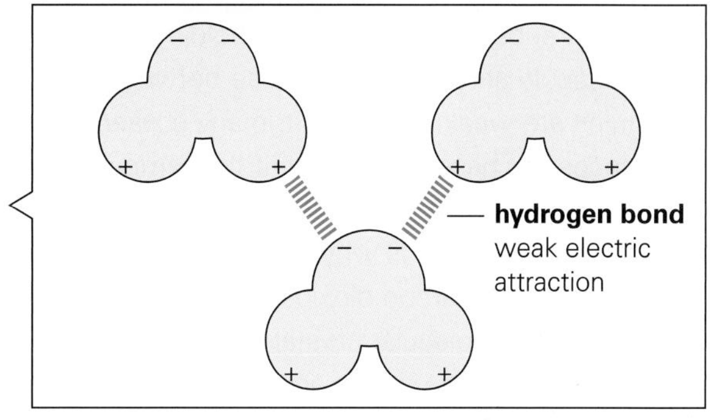

Attractions between the oppositely charged parts of adjacent molecules – ***hydrogen bonds*** – hold them together as a liquid at room temperature. Much energy (heat) must be supplied to overcome these forces and turn water to gas.

Methane has molecules of similar size and mass to water but no charged ends. In other words, it is non-polar. Because there are no attractions between the molecules of methane, it normally exists as a gas.

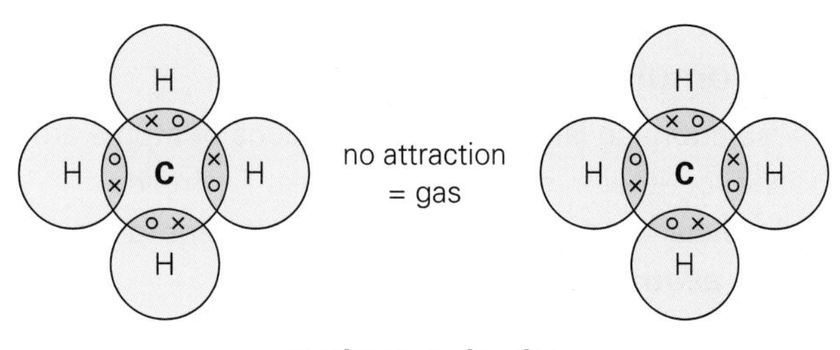

methane molecules

Dissolving

It is because its molecules are highly polar that water is such a good solvent. When sodium chloride dissolves, the negative ends of water molecules surround the positive sodium ions and they float away. Similarly, the negative chloride ions are attracted to the positive ends of other water molecules. These mobile ions will carry an electric current through the solution. The symbols for the ions are Na^+*(aq)* and Cl^-*(aq)* (see Spread 3.10).

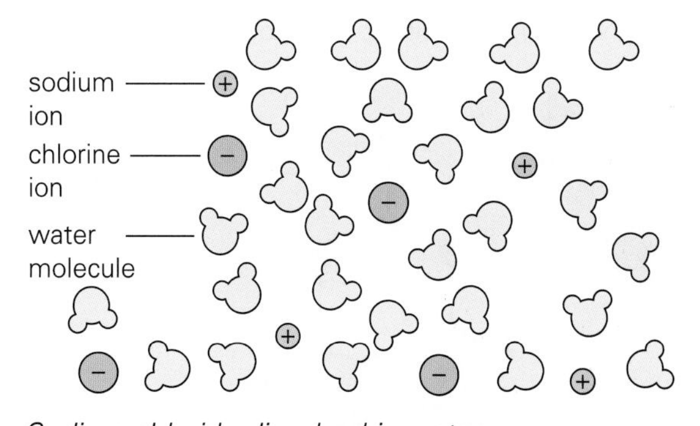

Sodium chloride dissolved in water

1 What holds the atoms together in a molecule of chlorine?

2 Why does oxygen exist as diatomic molecules O_2?

3 Draw the bonds present in the ammonia and carbon dioxide molecules using both dot–cross and line methods.

4 Why does the water molecule have slight electrical charges at different points?

5 Why is water a liquid at room temperature?

6 Describe what happens to its particles when sodium chloride dissolves in water. Why does this solution conduct a current?

3.09 Structures of solids

By the end of this spread, you should be able to:
- *describe ionic, molecular, and metallic crystals*
- *link the properties of solids to their structures.*

Ionic substances, such as sodium chloride, form ionic crystals – lattices in which positive and negative ions are bound tightly to each other (see Spread 3.07).

In covalent substances, the bonds holding atoms in each molecule are strong, but those between molecules are weak. As a result, many covalent substances are gases. However, if the temperature is low enough, the molecules may move slowly enough to bond together. That is why iodine (at room temperature) and carbon dioxide (cooled to –79 °C) both solidify as molecular crystals. But the weak molecule-to-molecule bonds are easily broken by heat, so these solids sublime (change to gas).

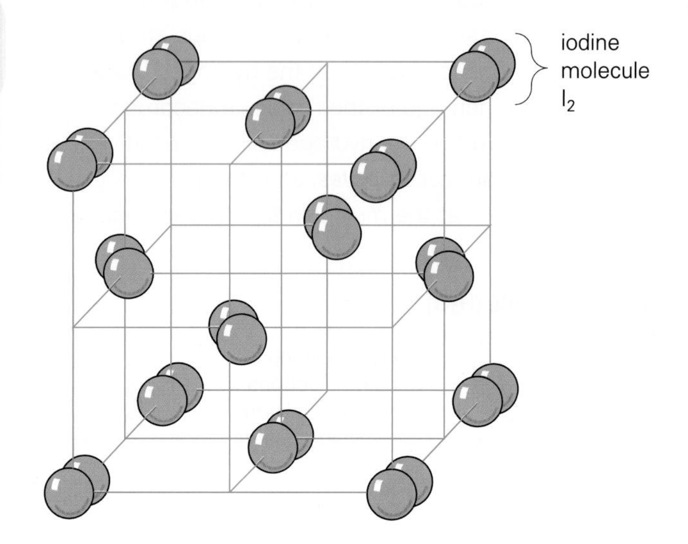

Structure of an iodine crystal. The blue lines show how the molecules are arranged. They are not bonds. In reality, the molecules would be much closer together.

Macromolecular crystals

These are huge lattices in which millions of atoms are bound together in one giant molecule. Examples include diamond and graphite – two different forms (called ***allotropes***) of carbon.

Diamond Each carbon atom forms covalent bonds with four others. The bonds are very strong, which is why diamond is one of the hardest substances known and has a very high melting point (3550 °C).

Graphite The carbon atoms are arranged in flat sheets in which each atom links with three others. Weaker forces hold these sheets loosely together in a layered structure. The layers slide over each other easily, making graphite suitable for the 'lead' in pencils and as a lubricant. The fourth outer electron of each carbon atom can move through the structure, allowing graphite to conduct electricity.

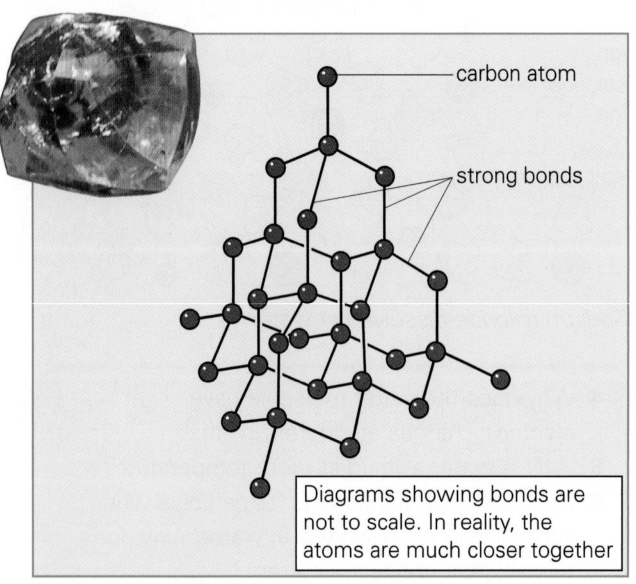

Structure of diamond

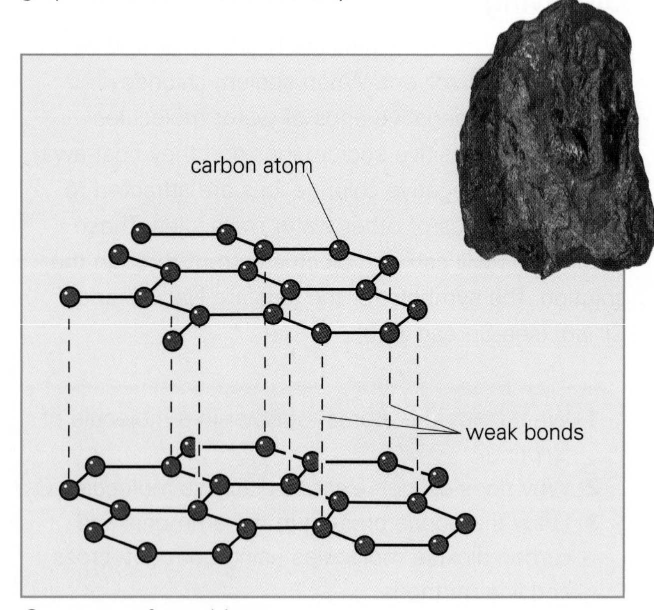

Structure of graphite

From quartz to ceramics

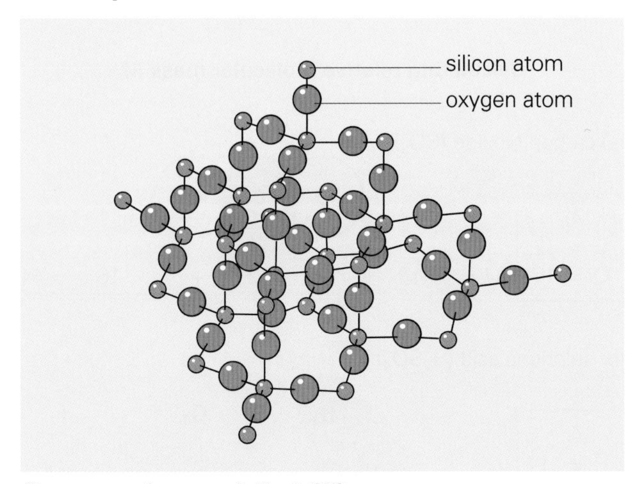

Structure of quartz (silica) SiO_2

Quartz (pure silica, SiO_2) has a diamond-like structure. It is very hard and has a high melting point. Many rocks are based on silica combined with metals.

Sand is ground-up quartz mixed with impurities.

Glass is made by strongly heating a mixture of sand, soda (sodium carbonate), and limestone (calcium carbonate) to produce a thick, sticky liquid. This gradually hardens as it cools, but it does not form crystals and remains transparent. Glasses are mixtures, not compounds, so their composition can vary and they do not have fixed melting points.

Clay is based on quartz combined with aluminium and hydrogen atoms. It has a giant layer structure something like graphite. Water molecules can get between the layers and allow them to slide over each other. Wet clay is therefore pliable and slippery.

When clay is fired (heated in a kiln), the atoms form new bonds with the layers above and below. The result is a hard, brittle material called a ceramic:

1 What holds the particles together in **(a)** ionic crystals **(b)** molecular crystals **(c)** metallic crystals?
2 Why is carbon dioxide a gas at room temperature?
3 Why is it so difficult to melt diamond?
4 Why does graphite feel slippery?
5 Glass does not have a fixed melting point. Why?
6 Why are alloys harder than pure metals?

Metallic crystals

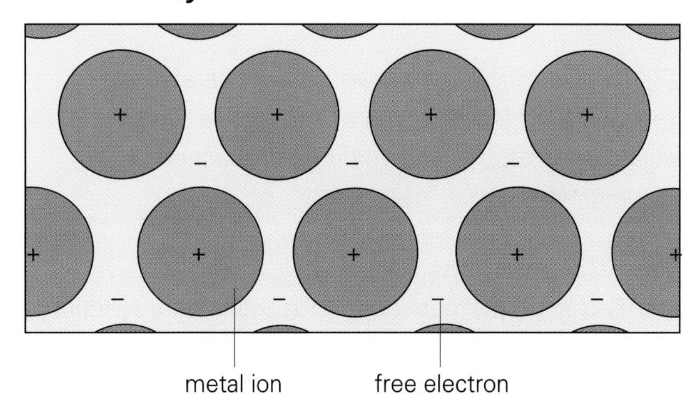

Metals are also made up of giant structures. The atoms are packed together closely – like oranges on display. Metals are malleable, so the bonds binding the atoms are both strong and flexible. These ***metallic bonds*** arise because the outer electrons become separated from their atoms. As a result, there is a lattice of positive ions surrounded by a 'sea' of free electrons. The mobile electrons bond the metal ions tightly into the lattice, making it difficult to melt. They also provide a means of conducting electricity.

Alloys

Metals bend easily because the layers of atoms in them can slide over each other. If a small amount of another metal is added to make an alloy, the atoms of the added metal are a different size. They make it more difficult for the layers to slide. That is why the alloy is harder than the pure metal.

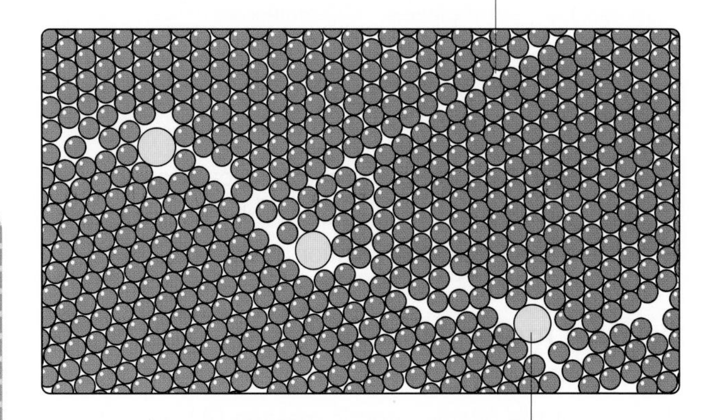

Structure of an alloy

3.10 Equations, calculations, and moles

By the end of this spread you should be able to:
- understand the meaning of chemical equations
- calculate masses and volumes of reacting substances.

Chemical equations use symbols to show how atoms change partners during reactions. In a reaction, no atoms are created or destroyed, so there must be the same number of each kind of atom in the *reactants* (starting materials) as there are in the *products* (new substances formed). In other words, the equation must *balance*. For example, a mixture of hydrogen and oxygen explodes when lit. Below you can see how the atoms regroup to make molecules of water.

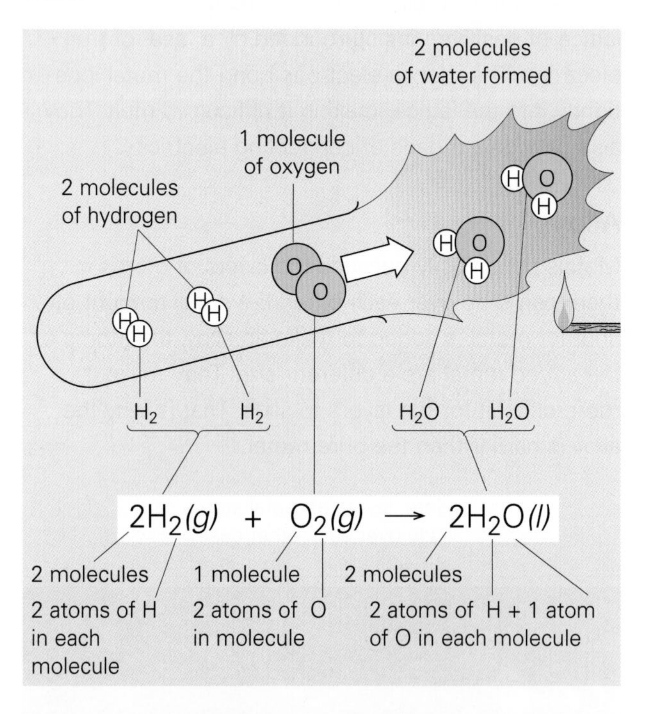

The symbols in brackets give the physical state of the substances: *(s)* is solid, *(l)* is liquid, *(g)* is gas, and *(aq)* is an aqueous solution (a solution in water).

Relative molecular mass (M_r)

The ***relative molecular mass*** (M_r) of a compound is the sum of the relative atomic masses (A_r) of all the atoms making up a molecule of the compound.

For ionic compounds, the term used is the ***relative formula mass***, but the calculation is the same.

Calculating relative molecular mass M_r

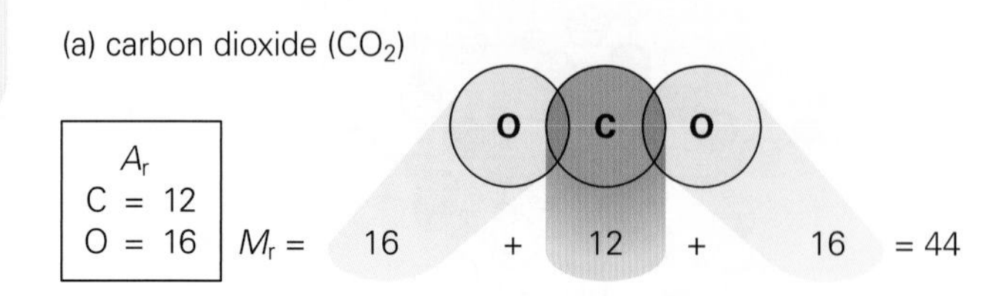

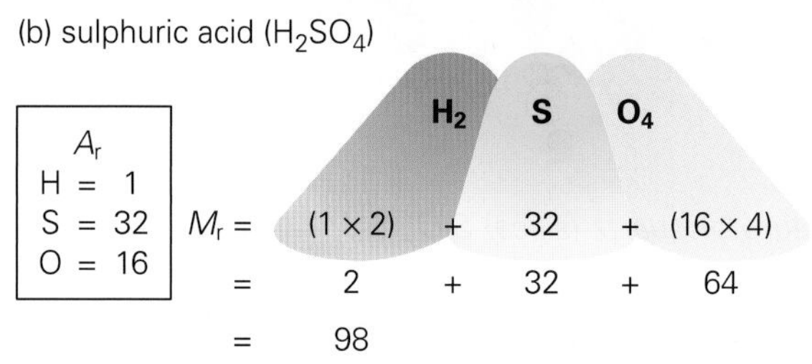

Ionic equations

In solution, sodium chloride exists as separate ions $Na^+(aq)$ and $Cl^-(aq)$ (see Spread 3.08). Dissolved in water, silver nitrate also separates – into silver ions $Ag^+(aq)$ and nitrate ions $NO_3^-(aq)$. When the two solutions are mixed, silver ions and chloride ions join together as insoluble silver chloride. This appears as a white solid. The equation for the reaction can be written:

$$\text{silver nitrate} + \text{sodium chloride} \rightarrow \text{sodium nitrate} + \text{silver chloride}$$

$$AgNO_3(aq) + NaCl(aq) \rightarrow NaNO_3(aq) + AgCl(s)$$

It is more accurate, however, to show the separate ions:

$$Ag^+(aq) + NO_3^-(aq) + Na^+(aq) + Cl^-(aq) \rightarrow Na^+(aq) + NO_3^-(aq) + AgCl(s)$$

The sodium and nitrate ions, highlighted in yellow, remain unchanged in solution. Taking no part in the action, they are ***spectator ions*** and can be left out of the equation. The ions which *do* change are shown as an ***ionic equation***:

$$Ag^+(aq) + Cl^-(aq) \rightarrow AgCl(s)$$

Ionic equations are used to illustrate many chemical changes, especially those occurring during electrolysis (see Spread 3.25).

The mole

In chemical reactions, *equal amounts* of different substances means *equal numbers of particles* (such as atoms or molecules), and not equal masses.

If you take the A_r or M_r of any substance (for example 12 for carbon) and write it as a mass in grams (12 g), then the mass contains 6×10^{23} particles. This amount is called one ***mole***. The number 6×10^{23} is known as the ***Avogadro constant***. The mass of one mole of a substance is called the ***molar mass***:

1 mole (6×10^{23}) of...	Molar mass
...atoms of helium ($A_r = 4$)	4 g
...molecules of oxygen ($M_r = 32$)	32 g
...molecules of carbon dioxide ($M_r = 44$)	44 g

Molar gas volume

One mole of any gas contains 6×10^{23} particles. At normal room temperature and pressure ***(r.t.p.)***, these always occupy the same volume, 24 dm^3 (24 litres). This ***molar gas volume*** is the same for all gases.

What is the mass of 48 dm^3 of methane (at r.t.p.)?

The M_r of methane (CH_4) = 12 + (1 × 4) = 16

1 mole of methane occupies 24 dm^3. Its mass is 16 g.

2 moles of methane occupy 48 dm^3. Its mass is 32 g.

Concentration of a solution

The ***concentration*** of a solution is the amount (in moles) of solute dissolved per dm^3 of solution.

If some dilute sulphuric acid has 196 g of pure acid in 1 dm^3 of solution, what is the concentration?

For sulphuric acid (H_2SO_4): M_r = 98

So, number of moles of H_2SO_4 = $^{196}/_{98}$ = 2

So, concentration = 2 mole/dm^3, written 2M.

Calculating from equations

The equation for a reaction is used when solving problems about masses and volumes:

What mass of calcium oxide could be produced by heating 500 g of calcium carbonate? What volume of carbon dioxide (at r.t.p.) would be given off?

The equation for this reaction is:

	calcium carbonate	→	calcium oxide	+	carbon dioxide
	$CaCO_3(s)$	→	$CaO(s)$	+	$CO_2(g)$
Amount:	1 mole	gives	1 mole	and	1 mole
M_r:	40 + 12 + (16 × 3)		40 + 16		12 + (16 × 2)
Molar mass:	100 g	gives	56 g	and	44 g

Since 100 g of calcium carbonate is 1 mole, 500 g is 5 moles, which form 5 moles of each product.

So, mass of calcium oxide = 5 × 56 = 280 g

As 1 mole of a gas occupies 24 dm^3 at r.t.p.:

volume of carbon dioxide = 5 × 24 = 120 dm^3

Finding a formula

From the masses of substances involved in a reaction it is possible to work out the formula of a compound:

When 8.0 g of copper(II) oxide is heated in hydrogen, the oxygen is removed, and 6.4 g of copper remains. What is the formula of copper(II) oxide?

To make copper(II) oxide, 6.4 g of copper was linked to 8.0 – 6.4 = 1.6 g of oxygen. So,

the ratio of copper to oxygen is	6.4 g : 1.6 g
Dividing by the A_r values	$^{6.4}/_{6.4}$: $^{1.6}/_{1.6}$
...changes the ratio to moles:	0.1 mole : 0.1 mole
Therefore the ratio of atoms is	1 : 1

So: formula for copper(II) oxide is Cu_1O_1, or just CuO.

1. $2Mg(s) + O_2(g) \rightarrow XMgO(s)$
   In this equation **(a)** what should the number X be? **(b)** what do *(s)* and *(g)* stand for?
2. What do the symbols A_r and M_r stand for? What is the M_r for water (H_2O)?
3. What is meant by a *mole* of methane molecules? What is their **(a)** mass **(b)** volume at r.t.p?
4. If a solution has *a concentration of 3M*, what does this mean?
5. Methane (CH_4) burns in oxygen, forming water and carbon dioxide. Write a blanced equation for the reaction. Starting with 48 dm^3 of methane (at r.t.p.), what mass of water and what volume of carbon dioxide (at r.t.p.) would be formed?

3.11 Breaking and making bonds

By the end of this spread, you should be able to:
- *describe chemical reactions as the breaking and making of bonds between atoms*
- *calculate the heat produced by a fuel burning.*

Before studying this spread, see:
3.08 on bonds
4.17 on energy
3.10 on chemical equations and moles

When things burn, they combine with oxygen and give out energy (heat). Methane (natural gas) is the fuel used in bunsen burners and gas cookers. The equation for the burning reaction is:

$CH_4(g)$ + $2O_2(g)$ → $CO_2(g)$ + $2H_2O(l)$ + *energy*

methane oxygen carbon dioxide water

To start the fuel burning, some energy must first be supplied by a match or a spark. This ***activation energy*** breaks down the bonds between the carbon and hydrogen atoms in the methane, and between the two atoms in the oxygen molecule. The separated atoms are free to form new bonds between carbon and oxygen (making carbon dioxide) and between hydrogen and oxygen (making water). Much more energy is released in this process than was taken in to break down the old bonds, so the reaction gives out heat – it is ***exothermic***. Many chemical changes, besides burning, are exothermic.

A covalent bond is usually shown as a line between the symbols for the atoms joined together.

A hydrogen molecule, H—H, has a *single bond* (two shared electrons).

An oxygen molecule, O═O, has a *double bond* (four shared electrons).

In a chemical reaction, the starting substances are the ***reactants***; the new substances formed are the ***products***.

In other reactions, a lot of energy is needed to break down the bonds in the reactants, and only a little is given out by the making of the new bonds of the products. Such reactions take in heat from the surroundings and feel cold – they are ***endothermic***.

Energy is *absorbed* when bonds are *broken*.

Energy is *released* when bonds are *formed.*

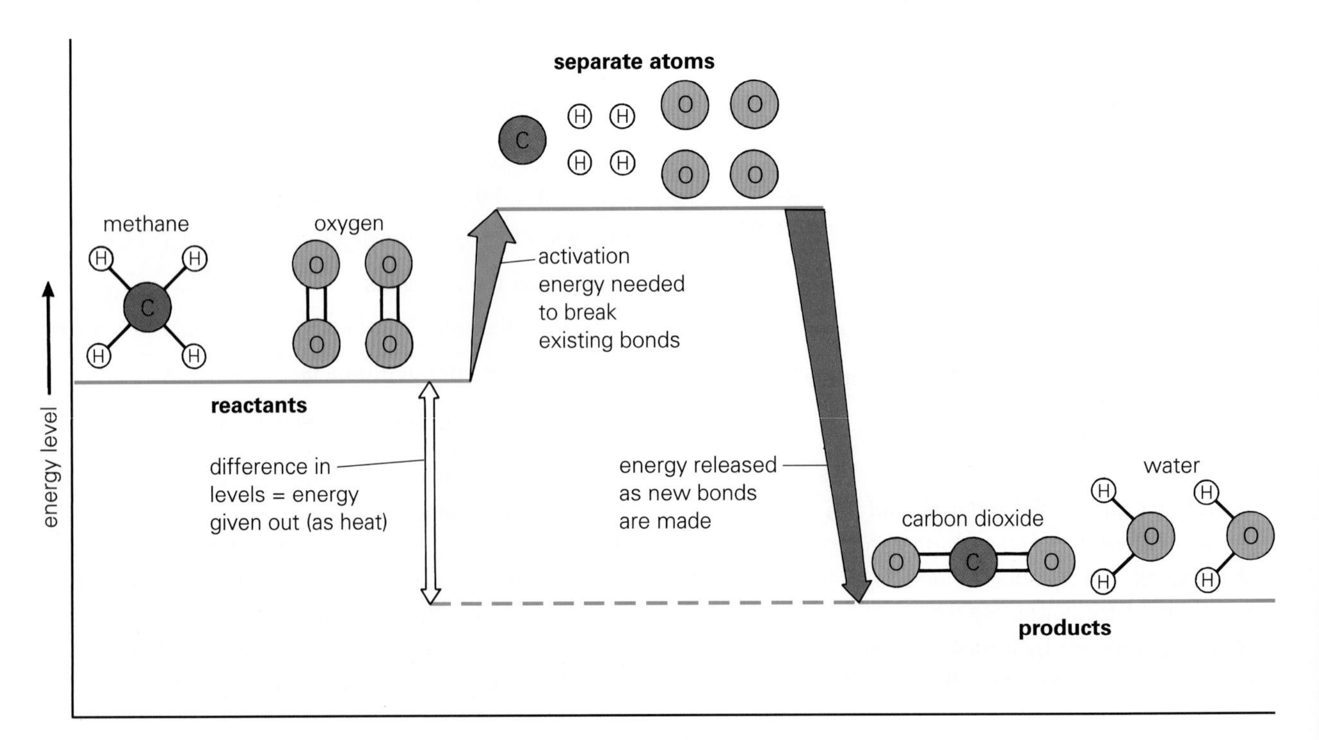

Bond energies

The amount of energy needed to break a particular bond is the same as the amount given out when the bond is made. This is the ***bond energy***. It is measured in kilojoules per mole (kJ/mol). The table on the right shows some typical values.

Bond		Bond energy in kJ/mol
C—C		347
C═C		612
C—H	in methane	435
C—H	in others	413
C═O		805
O—H		464
O═O		498
H—H		436

— single bond ═ double bond

Calculating energy changes

If you know the equation for a reaction and the structures of the substances taking part, you can use bond energies to calculate the overall heat change:

How much heat energy is given out when 24 dm^3 of methane is completely burned in oxygen? (24 dm^3 is the volume of 1 mole of gas.)

The chart below shows how to calculate the answer:

Bond breaking

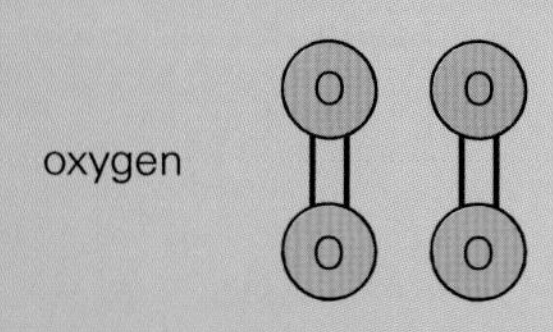

4 C—H bonds need 4 × 435 = 1740 kJ

oxygen

2 O—O bonds need 2 × 498 = 996 kJ

Total energy absorbed = 2736 kJ

Bond making

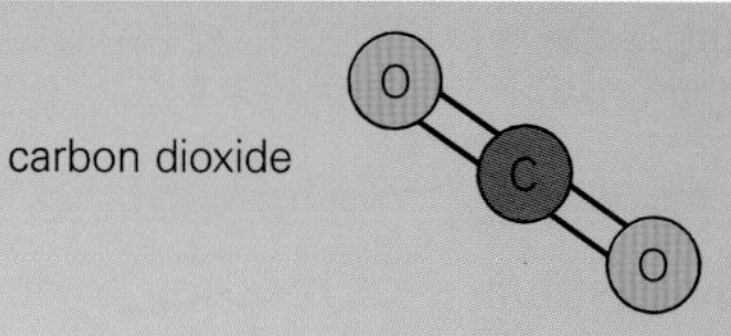

2 C═O bonds release 4 × 805 = 1610 kJ

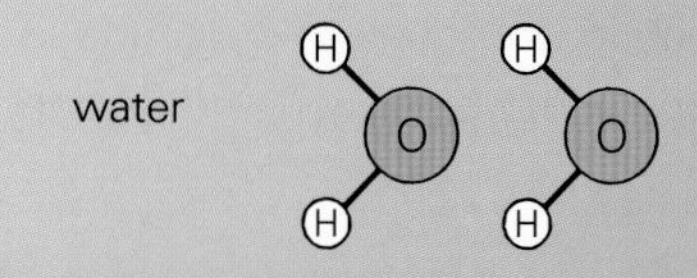

4 O—H bonds release 4 × 464 = 1856 kJ

Total energy released = 3466 kJ

Energy (heat) given out in reaction = total energy released – total energy absorbed
= 3466 kJ – 2736 kJ
= 730 kJ

1. Give an example of a molecule containing a double bond. What does this mean?
2. What is meant by activation energy?
3. Explain why heat is given out when methane burns.
4. Why are some reactions *endothermic?*
5. What is meant by *bond energy*?
6. Calculate the heat energy given out when 1 mole of propane (C_3H_8) is burnt in 5 moles of oxygen to give 3 moles of carbon dioxide and 4 moles of water. (The structure of propane is on the right.)

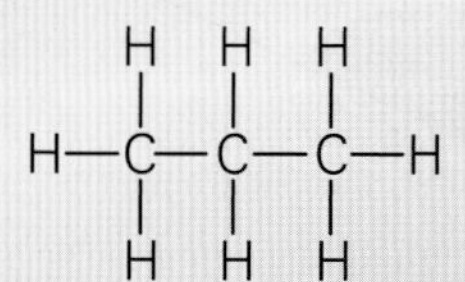

Structural formula of propane

3.12 Types of chemical reaction

By the end of this spread you should be able to:
- *recognize different types of chemical reaction*
- *apply the term redox to a range of reactions*

The different reactions by which new substances are made can be grouped into a few main types.

Synthesis is the building up of a new substance from two or more simpler ones. Iron and sulphur, heated together, join to form iron(II) sulphide.

iron	+	sulphur	→	iron(II) sulphide
$Fe(s)$	+	$S(s)$	→	$FeS(s)$

In photosynthesis (see Spread 2.02), green plants create glucose from carbon dioxide and water with the help of sunlight. Here, there is a by-product, oxygen, which renews the atmosphere.

carbon dioxide	+	water	→	glucose	+	oxygen
$6CO_2(g)$	+	$6H_2O(l)$	→	$C_6H_{12}O_6(aq)$	+	$6O_2(g)$

Thermal decomposition is the breaking down of a substance into simpler ones by the action of heat.

Quite gentle heating is enough to decompose green copper(II) carbonate into black copper(II) oxide and carbon dioxide.

copper(II) carbonate	→	copper(II) oxide	+	carbon dioxide
$CuCO_3(s)$	→	$CuO(s)$	+	$CO_2(g)$

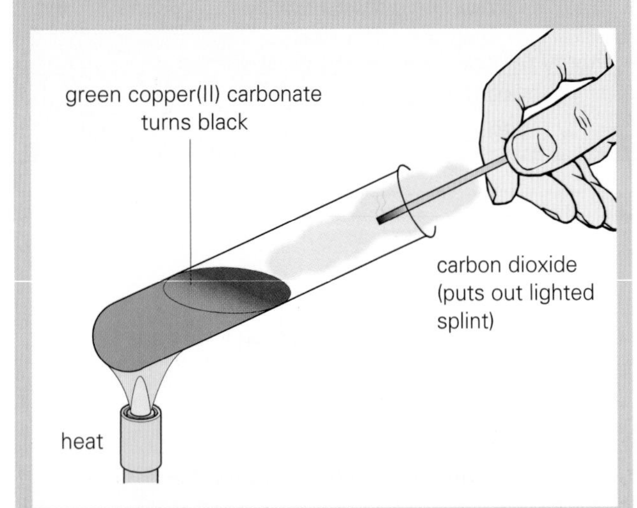

Substances can also decompose when acted upon by bacteria or enzymes (see Spread 3.16).

Magnesium burning in oxygen.

Precipitation is the formation of solid insoluble particles when two clear solutions are mixed. The bits appear as a suspension (cloudiness) called a ***precipitate***. Usually, the solutions are of ionic compounds such as silver nitrate and sodium chloride (as in Spread 3.04). Similarly, mixing solutions of barium nitrate and copper(II) sulphate produces a white precipitate of solid barium sulphate:

barium nitrate	+	copper(II) sulphate	→	barium sulphate	+	copper(II) nitrate
$Ba(NO_3)_2(aq)$	+	$CuSO_4(aq)$	→	$BaSO_4(s)$	+	$Cu(NO_3)_2(aq)$

The equation can also be written ionically:

$$Ba^{2+}(aq) + SO_4^{2-}(aq) \rightarrow BaSO_4(s)$$

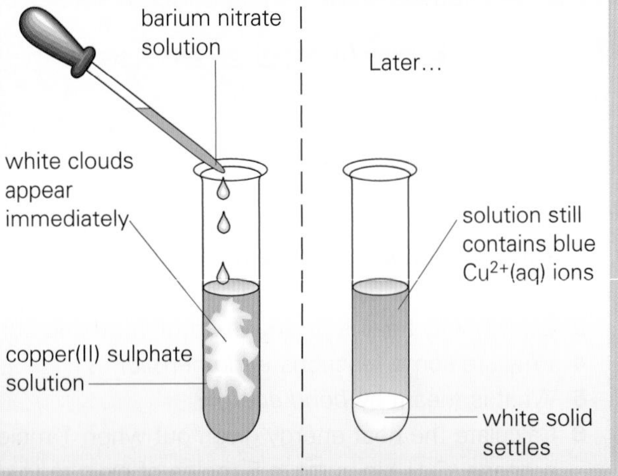

Neutralization is the reaction of an acid with a base to form a salt (see Spread 3.14).

Redox is described on the next page.

Redox

Black copper(II) oxide, heated in a stream of hydrogen, changes to red-brown copper metal. The hydrogen has removed the oxygen and ***reduced*** the copper(II) oxide. Gaining oxygen to become water, the hydrogen has been ***oxidized***.

copper(II) oxide + hydrogen → copper + water

$$CuO(s) + H_2(g) \rightarrow Cu(s) + H_2O(l)$$

Reduction and oxidation are two parts of the same process, called ***redox***.

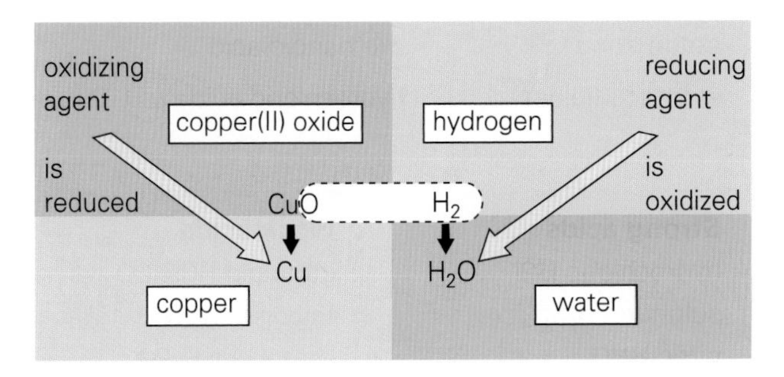

Magnesium, burning in oxygen, is oxidized to magnesium oxide. The two outer electrons of the magnesium atom are transferred to the oxygen atom.

$$Mg + O \rightarrow Mg^{2+}[O]^{2-}$$

Oxidation of a substance is the *addition of oxygen* to it or (more generally) the *removal of electrons* from it. ***Reduction*** is therefore the *removal of oxygen* or the *addition of electrons*.

Some metals can form more than one kind of ion. When iron is oxidized, its atoms lose either two electrons to become Fe^{2+} ions or three giving Fe^{3+} ions. The Roman number in the name of a compound shows the oxidation state of the ion. Hence, iron(II) sulphate is $Fe^{2+}\ SO_4^{2-}$ or, simply, $FeSO_4$.

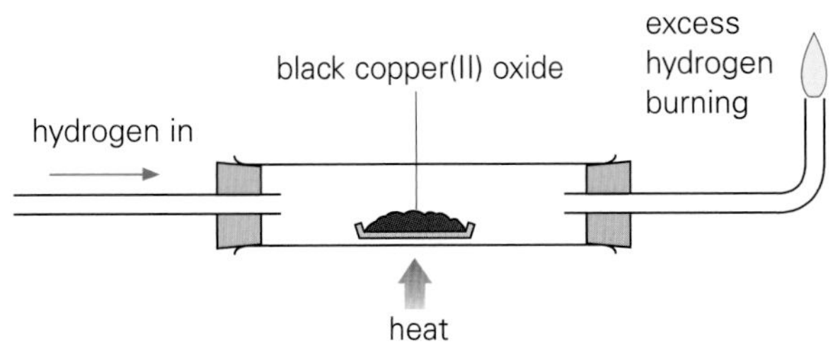

An ***oxidizing agent*** often contains oxygen, which it gives up. Hydrogen peroxide H_2O_2, for instance, will oxidize acidified iron(II) sulphate solution to brown iron(III) sulphate. The oxygen atom released from the peroxide does not join on to the iron(II) sulphate, however. Instead, it removes one electron from each of two iron(II) ions and joins with hydrogen ions from the acid (see Spread 3.13) to form water.

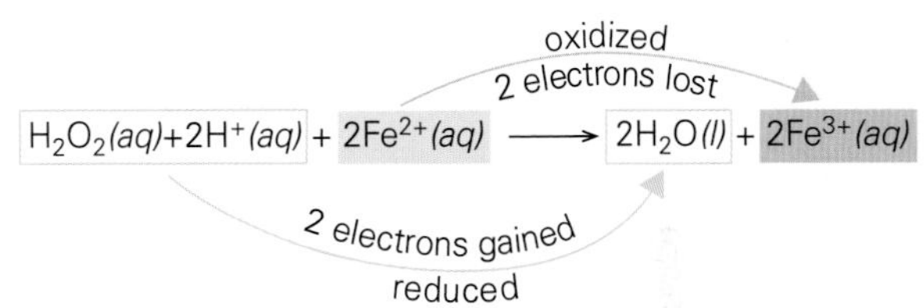

In redox reactions, electrons pass from a reducing agent to an oxidizing agent.

This definition can be applied to a range of reactions not involving oxygen, including some also known as displacement reactions.

Reaction	Electrons pass from reducing agent to oxidizing agent		
zinc dissolving in sulphuric acid	$Zn(s)$	2e ↷	$2H^+(aq)$
sodium dissolving in water	$Na(s)$	1e ↷	$H_2O(l)$
sodium combining with chlorine	$2Na(s)$	2e ↷	$Cl_2(g)$
silver plating (electrolysis)	cathode	1e ↷	$Ag^+(aq)$
iron displacing copper from copper (II) sulphate solution	2Fe(s)	2e ↷	$Cu^{2+}(aq)$

1. What is *thermal decomposition?*
2. What do you see when copper(II) carbonate is heated? Write the equation for the reaction.
3. Name the precipitate which forms when solutions of barium nitrate $Ba(NO_3)_2$ and copper(II) sulphate $CuSO_4$ are mixed. Write an equation for the reaction.
4. What do the terms *oxidation* and *reduction* mean?
5. What two particles can iron atoms become when they are oxidized?
6. Which substance is oxidized in the reaction between **(a)** zinc and sulphuric acid **(b)** water and sodium **(c)** iron and chlorine?

3.13 Acids and bases

By the end of this spread you should be able to:
- *give the main properties of acids and bases*
- *describe what indicators and the pH scale show*

Acids

Sour fruit, vinegar, nettles, and your stomach all contain natural acids. These, and the more powerful acids made industrially, are usually aqueous solutions. The more water there is, the more ***dilute*** the acid. Acids are ***corrosive***, eating into many materials. When ***concentrated*** (with little or no water), acids can be very dangerous and should never be handled.

Pure acids are covalent substances containing hydrogen. In water, the molecules split up, releasing some, or all, of the hydrogen into solution as positive hydrogen ions, ***H^+(aq)***. Dilute sulphuric acid, for example, ionizes as shown:

sulphuric acid → hydrogen ions + sulphate ions

$H_2SO_4(aq) \rightarrow 2H^+(aq) + SO_4^{2-}(aq)$

It is the hydrogen ions in dilute acids which attack and dissolve metals placed in them. The vigorous fizzing which occurs when magnesium dissolves in dilute sulphuric acid is hydrogen gas being given off:

sulphuric acid + magnesium → hydrogen + magnesium sulphate

$H_2SO_4(aq) + Mg(s) \rightarrow H_2(g) + MgSO_4(aq)$

$2H^+(aq) + Mg(s) \rightarrow H_2(g) + Mg^{2+}(aq)$

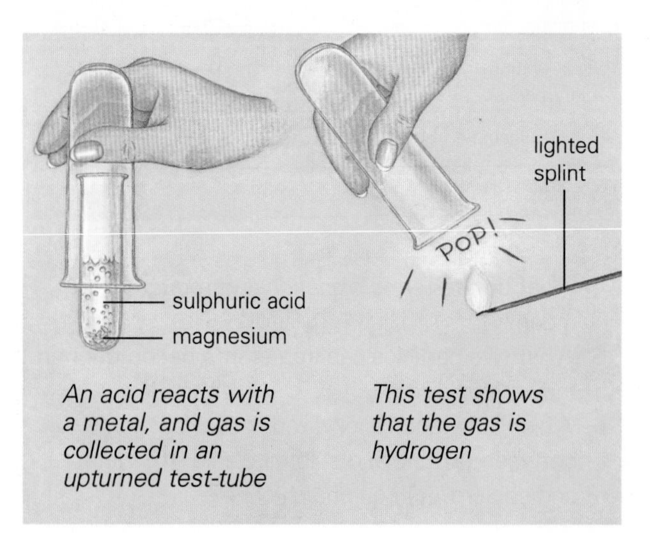

An acid reacts with a metal, and gas is collected in an upturned test-tube

This test shows that the gas is hydrogen

Some naturally occurring acids

	contains...
lemon juice	citric acid
vinegar	ethanoic acid (acetic acid)
tea	tannic acid
sour milk	lactic acid
grapes	tartaric acid
nettle sting	methanoic acid
stomach (juices)	hydrochloric acid

Strong acids
hydrochloric HCl
sulphuric H_2SO_4
nitric HNO_3

Weak acids
ethanoic HCH_3CO_2
carbonic H_2CO_3
phosphoric H_3PO_4

Acids which ionize completely and have lots of hydrogen available are called ***strong acids***. Those which only partly ionize, giving fewer hydrogen ions, are ***weak acids***.

Bases

Bases are the chemical 'opposites' of acids. They react strongly with acids, ***neutralizing*** them by removing their hydrogen ions.

Commonly, ***bases*** are oxides and hydroxides of metals. The few bases which dissolve in water are called ***alkalis***. Alkaline solutions contain negative hydroxide ions, ***OH^-(aq)***. Like acids, alkalis can be strong or weak.

sodium hydroxide → sodium ions + hydroxide ions

$NaOH(aq) \rightarrow Na^+(aq) + OH^-(aq)$

Strong alkalis
sodium hydroxide
potassium hydroxide

Weak alkali
ammonia
sodium hydrogen carbonate

Alkalis can be just as corrosive as acids. Their powerful chemical action is often used in bath, sink, and oven cleaners.

Indicators

There are many natural dyes which have a different colour depending on whether they are in an acidic or alkaline solution. Beetroot juice is just one example. Dyes like this are called ***indicators***. Litmus (extracted from lichen) is commonly used to distinguish between acids and alkalis.

Acids turn litmus red.

Alkalis turn litmus blue.

Laboratory indicators come as solutions or test papers.

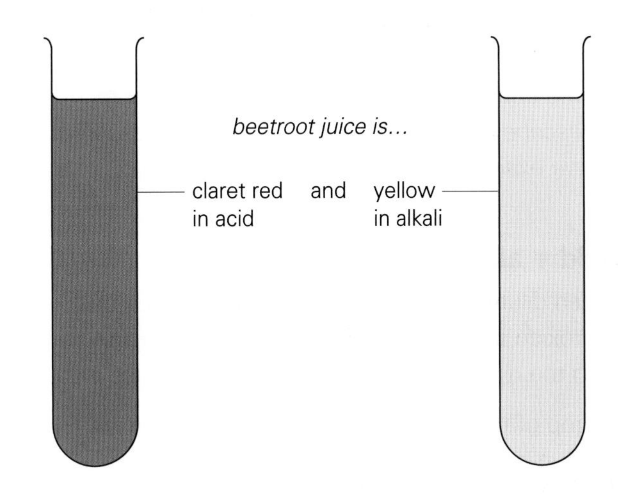

The pH scale

This is a scale of numbers from 0 to 14, which shows how strong or weak an acid or alkali is. The strongest acids have a pH of 0 or 1. The strongest alkalis have a pH of 13–14.

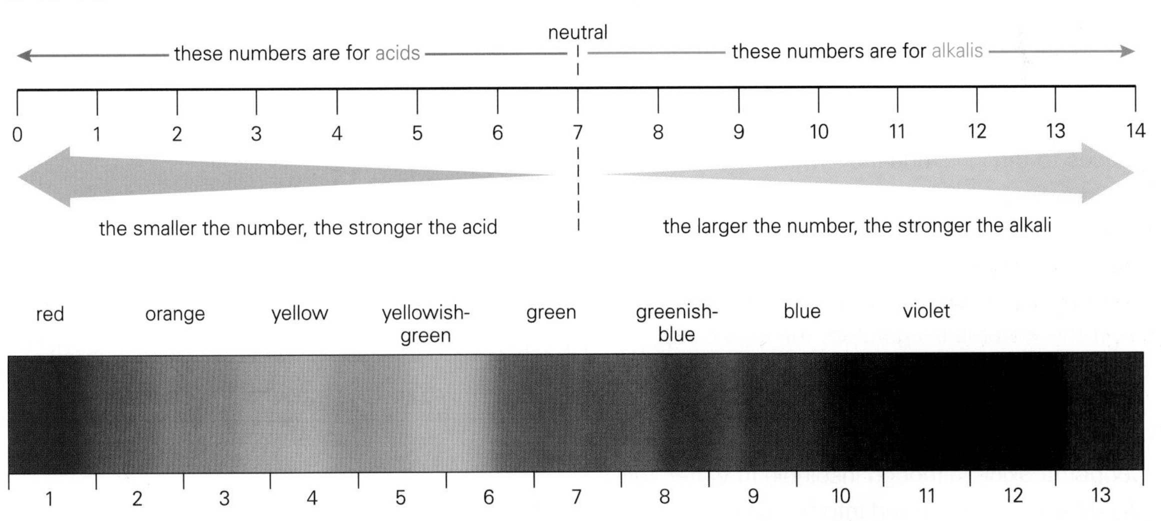

An *acid* has a *pH* value *less than 7*.

An *alkali* has a *pH* value *greater than 7*.

Pure water and *neutral* solutions have a *pH* value of *exactly 7*.

You can measure pH with ***universal indicator***. This contains a mixture of dyes. It goes a different colour at different pH values.

Protect your eyes! always wear safety glasses when experimenting with acids and alkalis.

1. What element is found in all acids?
2. What happens to the molecules of an acid when it is diluted with water?
3. What is the difference between a *strong* acid and a *weak* acid?
4. Explain the terms *base* and *alkali*.
5. What causes the fizzing when magnesium is put into sulphuric acid? How would you test for the gas?
6. What is the chemical meaning of the term *indicator*?
7. Universal indicator paper dipped into vinegar shows a pH of 3. What does this tell you about the vinegar?
8. What is the pH value of pure water?

3.14 Neutralization

By the end of this spread you should be able to:
- describe what happens in neutralization reactions
- give examples of everyday neutralizations.

Acid + alkali

When dilute hydrochloric acid is added to sodium hydroxide solution, the hydrogen ions in the acid join with the hydroxide ions in the alkali to produce water.

A drop of the mixture dabbed onto pH paper will show when it has become ***neutral***.

hydrochloric acid + sodium hydroxide → water + sodium chloride

$HCl(aq) + NaOH(aq) \rightarrow H_2O(l) + NaCl(aq)$

Showing the ions present:

$H^+(aq) + Cl^-(aq) + Na^+(aq) + OH^-(aq) \rightarrow H_2O(l) + Na^+(aq) + Cl^-(aq)$

This reduces to the ionic equation which represents all acid/alkali reactions:

$$H^+(aq) + OH^-(aq) \rightarrow H_2O(l)$$

The sodium and chloride ions stay unchanged in solution. If the water is evaporated, the ions cling together as solid sodium chloride. This is a ***salt***.

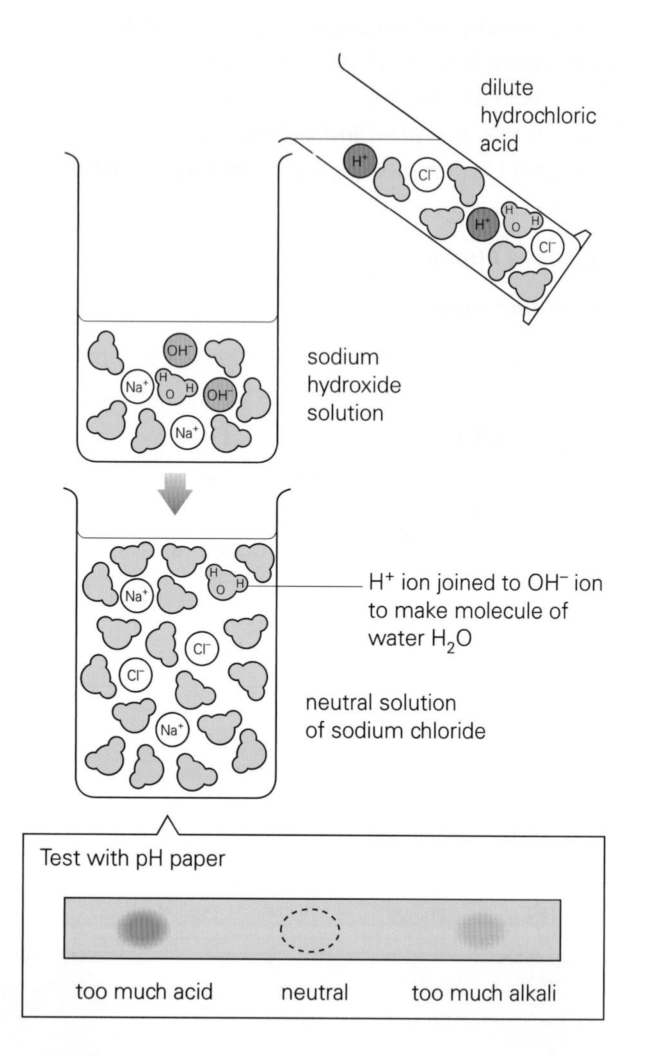

Acid + base

Black copper(II) oxide, although insoluble in water, will gradually dissolve when stirred into hot dilute sulphuric acid. The oxygen from the base combines with the hydrogen ions from the acid to form neutral water. Copper(II) ions pass into solution, turning it blue.

sulphuric acid + copper(II) oxide → water + copper(II) sulphate

$H_2SO_4(aq) + CuO(s) \rightarrow H_2O(l) + CuSO_4(aq)$

$2H^+(aq) + CuO(s) \rightarrow H_2O(l) + Cu^{2+}(aq)$

colourless   black   blue

Evaporating the solution will again produce a solid salt – blue crystals of copper(II) sulphate in this case. This is an example of the general reaction

acid + base → water + salt

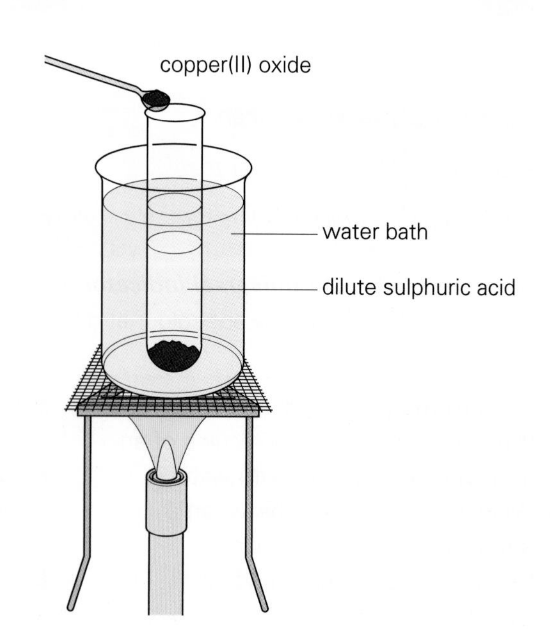

Acid + carbonate

If you put copper(II) carbonate into dilute nitric acid, a violent fizzing occurs as the green powder dissolves. The fizzing is the release of carbon dioxide which accompanies the neutralization of the acid. The remaining blue solution will yield the solid salt copper(II) nitrate if left to crystallize.

nitric acid	+	copper(II) carbonate	→	water	+	carbon dioxide	+	copper(II) nitrate
$2HNO_3(aq)$	+	$CuCO_3(s)$	→	$H_2O(l)$	+	$CO_2(g)$	+	$Cu(NO_3)_2(aq)$
$2H^+(aq)$	+	$CuCO_3(s)$	→	$H_2O(l)$	+	$CO_2(g)$	+	$Cu^{2+}(aq)$
colourless		green						blue

All carbonates react with acids in this way. Rain water contains the very weak carbonic acid (from dissolved carbon dioxide) and it slowly eats away limestone (calcium carbonate) rock. This erosion is greatly accelerated by ***acid rain*** – rain which has dissolved the polluting gases, sulphur dioxide and nitrogen dioxide.

One effect of acid rain

Salts

These are the substances remaining in solution when acids are neutralized. When solid, they have crystal lattices of positive metal (or ammonium) ions and negative ions, called ***acid radicals***, from the acids.

Acid	Salt	Use
nitric	ammonium nitrate	fertilizer
sulphuric	magnesium sulphate	laxative
phosphoric	sodium phosphate	water-softening
carbonic	sodium carbonate	glass-making

Acids which have two ionizable hydrogen atoms in their molecules can produce two sets of salts. So, carbonic acid and sodium hydroxide can make either sodium carbonate, Na_2CO_3 (washing soda), or sodium hydrogencarbonate, $NaHCO_3$ (baking soda or bicarb).

Counteracting overacidity

Acid soils contain substances which dissolve in rain water to produce acids. Most plants grow best in soil with a pH of around 7, so farmers neutralize acids by spreading slaked lime (calcium hydroxide) on fields:

Indigestion pains result when our stomachs produce too much hydrochloric acid. The excess acid can be neutralized by swallowing tablets containing weak bases such as magnesia (magnesium hydroxide) or baking soda (sodium hydrogencarbonate). Baking soda is also used in toothpastes to prevent tooth decay caused by excess mouth acids.

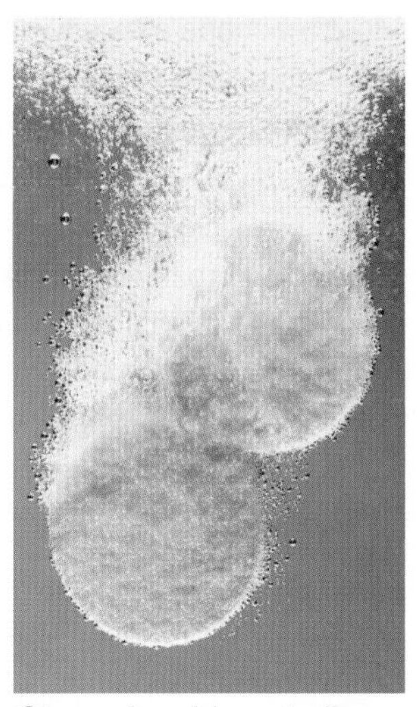

Stomach acid neutralizers

When an acid is neutralized, its hydrogen ions are converted into water, and a salt is formed.

1 What substances are formed when hydrochloric acid is neutralized by sodium hydroxide? Write the equation.

2 Write an equation which represents all acid and alkali neutralizations.

3 Write the word equation for the general reaction between an acid and a base.

4 How can copper(II) sulphate crystals be made?

5 Write a word equation for the eroding of limestone by acid rain (containing nitric acid).

6 Give *two* examples of the application of neutralization in everyday life.

3.15 Rate of reaction

By the end of this spread you should be able to:
- *describe the factors which affect reaction rates*
- *explain that particles need to collide often and with sufficient energy if they are to react.*

All reactions proceed at measurable rates. Some, like fireworks exploding, take only fractions of a second. Others, like sugar fermenting to alcohol, may take days. Industrial chemists usually want to speed reactions up in order to make their products economically – heating is the most usual way.

Effective collisions

Most reactions are between substances in liquid or gas states, where their particles are moving around at different speeds. When slow-moving particles collide, they simply bounce off each other. Only very fast-moving particles have sufficient energy for the collision to break down their existing bonds and allow them to react. This minimum energy is the ***activation energy*** (Spread 3.11). If reacting substances are heated, their particles move faster and more of them will possess the necessary activation energy. A rise of 10 °C often doubles the rate of a reaction.

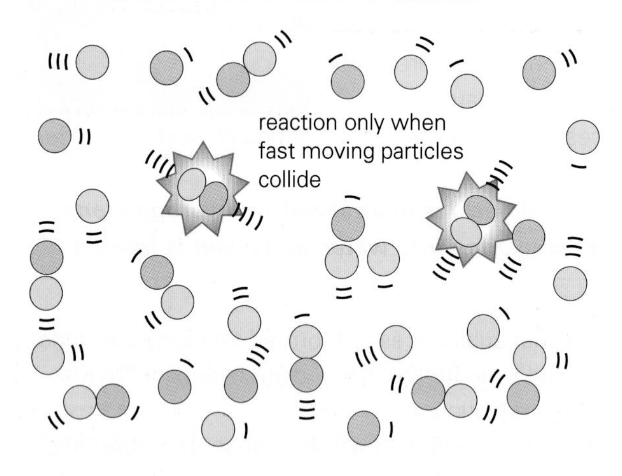

Increasing the surface area of a solid

In reactions involving a solid, colliding particles can only meet at the surface of the solid. Breaking up the solid into smaller pieces exposes more surface and the reaction is faster. The dissolving of marble chips (a form of calcium carbonate) in dilute hydrochloric acid illustrates this.

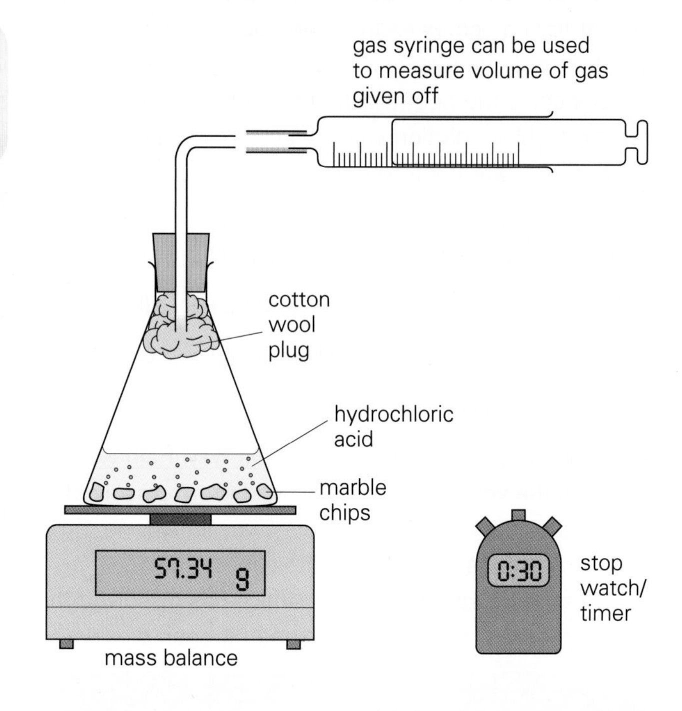

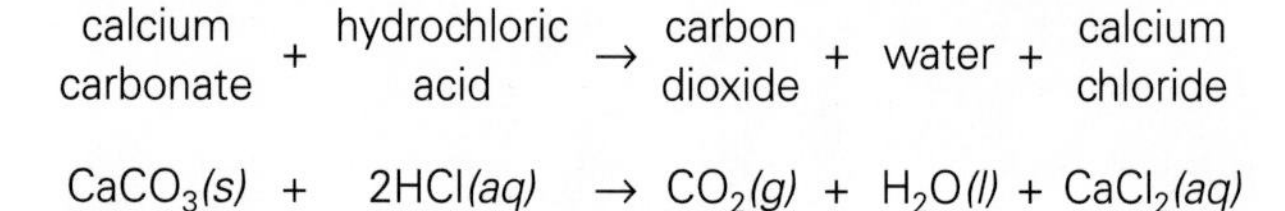

$$CaCO_3(s) + 2HCl(aq) \rightarrow CO_2(g) + H_2O(l) + CaCl_2(aq)$$

The rate of this reaction can be measured by recording the loss in mass in the reaction flask every 10 seconds. Alternatively, you could measure the volume of carbon dioxide given off. Line graphs of the results will level off when the reaction is complete.

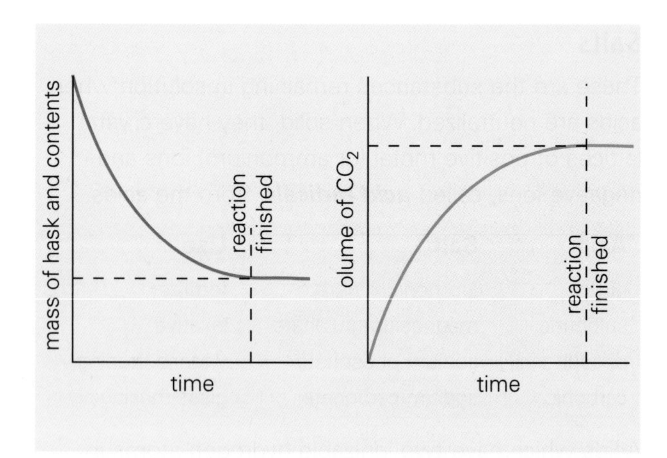

If the experiment is repeated using the same starting mass of marble but now crushed into powder, it is over much more quickly. The acid is in contact with a much larger surface area of calcium carbonate.

Increasing the concentration of solutions

Particles in solution move around and meet frequently. The more there are in a given volume, the greater chance there is of effective collisions occurring. The fewer particles in dilute solutions react more slowly.

Solutions of sodium thiosulphate and hydrochloric acid, added together, react to form a precipitate of sulphur which gradually thickens. The reaction rate can be found by timing how long it takes for the precipitate to obscure a cross drawn on paper beneath the reaction flask.

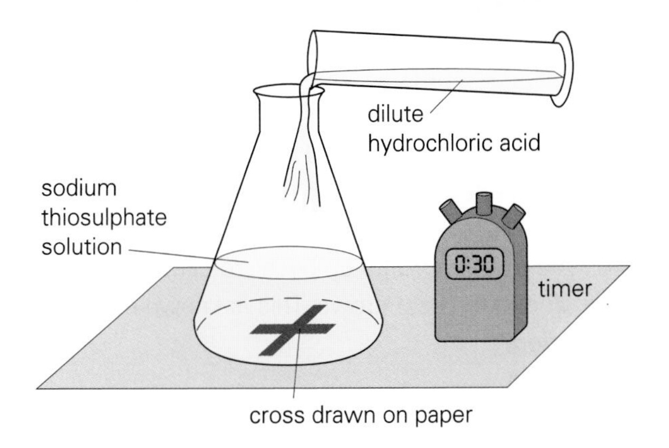

If the process is repeated using thiosulphate diluted to half its original concentration, it takes twice as long for the cross to disappear. In this reaction, the rate is directly proportional to the concentration of the sodium thiosulphate.

This experiment can also be used to illustrate the temperature effect. If the two original solutions are heated by 10 °C and then mixed, the cross disappears in half the time.

Typical experimental results

Concentration of thiosulphate	Temperature	Time taken
1M	25 °C	3 mins
0.5M	25 °C	6 mins
1M	35 °C	1.5 mins

In other reactions, doubling the concentration of one reactant can increase the rate by four times.

Increasing the pressure on gases has the same effect as increasing the concentration of solutions. Because more molecules are squeezed into the same volume, they meet more often and react more quickly.

Using catalysts

Catalysts are substances which speed up reactions without being used up themselves. Hydrogen peroxide, by itself, decomposes very slowly releasing oxygen and becoming water. If a small amount of manganese dioxide is added, however, the liquid fizzes as the oxygen escapes quickly. When the reaction is finished, the manganese dioxide can be recovered and shown to have the same mass as before.

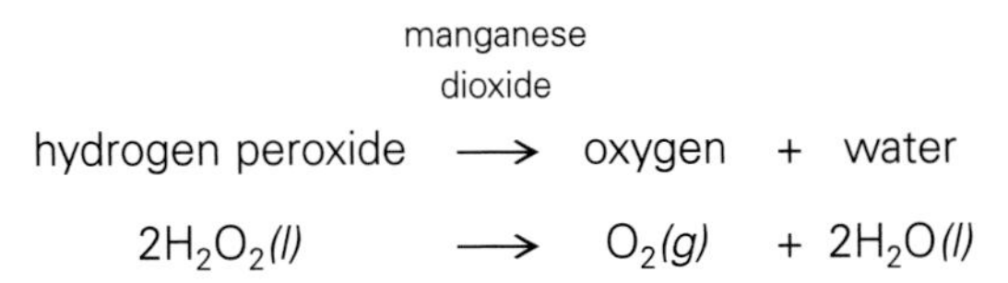

$$\text{hydrogen peroxide} \xrightarrow{\text{manganese dioxide}} \text{oxygen} + \text{water}$$

$$2H_2O_2(l) \longrightarrow O_2(g) + 2H_2O(l)$$

A catalyst lowers the activation energy needed for a reaction by forming loose, temporary bonds with the reactants. Transition metals and their compounds form such bonds readily and are good catalysts. Each one works best on only one particular reaction.

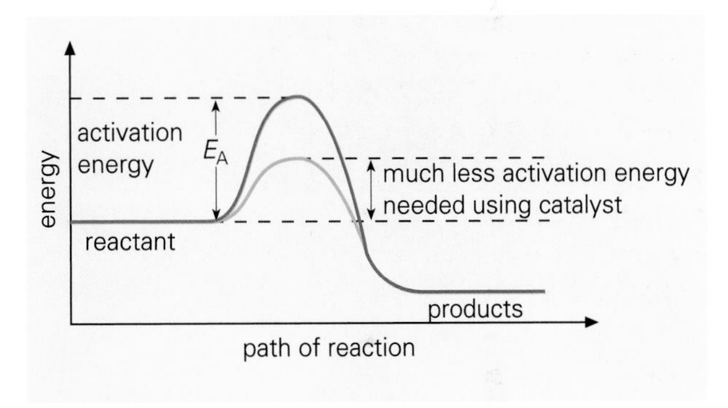

When a metal is used as a catalyst, the action takes place at the surface of the metal. The larger the surface area, the more effective it is. Catalysts are often finely divided powders, sometimes deposited on porous ceramic material.

1. Explain the term *effective collisions*.
2. **(a)** List *three* ways of speeding up the reaction between marble chips and dilute hydrochloric acid. **(b)** How can you tell when this reaction is finished?
3. What happens when solutions of sodium thiosulphate and hydrochloric acid are mixed?
4. What is a catalyst and how does it work?
5. What catalyst will speed up the decomposition of hydrogen peroxide? Write the equation.
6. Name the catalyst and the reaction in one important industrial process (see Spread 3.21).

3.16 Reactions involving enzymes

By the end of this spread you should be able to:
- *describe how enzymes work*
- *appreciate the impact of biotechnology on today's world.*

When you chew a sandwich, the starches in the bread begin to break down into glucose. The reaction is brought about by the amylase in saliva.

Amylase is just one of thousands of biological catalysts called ***enzymes*** – each one doing a specific job. They control reactions in our bodies and in food production.

Dairy food

Cheese is made by adding ***rennin***, an enzyme originally obtained from calves' stomachs, to milk. The milk separates into solid curds and a watery liquid – whey. The curds are pressed and wrapped in cloths to dry. Addition of other, different enzymes gives cheeses their different tastes; for example, lipase gives Danish Blue.

Yogurt Milk is heat treated to begin thickening, and two bacteria are added. Enzymes present produce lactic acid from the milk sugar, cause further thickening, and give a creamy taste.

Many yogurts are heated again to kill the bacteria, but you can add a carton of 'live' yogurt to warm milk to make a lot more.

Made using yeast

Yeast products

Alcohol Yeast is a single-celled fungus which contains several useful enzymes. One of these, ***zymase***, turns a solution of glucose into ethanol – usually called alcohol. The solution froths up because carbon dioxide is given off as the yeast grows. This is known as ***fermentation***.

$$\text{glucose} \xrightarrow{\text{zymase}} \text{ethanol} + \text{carbon dioxide}$$

$$C_6H_{12}O_6(aq) \longrightarrow 2C_2H_5OH(aq) + 2CO_2(g)$$

Fermentation stops when all the glucose has gone or when enough alcohol (about 12%) to kill the yeast has been produced.

Wine is made from grapes, which contain a lot of natural glucose and yeast on their skins. Beer is made from the malt sugar in partly germinated barley grain. The maltose in yeast added to this changes it first to glucose and then to alcohol.

Bread is made from dough, a mixture of flour, water, and yeast. The yeast causes slight fermentation of the starches in the flour and the dough swells up with carbon dioxide bubbles. Baking kills the yeast and dries the dough.

Enzyme action

Enzymes work very quickly and efficiently owing to the shapes of the enzyme and the molecules it affects. The complex, three-dimensional protein structure of an enzyme has a uniquely shaped hole – or ***active site*** – in it. Only molecules with a corresponding shape – ***substrates*** – will fit into the hole. The enzyme and substrate form a short-lived complex which weakens the bonds in the substrate. It breaks down easily and the products are released quickly, leaving the enzyme free to accept another substrate.

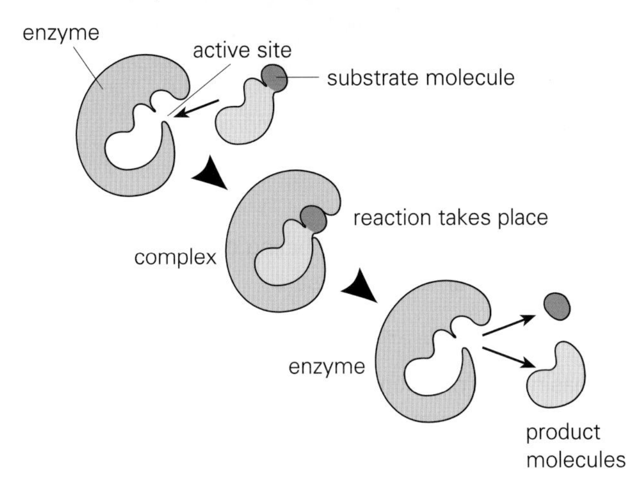

Enzymes are very sensitive to both acid/alkali conditions and to heat. Digestive pepsins in the stomach work well in the very acidic conditions there (pH 1.5–2.0). Lipase and other pancreas enzymes need alkaline conditions (pH 7.5–8.8).

There is an optimum temperature for enzymes to work well. Too cold, and the enzyme shuts down; too hot, and the enzyme can be destroyed (denatured).

How the rate of an enzyme catalyzed reaction varies with temperature

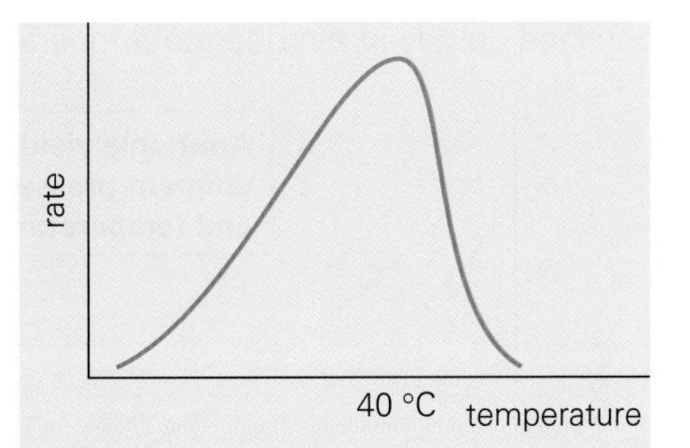

Biotechnology

Today, most fuels, most plastics, and many essential chemicals are obtained from crude oil. Alternatives will have to be found when the oil runs out. Already there are processes which use enzymes to make fuels (ethanol and hydrogen) from easily grown plants such as sugar cane. Animal dung and domestic waste are converted to methane gas in villages throughout India and China. Some other processes that depend on enzymes are shown below.

Application	Enzyme used
biological washing powder	proteinase
leather making – removing hair and improving pliability	amylase
medicine – dissolving blood clots	streptokinase
baby foods (predigested)	trypsin
clearing fruit juices	pectinase
manufacture of glucose syrup	amyloglucosidase

Meat substitute Meat is an excellent source of protein – a vital part of our diet. However, raising livestock can be difficult and expensive. Substitute protein can now be produced quickly and cheaply from potato or wheat starch using enzymes in the mould *Fusarium graminearum*.

Myco-protein meat substitute

1. What is an enzyme? What does amylase do?
2. How is cheese made?
3. Describe the fermentation of glucose.
4. Why is bread full of tiny holes?
5. Describe how enzymes work.
6. How does the temperature affect enzyme action?
7. How can meat-substitute protein be made?
8. List *three* industrial uses of enzyme action.

3.17 Reversible reactions

By the end of this spread you should be able to:
- *understand that many reactions are reversible*
- *understand that both yields and costs are important factors in manufacturing processes.*

A ***reversible*** reaction is one in which the products can recombine to form the original substances again. It can be shown as a single equation using the symbol

When calcium carbonate (limestone) is heated strongly, it splits up to form calcium oxide (quicklime) and carbon dioxide gas:

$$CaCO_3(s) \rightarrow CaO(s) + CO_2(g)$$

If the container is sealed, as below, the gas cannot escape, and some of it reacts with the calcium oxide, changing back to calcium carbonate:

$$CaO(s) + CO_2(g) \rightarrow CaCO_3(s)$$

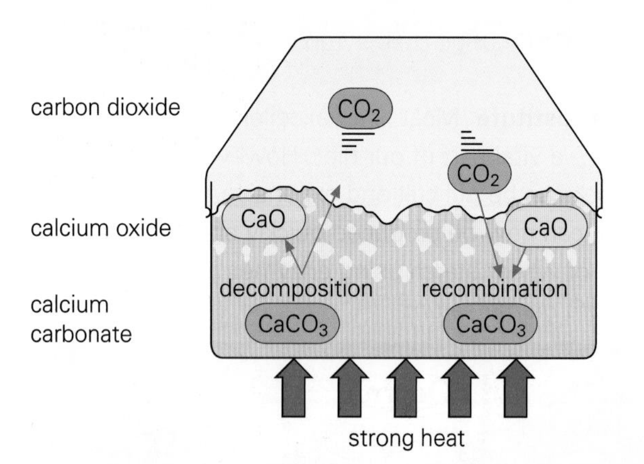

A time will come when the rate of decomposition of carbonate is balanced by the rate of recombination of the products. An ***equilibrium*** will exist:

$$CaCO_3(s) \rightleftharpoons CaO(s) + CO_2(g)$$

In the industrial manufacture of quicklime, the carbon dioxide is allowed to escape, moving the equilibrium to the right, and the reaction proceeds to completion.

Many of the reactions for making industrial chemicals are reversible. Some of the product splits up again, and only a small percentage may remain when equilibrium is reached. The percentage remaining is called the ***yield***. Producers must find ways of getting the best possible yield without too much cost.

Iron catalyst pellets from an ammonia plant

Ammonia

This is manufactured by combining nitrogen with hydrogen over an iron catalyst:

$$N_2(g) + 3H_2(g) \overset{Fe}{\rightleftharpoons} 2NH_3(g)$$

Increasing the pressure gives a better yield because the ammonia occupies a smaller volume than the reacting gases. It is more difficult for the ammonia to split up again. But a higher pressure also means that the cost and maintenance of equipment is greater. The favoured compromise is a pressure of about 150 atmospheres. Decreasing the temperature gives a better yield because the reverse reaction (the splitting up of ammonia) is endothermic and favoured by higher temperatures. However, the process is uneconomically slow at low temperatures. The use of an iron catalyst enables a satisfactory yield to be obtained quickly at a moderate temperature (450 °C).

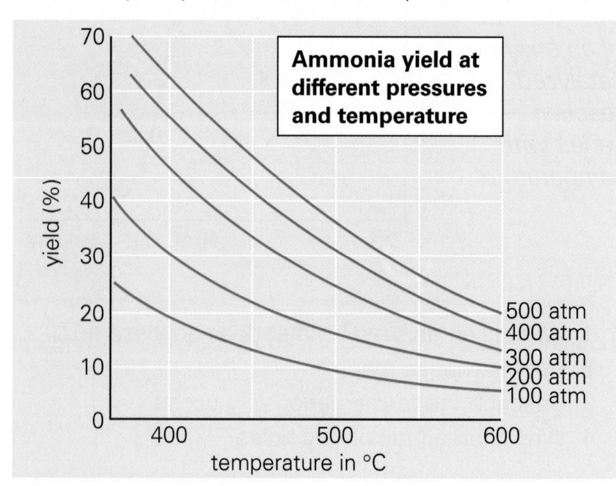

Upsetting the balance

The position of equilibrium of a reversible reaction depends upon the conditions prevailing. If you alter the conditions, the balance is upset and the reaction adjusts to reduce the effect of the change. The table below shows the effects of altering conditions on the general reaction

$$2A(g) + B(g) \rightleftharpoons C(g) + D(g) + \textit{energy given out}$$

3 volumes 2 volumes

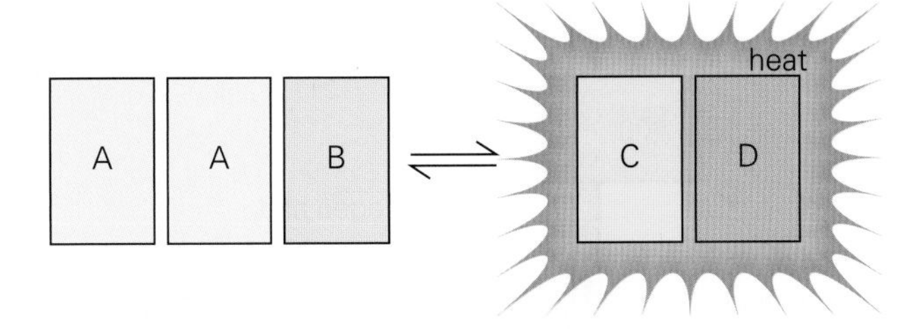

Change	Effect on equilibrium
removal of C	moves to right to make more C
increase in pressure	moves to right – smaller volume
increase in temperature	moves to left
adding catalyst	unchanged but reached more quickly
increase in concentration of A (if in solution)	moves to right

1 **(a)** What is a *reversible reaction*? **(b)** How can the decomposition of calcium carbonate be reversed? Write the equation for this.

2 When ammonia is manufactured **(a)** what pressure and temperature are used? **(b)** why are these conditions chosen? **(c)** Use the graph above to estimate the yield under these conditions. **(d)** What conditions would give a yield of 70%?

3 What is the effect of a catalyst on the equilibrium of a reversible reaction?

4 **(a)** Give *two* reasons why no sulphur dioxide should be allowed to escape unchanged in Stage 2 of the Contact process.
(b) How is a yield of over 99.5% of sulphur trioxide achieved?

Sulphuric acid

The most widely used chemical in the world, sulphuric acid, is manufactured by the three-stage ***Contact process***.

Stage 1 Liquid sulphur is burnt in a stream of cleaned and dried air to produce sulphur dioxide gas.

sulphur + oxygen → sulphur dioxide

$$S(l) + O_2(g) \rightarrow SO_2(g)$$

Stage 2 The sulphur dioxide is combined with more oxygen (in air) to convert it to sulphur trioxide. Layers of vanadium(V) oxide catalyst at 450°C bring this about.

sulphur dioxide + oxygen ⇌ sulphur trioxide

$$2SO_2(g) + O_2(g) \rightleftharpoons 2SO_3(g)$$

2 volumes 1 volume 2 volumes

The reaction is reversible and the forward direction gives out heat. It is important both economically and environmentally that no sulphur dioxide escapes unchanged to pollute the atmosphere. The equilibrium will be pushed to the right by increasing the pressure and lowering the temperature. In practice, a yield of over 99.5% is achieved at normal pressure and 450 °C by using excess oxygen, and by cooling and removing the sulphur trioxide as soon as it is formed.

Stage 3 The sulphur trioxide is dissolved in water to produce sulphuric acid.

sulphur trioxide + water → sulphuric acid

$$SO_3(g) + H_2O(l) \rightarrow H_2SO_4(aq)$$

The sulphur trioxide dissolves best in water which already has sulphuric acid in it. The very concentrated acid can be diluted as needed.

A sulphuric acid plant

3.18 Ammonia and fertilizers

By the end of this spread, you should be able to:
- *explain how ammonia is made*
- *describe how fertilizers are made and the problems in siting a suitable factory.*

Plants constantly take in nitrates from the soil. The nitrates must be replaced, either naturally or by adding fertilizers usually made from ammonia (NH_3).

Making ammonia: the Haber process

To make ammonia, nitrogen (taken from air) is combined with hydrogen. The hydrogen is made from methane heated with steam. The gases are then pumped at high pressure over hot iron catalyst, and this reaction takes place:

$$N_2(g) + 3H_2(g) \rightleftharpoons 2NH_3(g)$$

The reaction is reversible (see Spread 3.17) and only 10–15% of the gases become ammonia. The ammonia is removed as liquid by cooling the mixture down. The unchanged gases are recirculated until conversion is complete.

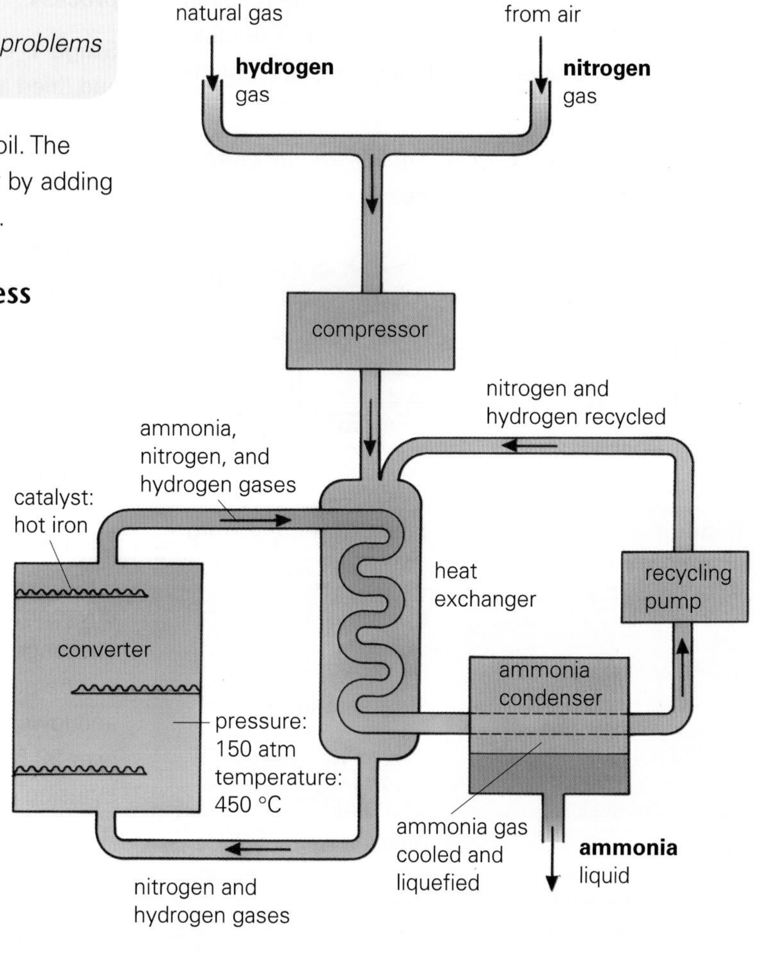

Nitric acid and Nitram

Ammonia can be combined with oxygen to make nitric acid in a series of reactions as shown below:

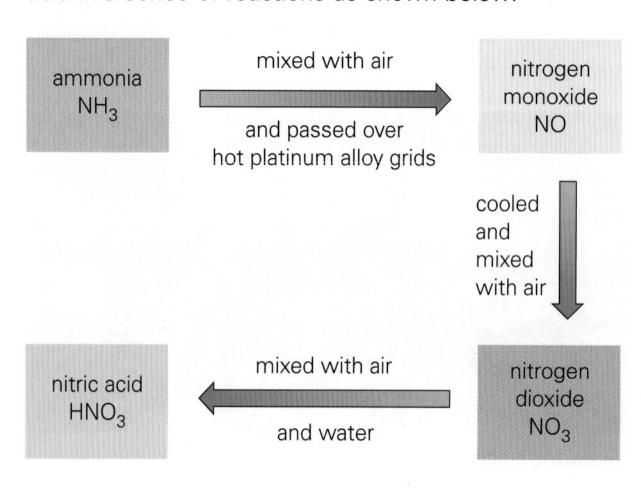

Nitric acid is used to make explosives, plastics, and drugs, but much of it is converted into ***Nitram*** fertilizer. This is done by neutralizing nitric acid with ammonia to produce ammonium nitrate solution:

$$HNO_3(aq) + NH_3(g) \rightarrow NH_4NO_3(aq)$$

nitric acid ammonia ammonium nitrate (solution)

Evaporating the solution leaves crystals of ammonium nitrate – Nitram.

Nitram is often mixed with ammonium phosphate and potassium chloride to produce the ***NPK compound fertilizers***. These provide nitrogen, phosphorus, and potassium – the three essential elements for good plant growth.

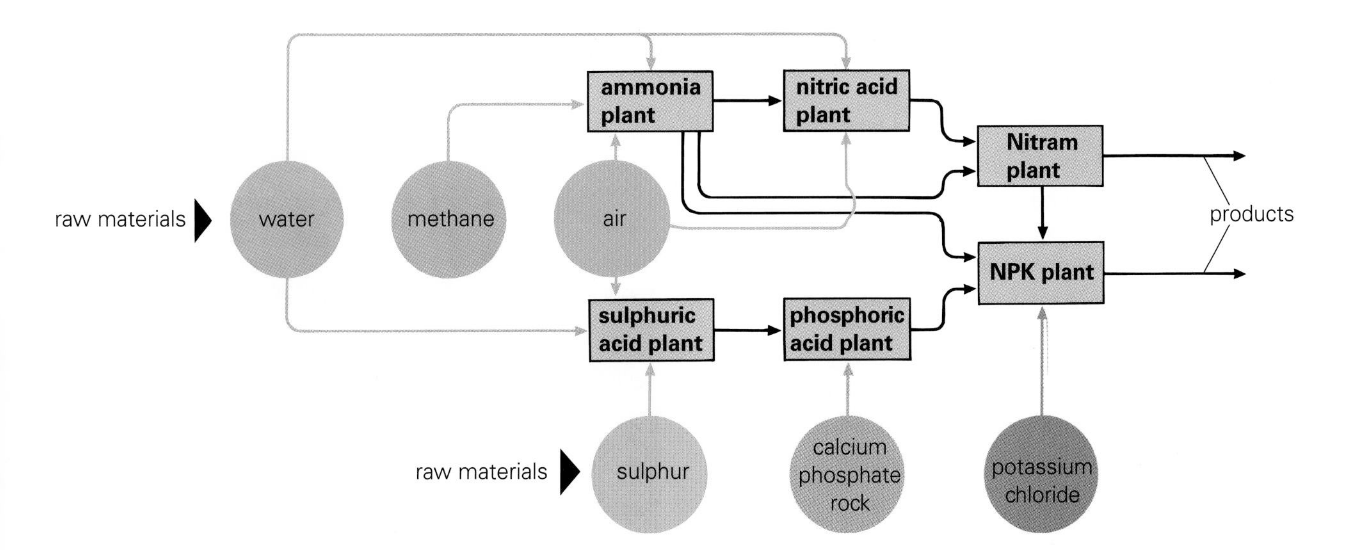

Siting a fertilizer factory

It is usual for all the chemical processes for making fertilizers to be carried out on one site. The diagram above shows how the six plants are interlinked.

In choosing a site for the factory both economic and environmental costs have to be considered. To produce fertilizers competitively, the factory should be near:

- a river for a water supply
- a port for sulphur and phosphate supplies
- a natural gas (methane) supply
- a town to supply the workforce
- motorways and railways to transport the product.

The huge area of land needed for the plant will affect the environment greatly. The site should not be in:

- good farmland
- a conservation area
- a built-up area, because of the possibility of accidental leakage of dangerous chemicals.

Wherever the site is situated, emissions of polluting gases (especially nitrogen oxides) must be strictly controlled.

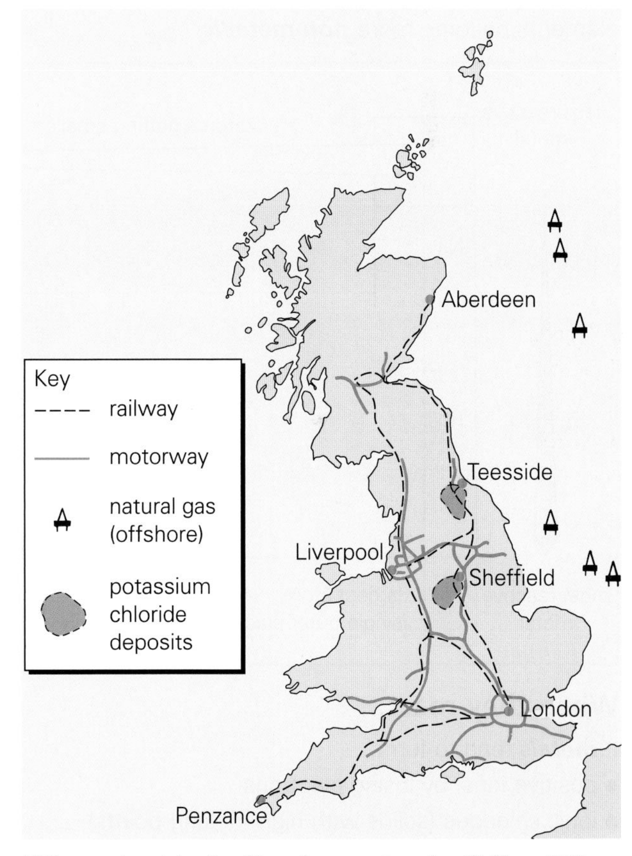

Where should a fertilizer factory be sited? (See Q6)

1 What gases are needed to make ammonia? How are they obtained?
2 What is the catalyst in the Haber process?
3 What is *Nitram* and how is it made?
4 Name *three* elements needed for good crop growth.
5 How many separate plants are there in a fertilizer factory? What does each one do?
6 Look at the map above. Decide where in the UK you would site a new fertilizer factory. Give the reasons for your choice.

3.19 Trends in the periodic table

By the end of this spread you should be able to:
- *describe the differences in properties of metals and non-metals, and some of their compounds.*

The nature of an element and the way it reacts with other substances are related to its position in the periodic table:

Down any one group, the atoms get *bigger*. They lose their outer electrons more easily to become positive ions. The elements become more ***metallic***.

Across a period, the atoms get *smaller*, and their nuclei more positive. The attraction for extra electrons increases, so negative ions form more easily. The elements become more ***non-metallic***.

Changes across period 3

Element	Group	Atomic radius (nm)*	Ion formed	Nature of oxide
sodium	1	0.191	Na^+	alkaline
magnesium	2	0.16	Mg^{2+}	basic
aluminium	3	0.13	Al^{3+}	basic/acidic
silicon	4	0.118	–	acidic
phosphorous	5	0.110	P^{3-}	acidic
sulphur	6	0.102	S^{2-}	acidic
chlorine	7	0.099	Cl^-	acidic

*1 nm = 0. 000 000 001 m

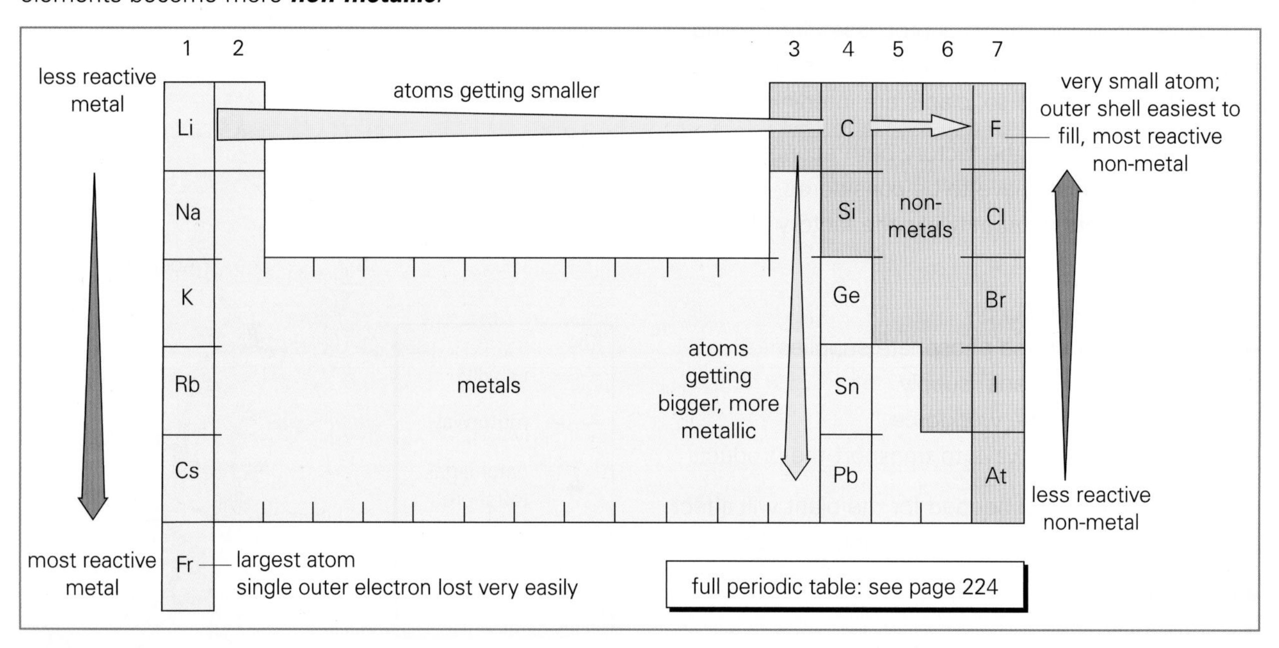

When they react...

...metals tend to form:
- positive ions, by losing electrons
- ionic chlorides (solids with high melting points)
- ionic oxides (solids which are bases).

...non-metals tend to form:
- negative ions by gaining electrons
- covalent chlorides (liquids with low boiling points)
- covalent oxides (gases which form acids with water, as in the following examples):

  nitrogen dioxide + water → nitric acid

  phosphorus(V) oxide + water → phosphoric acid

Group 4

These elements illustrate the change of chemical character down a group. At the top of the group, carbon is a non-metal: its oxide reacts with water to form carbonic acid. Silicon, the next one down, also has an acidic oxide which forms silicates – the basis of 75% of the Earth's rocks. However, silicon is slightly metallic because it is a semiconductor of electricity.

At the bottom of the group, tin and lead appear to be typical metals and are good conductors. Their oxides will neutralize acids but they also react with alkalis – non-metallic behaviour. Such oxides are ***amphoteric***.

Group 1: the alkali metals

Physically, the elements in Group 1 do not appear to be typical metals. They are soft enough to be cut with a knife, many are less dense than water, and they melt very easily. Each has one electron in its outer shell (see Spreads 3.06 and 3.07) and this is lost to form a positive ion. Because they form positive ions more easily than all other elements, they are the most metallic in their chemical behaviour.

Reaction with air

The Group 1 metals react so quickly with air that they have to be stored in jars of oil. Lithium, freshly cut, is silvery but becomes tarnished by a layer of oxide within minutes.

Lithium metal

lithium + oxygen → lithium oxide

$4Li(s) + O_2(g) \rightarrow 2Li_2O(s)$

The compound lithium oxide is an ionic solid consisting of two Li^+ ions balancing one O^{2-} ion.

Reaction with water If you drop a small piece of sodium into a beaker of water in a fume cupboard, it fizzes around on the surface, dissolving and giving off hydrogen.

sodium + water → sodium hydroxide + hydrogen

$2Na(s) + 2H_2O(l) \rightarrow 2NaOH(aq) + H_2(g)$

Lithium reacts more steadily and potassium more violently, often bursting into flames. In every case ,the remaining solution is strongly alkaline – hence the name *alkali metals.*

Potassium reacting with water

Alkali metals

Element	Atomic number	Melting point (°C)	Atomic radius (nm)
lithium (Li)	3	181	0.157
sodium (Na)	11	98	0.191
potassium (K)	19	63	0.203
rubidium (Rb)	37	39	0.216
caesium (Cs)	55	29	0.235

Reaction with chlorine All the alkali metals burn in chlorine, giving white salts called chlorides. These are highly ionic (see Spread 3.07). Sodium forms sodium chloride (common salt).

sodium + chlorine → sodium chloride

$2Na(s) + Cl_2(g) \rightarrow 2NaCl(s)$

It is possible to reverse this reaction. To do so, the sodium chloride must be melted to allow the ions to move. A large current of electricity is then passed through the hot liquid (see Spread 3.25). Chloride ions move to the positive terminal, release their extra electrons, and become chlorine gas. The sodium ions move to the negative terminal where they pick up electrons and become sodium metal. All the alkali metals can be isolated in this way.

Sodium chloride is an essential part of our diet. Unfortunately, too much of it can lead to high blood pressure. Potassium chloride tastes much the same without the possible harm and is used in 'low salt'.

1. How does the size of the atoms change **(a)** down a Group **(b)** across a period of the periodic table?
2. Give *three* chemical properties of metals.
3. The elements in Group 4 change in nature from non-metallic carbon at the top to metallic lead at the bottom. List the properties of germanium. Why do you think it is called a *metalloid*?
4. What type of compound is sodium oxide? How would its solution affect universal indicator?
5. How does potassium react with water? Write an equation for the reaction.
6. What is 'low salt' and what effect does it have on blood pressure?

3.20 The halogens: Group 7

By the end of this spread you should be able to:
- *list the physical properties of the halogens*
- *explain and illustrate the decrease in reactivity of the halogens down the group.*

All the elements in Group 7 combine with metals to form salts (halogen means 'salt producer'). Chlorine, bromine, and iodine, the three common halogens, react with hot iron wool as shown below.

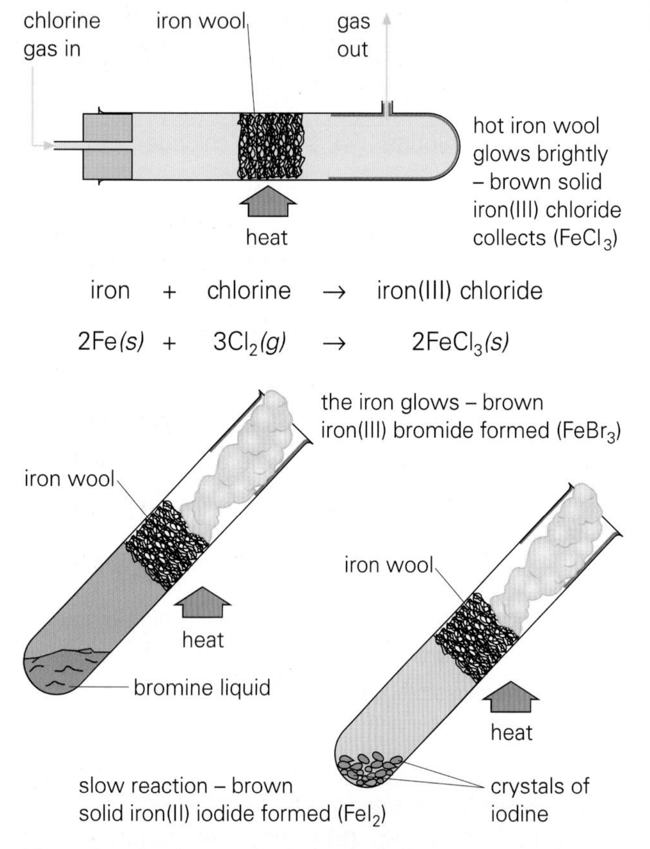

Note that the lower the halogen is in the group, the less readily it reacts with the metal.

Chlorine is used to keep this pool germ free.

Fluorine, at the top of Group 7, is the most reactive of all non-metals and is too dangerous to handle. The fluorine atom is very small and its positive nucleus strongly attracts one extra electron to become a negative ion, F^-, with a noble gas structure. The other halogens also form ions with a 1– charge, but less easily down the group. This is because their atoms are bigger and bigger and their nuclei are shielded increasingly by more full electron shells.

Bleaching action

When chlorine dissolves in water, it reacts to form a mild bleach which removes the colour from indicator paper. This bleach also kills germs and is used to sterilize drinking water and swimming pools.

Bromine also dissolves in water but bleaches more slowly. Iodine dissolves and bleaches hardly at all. Again, the halogens are steadily less reactive down the group.

Name of halogen	Molecule	Appearance	Atomic number	Shells of electrons	Radius of atom (nm)	Boiling point
fluorine	F_2	pale yellow gas	9	2e, 7e	0.071	–144 °C
chlorine	Cl_2	green gas	17	2e, 8e, 7e	0.099	–35 °C
bromine	Br_2	dark red liquid – changes to red-brown gas on warming	35	2e, 8e, 18e, 7e	0.114	+59 °C
iodine	I_2	shiny grey crystals – changes to purple gas on heating	53	2e, 8e, 18e, 18e, 7e	0.133	+184 °C

Uses of the halogens

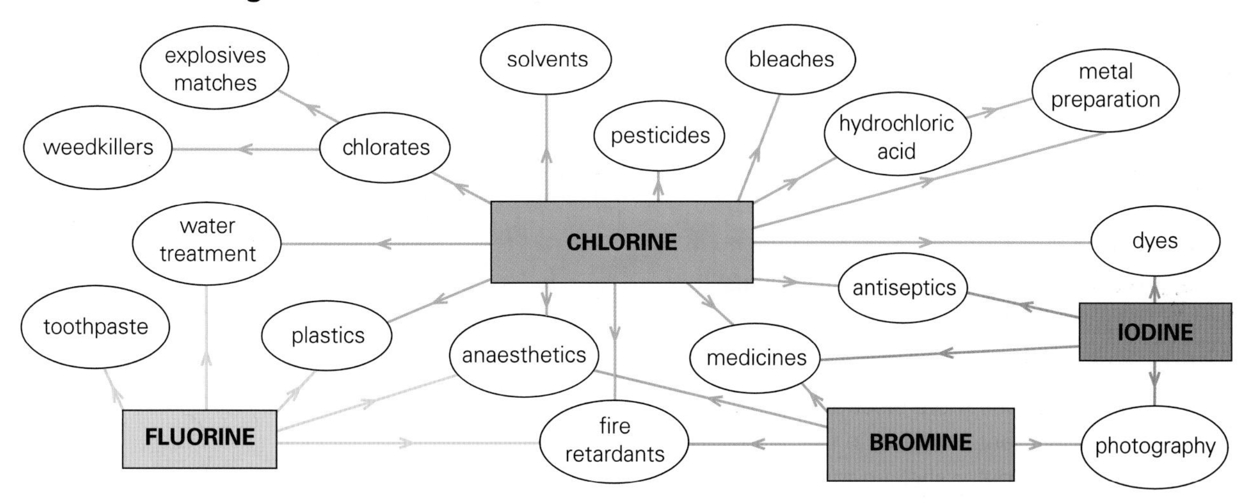

Reaction with hydrogen

A jet of hydrogen will burn in a jar of chlorine with a silvery flame forming hydrogen chloride gas. This is a covalent compound (see Spread 3.08).

hydrogen + chlorine → hydrogen chloride

$H_2(g) + Cl_2(g) \rightarrow 2HCl(g)$

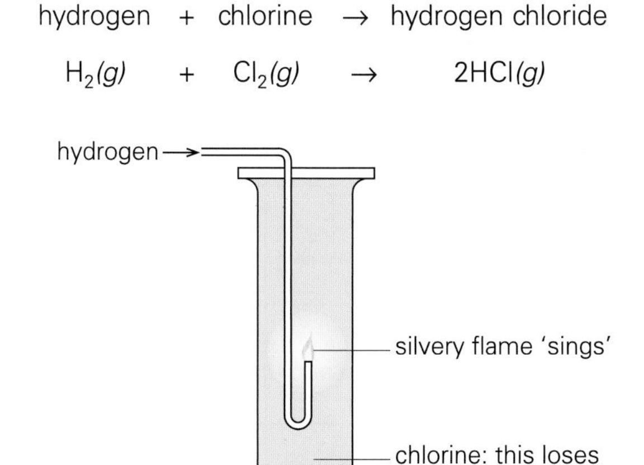

Bromine combines with hydrogen less readily, needing a heated platinum catalyst to form hydrogen bromide gas. The same catalyst is needed for the partial conversion of hydrogen and iodine vapour to hydrogen iodide gas.

All three gases dissolve very easily in water and ionize to form hydrochloric, hydrobromic, and hydriodic acids. The salts formed when these strong acids react with metals, bases, and carbonates are called ***chlorides***, ***bromides***, and ***iodides***.

Solutions of these salts all produce precipitates when mixed with silver nitrate solution (Spreads 3.10, 3.12).

Displacement

A more reactive halogen will displace a less reactive one from its compound. Chlorine solution, added to colourless potassium bromide solution, becomes chloride ions and converts the bromide ions into red-brown bromine:

chlorine + potassium bromide → bromine + potassium chloride

$Cl_2(aq) + 2KBr(aq) \rightarrow Br_2(aq) + 2KCl(aq)$

$Cl_2(aq) + 2Br^-(aq) \rightarrow Br_2(aq) + 2Cl^-(aq)$

Similarly, both chlorine and bromine will displace iodine from potassium iodide solution.

1 What does the name *halogen* mean?
2 When chlorine, bromine, and iodine react with hot iron, how do their reactions compare?
3 How do the **(a)** atomic radii **(b)** boiling points **(c)** reactivity of the elements change down Group 7?
4 What effect does chlorine have on indicator paper? Why is chlorine added to water?
5 When chlorine and bromine react with hydrogen, how do their reactions compare?.
6 Describe the reaction between solutions of potassium bromide and silver nitrate.
7 What happens when chlorine solution is added to aquous potassium bromide?

3.21 Transition metals

By the end of this spread you should be able to:
- *describe what transition metals have in common*
- *give examples of the wide range of uses of transition metals and their compounds.*

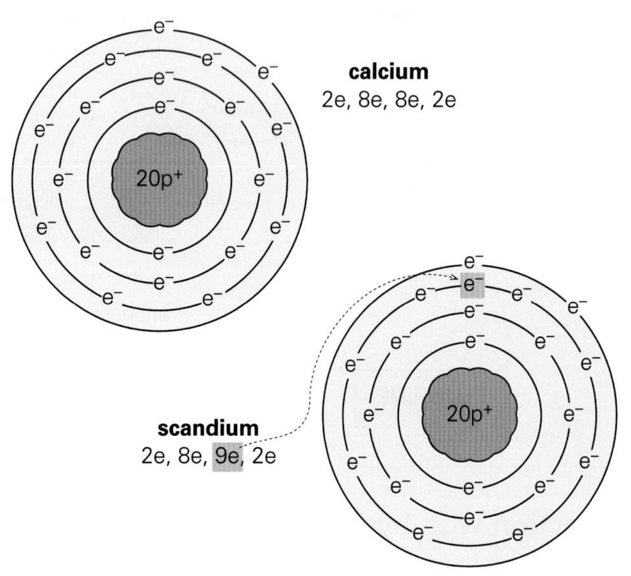

e = electron; p = proton; neutrons not shown

Between Groups 2 and 3 of the periodic table, there is a block of elements called the ***transition metals***. It includes many of the commonly known metals.

Scandium, atomic number 21, is the first metal in the block. Its atoms have one more electron than calcium (number 20), the element it follows in the table. Instead of making the outer shell up to three, this extra electron exists in the next inner shell – which then holds 9 electrons. The same sort of thing happes with the next nine metals until, at zinc, the inner shell holds 18 electrons.

Group 2 | ← transition metals → | Group 3

← these have similar properties → (titanium to zinc)

Group 2											Group 3
Ca calcium 20	Sc scandium 21	Ti titanium 22	V vanadium 23	Cr chromium 24	Mn manganese 25	Fe iron 26	Co cobalt 27	Ni nickel 28	Cu copper 29	Zn zinc 30	Ga gallium 31
Sr strontium 38	Y yttrium 39	Zr zirconium 40	Nb niobium 41	Mo molybdenum 42	Tc technetium 43	Ru ruthenium 44	Rh rhodium 45	Pd palladium 46	Ag silver 47	Cd cadmium 48	In indium 49
Ba barium 56	La lanthanum 57	Hf hafnium 72	Ta tantalum 73	W tungsten 74	Re rhenium 75	Os osmium 76	Ir iridium 77	Pt platinum 78	Au gold 79	Hg mercury 80	Tl thallium 81

In each row of the block, an inner electron shell builds up from 8e to 18e with 2e (or sometimes 1e) in the outer shell. The atoms in any one row are of similar size and have many properties in common.

Transition metals:
- have high melting points
- have high densities
- are hard and strong
- can be beaten into shape and drawn into wires
- can be mixed as alloys
- are often found in the ground as oxides or sulphides and can be extracted using heat and carbon.

Some uses of transition metals

iron	machinery, vehicles, structures
chromium	electro-plating, stainless steel
manganese	special steel alloys
nickel	alloys, catalyst, coins
titanium	aircraft parts
tungsten	electric bulb filaments
copper	plumbing, wiring
platinum	catalyst, jewellery

Coloured compounds

Many of the compounds of transition metals are highly coloured. Any one metal can form differently coloured compounds because it combines in several oxidation states (see Spread 3.12). The following table shows some manganese compounds.

Compound	Formula	Oxidation state	Colour
manganese carbonate	$MnCO_3$	+2	pink
manganese dioxide	MnO_2	+4	dark brown
sodium manganate(VI)	Na_2MnO_4	+6	dark green
potassium manganate(VII)	$KMnO_4$	+7	purple

Note that besides forming simple *positive* ions, the transition metals also form *negative* ions in combination with oxygen.

Because of their colours, transition metal compounds are used in dyes and pigments. Traces of them produce the colours in glasses and gemstones.

Oxidizing agents

It is relatively easy for transiton metals to change from one oxidation state to another. Their compounds are active redox agents.

Two of the most common oxidizing agents are potassium manganate(VII) and potassium dichromate(VI). In acid solution, they both remove electrons from other substances very readily and have distinctive colour changes.

manganate(VII) +	acid	+ electrons →	manganese(II) +	water
$MnO_4^-(aq)$	$+ 8H^+(aq) +$	$5e^- \rightarrow$	$Mn^{2+}(aq)$	$+ 4H_2O(l)$
purple			colourless	

dichromate(VI) +	acid	+ electrons →	chromium(III) +	water
$Cr_2O_7^{2-}(aq)$	$+ 14H^+(aq) +$	$6e^- \rightarrow$	$2Cr^{3+}(aq)$	$+ 7H_2O(l)$
orange			green	

The colour of a ruby comes from traces of chromium

Catalytic action

Transition metals and some of their compounds are widely used in industrial processes as catalysts (see Spread 3.15). Some of the more common applications are listed in the table below.

Catalyst	Process
iron	converting nitrogen and hydrogen to ammonia
platinum	oxidation of ammonia to make nitric acid
vanadium(V) oxide	oxidation of sulphur dioxide to make sulphuric acid
nickel	hardening of vegetable oils to make margarine

Each reaction is speeded up most by one particular catalyst. Sometimes the best catalyst is a combination of transition metals. Platinum–rhodium alloys, for example, are used in the ***catalytic converters*** now fitted to car exhausts. The alloys change the polluting gases carbon monoxide and nitrogen dioxide into harmless carbon dioxide and nitrogen.

Catalytic converter from a car

1 Name *five* transition metals and give a use of each.
2 Give *five* properties which most of the transition metals have in common.
3 Give the electron arrangements of:
(a) calcium **(b)** scandium **(c)** zinc **(d)** gallium.
4 What is the colour of rubies due to?
5 Name one transition metal compound used as an oxidizing agent. What colour change does it undergo?
6 What metals are used in catalytic converters? What do they do?

3.22 Reactivity of metals

By the end of this spread you should be able to:
- *describe how different metals react with air, water, and dilute acids*
- *describe metal corrosion and its prevention.*

Despite having similar physical properties, metals differ greatly in their ability to become positive ions in combination with other substances. The metals can be arranged in a ***reactivity series*** by observing how well they react with air, water, and dilute acid.

Reaction with air

The alkali metals are so reactive that they combine as soon as they contact air (see Spread 3.19).

Magnesium has to be heated strongly before it bursts into flames and burns to form magnesium oxide.

$$2Mg(s) + O_2(g) \rightarrow 2MgO(s)$$

The less reactive metals lead and copper do not burn when heated but become coated with a layer of oxide.

Gold does not react at all.

Reaction with water

Spread 3.19 describes the violent reactions of potassium and sodium with water. The diagram below shows how calcium reacts steadily in water.

calcium + water → hydrogen + calcium hydroxide

$$Ca(s) + 2H_2O(l) \rightarrow H_2(g) + Ca(OH)_2(s)$$

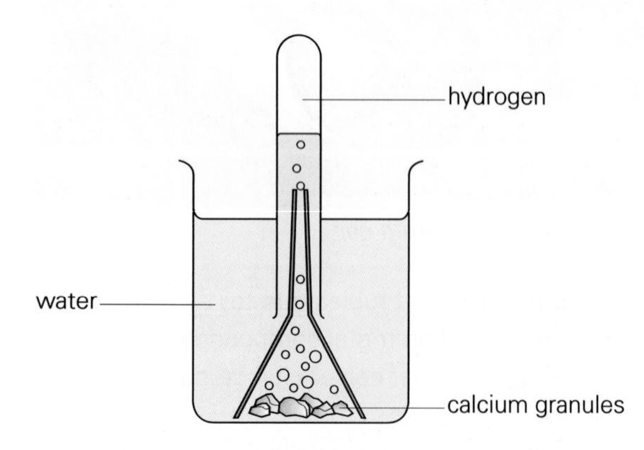

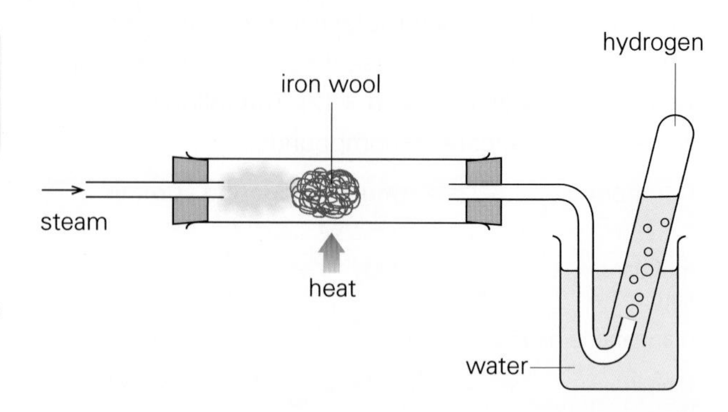

Many metals do not react with cold water, but some, like iron, will combine if heated in steam.

iron + steam → hydrogen + iron oxide

$$3Fe(s) + 4H_2O(g) \rightarrow 4H_2(g) + Fe_3O_4(s)$$

Reaction with dilute hydrochloric acid

Most of the metals will react with dilute acid to a greater or lesser degree. Those which do not are placed at the bottom of the 'activity league'.

Potassium and sodium are not tested in this way because it is too dangerous.

Zinc is in the middle of the reactivity series and it reacts steadily with moderately dilute hydrochloric acid. The reaction is used as a method of producing a steady supply of hydrogen.

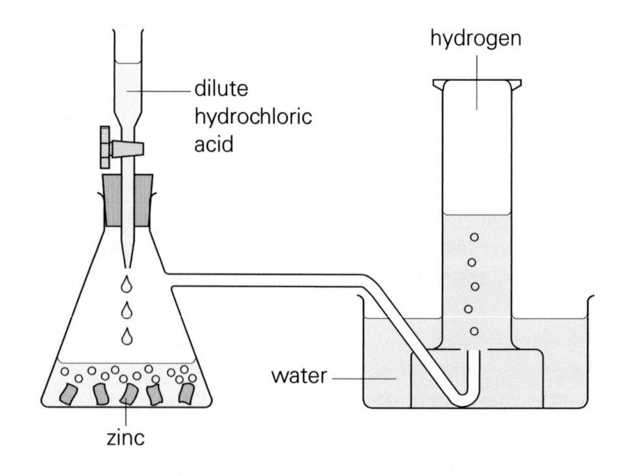

zinc + hydrochloric acid → hydrogen + zinc chloride

$$Zn(s) + 2HCl(aq) \rightarrow H_2(g) + ZnCl_2(aq)$$

$$Zn(s) + 2H^+(aq) \rightarrow H_2(g) + Zn^{2+}(aq)$$

Metal	Reaction with air	Reaction with water	Reaction with hydrochloric acid	Order of reactivity
potassium	oxidize very quickly (kept under oil)	violent – catches fire	explosive (not tested)	MOST
sodium		very vigorous		
calcium	oxide forms quite quickly	steady – forms hydrogen	very vigorous – gives off hydrogen freely	
magnesium		slow when cold, quick when heated in steam		
aluminium	protective layer of oxide forms steadily	moderate when heated in steam	good reaction when warmed	
zinc			steady	
iron	rusts slowly	slow when heated in steam	steady when warmed	
lead	oxide coating forms on strong heating	very little reaction	very slow	
copper		no reaction	no reaction	
silver	tarnishes very slowly			
gold	no reaction			LEAST

Corrosion

This is the formation of compounds on the surface of metals in contact with air and/or water.

Calcium, high in the reactivity series, will corrode away to a white powder if left exposed to air. Gold, at the bottom, never corrodes, and ornaments made from it remain bright and attractive. Copper and lead react so little with air and water that they can be used for water pipes and roofing. Aluminium and zinc react slowly with air, but they do not corrode away because they form a layer of oxide which protects the rest of the metal underneath. A protective layer of oxide (sometimes coloured) is often put on to aluminium articles by ***anodizing*** (see Spread 3.25).

Although iron reacts with air more slowly than aluminium or zinc, its corrosion is worse. This is because the rust which forms on the surface flakes off and exposes the underlying metal. Eventually iron will rust right through. Rust is hydrated iron oxide, and it forms only when both oxygen and water are present. Traces of salt or acid in the water accelerate rusting.

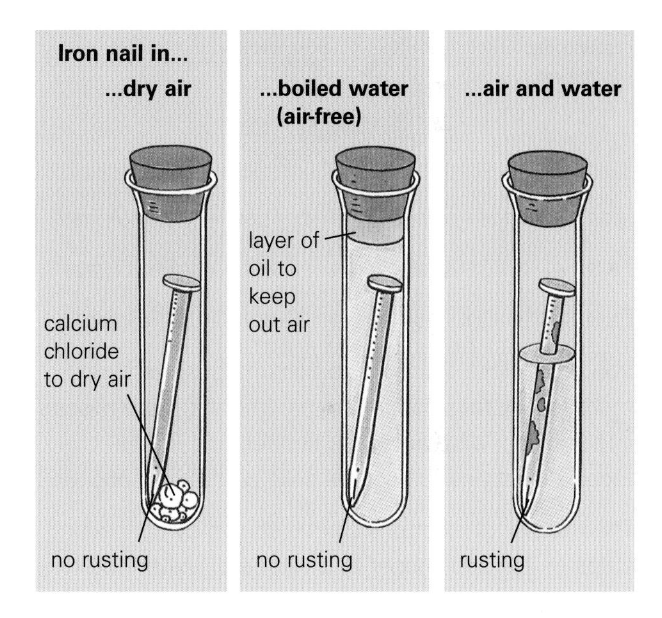

The experiment above shows that both air *and* water are needed for rusting. To stop rusting, iron and steel can be coated with paint, grease, plastic, or another metal, to keep out air or water. Galvanized screws are made of iron coated with zinc (see Spread 3.23).

1 The metals in the reactivity series are arranged in order. How can this order be worked out?

2 What gas is given off when a metal reacts with a dilute acid?

3 Tin lies between iron and lead in the reactivity series. How would you expect tin to react with **(a)** air? **(b)** water? **(c)** dilute hydrochloric acid?

4 Name a metal that corrodes quickly, and another that never corrodes.

5 What prevents aluminium from corroding right through?

6 How could you show that both air and water are needed for iron to rust?

7 What is *galvanized iron?*

3.23 Displacement of metals

By the end of this spread you should be able to:
- *describe what happens in a displacement reaction*
- *place an unknown metal in the reactivity series by observing its displacement reactions.*

If an iron nail is placed in copper(II) sulphate solution, the iron slowly dissolves, while copper is deposited on the nail as a brown coating:

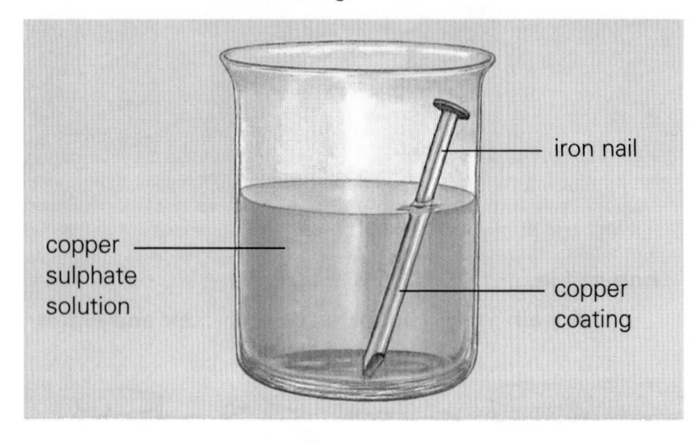

iron + copper sulphate → copper + iron sulphate

$Fe(s) + CuSO_4(aq) \rightarrow Cu(s) + FeSO_4(aq)$

$Fe(s) + Cu^{2+}(aq) \rightarrow Cu(s) + Fe^{2+}(aq)$

Iron atoms lose electrons to become positive ions more readily than copper atoms. Iron therefore passes into solution, ***displacing*** the copper ions. These receive electrons to become solid particles of copper.

The transfer of electrons can be shown by connecting strips of iron and copper to a voltmeter. When the two strips are dipped into a solution of salt, electrons flow from the iron to the copper and give a reading on the meter. This is one type of electric cell.

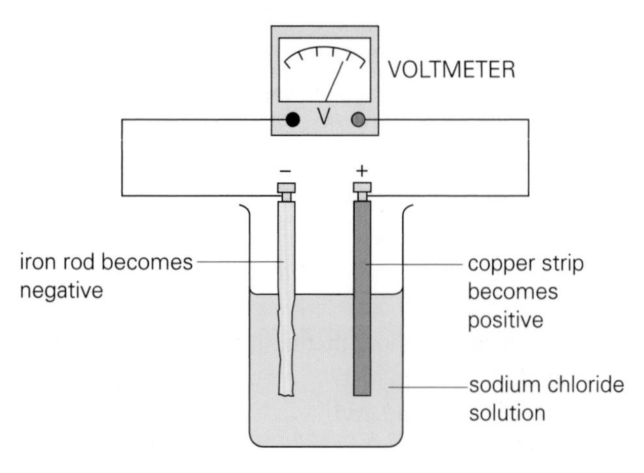

This block of zinc on the hull of a boat will prevent corrosion of the steel

Sacrificial Protection

Underground pipes, oil rigs, and ships made of iron (as steel) are protected from rusting by wiring blocks of magnesium or zinc to them. These metals are higher in the reactivity series than iron and will become the negative pole (cathode) in the cell made with iron in water. They corrode away leaving the iron untouched. The blocks are replaced when they are used up.

These galvanized screws are made of steel (mainly iron) with a protective zinc coating

Any metal will displace a less reactive metal (lower in the series) from a solution of one of its salts.

An unknown metal can be placed in the reactivity series by putting small pieces of it into a range of solutions of other metal salts. The metal might have no reaction with iron(II) sulphate solution but slowly displace lead from lead nitrate solution. In this case, the metal lies between iron and lead in the series.

Displacement of hydrogen

Hydrogen is obviously not a metal, but it does form positive ions when acids dissolve in water. Reactive metals dissolve in dilute acids by displacing hydrogen ions and releasing them as hydrogen gas molecules.

iron metal + dilute acid → iron ions + hydrogen gas

$$Fe(s) + 2H^+(aq) \rightarrow Fe^{2+}(aq) + H_2(g)$$

Copper and the other unreactive metals below it in the series form positive ions less easily than hydrogen and will not displace it from dilute acids (see Spread 3.22).

Hydrogen, therefore, is often included in the reactivity series between lead and copper as a reference point.

Metal	Ion formed	Order of displacement	Reaction with acid
potassium	K^+	higher metal displaces lower metal from a solution of one of its salts	displace hydrogen less easily
sodium	Na^+		
calcium	Ca^{2+}		
magnesium	Mg^{2+}		
aluminium	Al^{3+}		
zinc	Zn^{2+}		
iron	Fe^{2+}		
lead	Pb^{2+}		
hydrogen	H^+		
copper	Cu^{2+}		do not displace hydrogen
silver	Ag^+		
gold			

Competing for oxygen

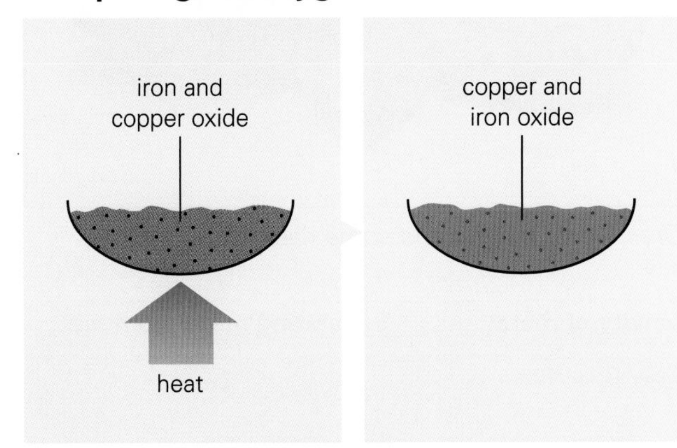

If a mixture of powdered iron and copper(II) oxide is heated in a crucible, it begins to glow as a reaction starts. It will continue to glow when the bunsen burner is removed because of the heat given out by the exothermic reaction. The products are iron(II) oxide and copper.

iron + copper(II) oxide → iron(II) oxide + copper

$$Fe(s) + CuO(s) \rightarrow FeO(s) + Cu(s)$$

Iron, being higher in the series than copper, combines with oxygen more readily and takes the oxygen away from the copper(II) oxide. The displaced copper has lost oxygen and has been *reduced*. Displacements are examples of *redox reactions* (see Spread 3.12).

The experiment is more spectacular when a mixture of powdered aluminium and iron(III) oxide is used. It is called the ***Thermit process*** and is used to produce small amounts of iron as in the joining of iron rails.

1 What happens when iron is placed in solutions of **(a)** copper(II) sulphate? **(b)** silver nitrate?

2 Why are zinc blocks wired to the oil rig legs?

3 Why is hydrogen often included in the reactivity series of metals?

4 What are the products of heating mixtures of
(a) iron and copper(II) oxide
(b) aluminium and iron(III) oxide?

5 The diagram shows what happens when pieces of metal M are placed in different solutions.

metal M in solution of...

magnesium sulphate	zinc sulphate	iron(II) sulphate	copper(II) sulphate
no reaction	no reaction	slight reaction	copper deposited

metal M

(a) With which solutions does M react?
(b) Where is metal M in the reactivity series.
(c) How would you expect M to react with
(i) water? **(ii)** dilute hydrochloric acid?
(iii) copper(II) oxide?

3.24 Metal ores

By the end of this spread you should be able to:
- *explain why different methods are needed to extract metals from their ores*
- *describe how iron and steel are made.*

Ores

Minerals are naturally occurring compounds which make up the Earth's rocks. Some minerals can be treated to produce metals (and other substances), and these are called ***ores***. A workable ore should contain a high proportion of metal and a minimum of impurity.

Very unreactive metals, such as gold and silver, are found uncombined and are simply dug out – mined – from the ground. More reactive metals are found in combination with oxygen, sulphur, carbon dioxide, or chlorine. Compounds which form readily are hard to decompose. The more reactive the metal, the more difficult it is to separate it from its ore.

Ores – gold nugget and some haematite (iron ore)

Metal	Ore/how found	Reactivity of metal	Separating metal from ore
potassium (K)	silvine (KCl)	very reactive	electrolysis (using electricity)
sodium (Na)	rock salt (NaCl)	getting more difficult to separate from ore (arrow)	
calcium (Ca)	limestone ($CaCO_3$)		
magnesium (Mg)	magnesite ($MgCO_3$)		
aluminium (Al)	bauxite (Al_2O_3)		
zinc (Zn)	calamine ($ZnCO_3$)		heating with carbon or carbon monoxide
iron (Fe)	haematite (Fe_2O_3)		
tin (Sn)	cassiterite (SnO_2)		
lead (Pb)	galena (PbS)		heating in air
copper (Cu)	chalcopyrite ($CuFeS_2$) and as an element		
silver (Ag)	argentite (Ag_2S) and as an element		
gold (Au)	as an element	unreactive	

Extraction of copper

Copper pyrites contains copper(I) sulphide. When this is separated from the ore, the copper can be obtained by simply roasting in air.

copper(I) sulphide + oxygen → copper + sulphur dioxide

$$Cu_2S(s) + O_2(g) \rightarrow 2Cu(s) + SO_2(g)$$

Smelting

Ores of metals mid-way in the reactivity series decompose less easily. To release the metals, the ores have to be mixed with carbon in the form of coke or anthracite and heated to very high temperatures. This is called ***smelting***. Iron is extracted from haematite in this way.

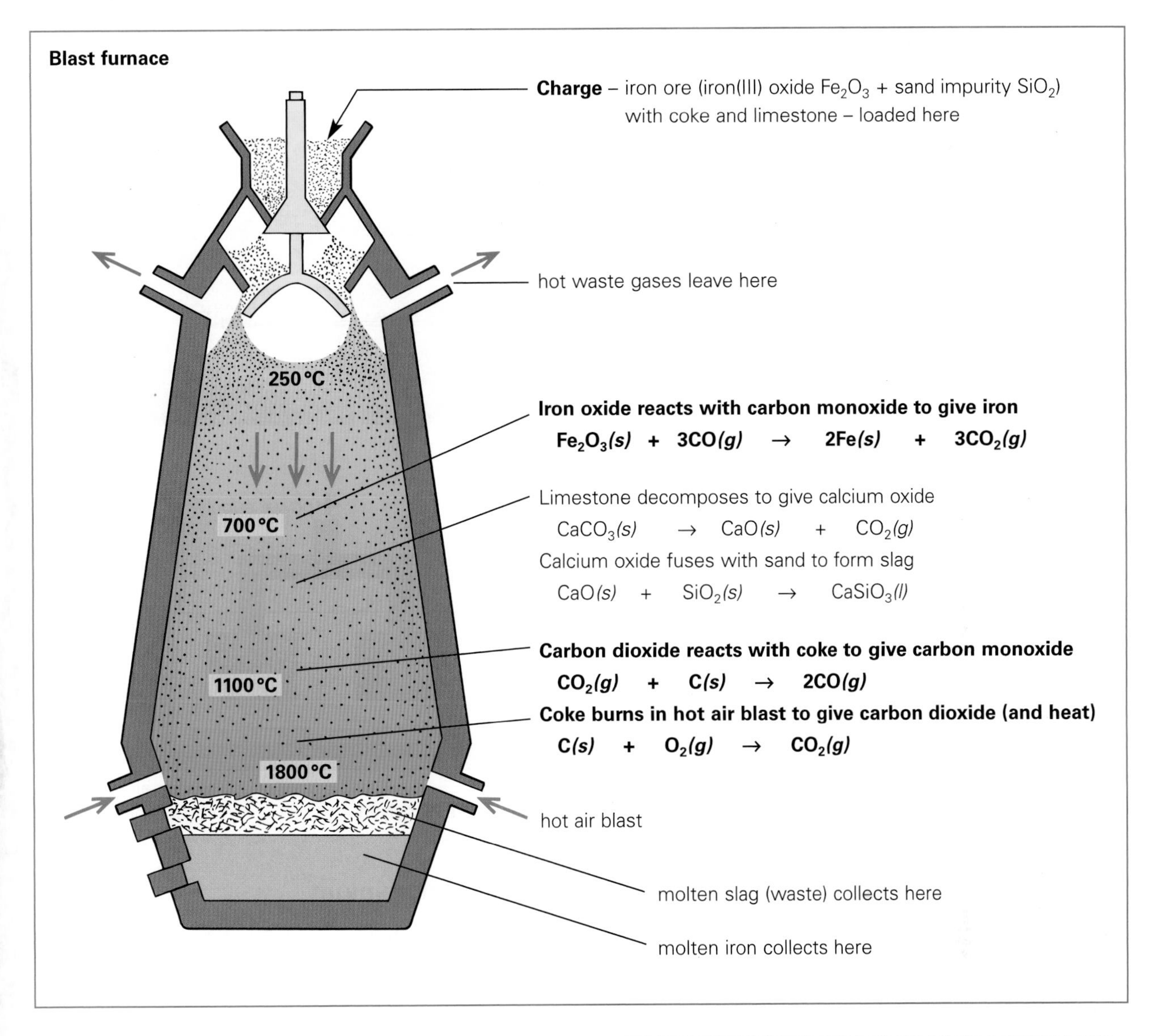

The iron tapped off from a blast furnace contains a lot of carbon (about 4%). If allowed to cool, it takes up the exact shape of the mould, becoming hard but brittle. This is ***cast iron***.

Most of the furnace iron is poured into a ***converter*** to be made into ***steel***. Here, oxygen and powdered lime (calcium oxide) are blown onto the surface of the molten iron. The carbon and other impurities oxidize and combine with the lime to form slag, which is poured off. Only a small amount of carbon (up to 1.5%) remains in the steel, making it tough but also springy.

Other metals are sometimes mixed with molten steel to produce alloys. ***Stainless steel*** has chromium and nickel added. Chisels and drills are made of steel toughened with manganese or tungsten.

1 What is an *ore?*
2 Name an ore of **(a)** lead **(b)** iron **(c)** aluminium.
3 Why can pure gold be found in the ground but not aluminium?
4 Which metal in the series is the most difficult to extract from its ore and what method is used?
5 Describe what is produced in a blast furnace by the reaction between: **(a)** coke and air **(b)** iron ore and carbon monoxide **(c)** limestone and sand impurities.
6 How is cast iron converted into steel, and what is the main difference between them?
7 What metals are mixed with steel to make **(a)** stainless steel? **(b)** tool steel?

3.25 Electrolytic reduction and refining

By the end of this spread you should be able to:
- *describe how aluminium is extracted*
- *describe how metals can be refined by electrolysis.*

Aluminium

Aluminium is the most abundant metal, occurring in many minerals, clay, and gemstones. Being a reactive metal, its compounds are hard to decompose, and the only ore worth treating is ***bauxite***.

After a number of processes to remove unwanted iron(III) oxide and sand, bauxite yields pure aluminium oxide Al_2O_3. Large currents of electricity are needed to extract the aluminium from this in a process called ***electrolysis***.

Electrolysis is the splitting up of an ionic liquid – the ***electrolyte*** – by passing a current between two ***electrodes*** dipping into it. The ***anode*** is connected to the positive of the power supply and the ***cathode*** to the negative. The process takes place in an ***electrolytic cell***.

In the cell on the right, the electrolyte is aluminium oxide mixed with another aluminium compound called cryolite, liquified by the heat from the huge electric current flowing. The cathode is the graphite lining of the steel tank, and the anode consists of blocks of graphite (a form of carbon which conducts well) dipping into the melt.

The aluminium oxide splits into positive aluminium ions, $Al^{3+}(l)$ and negative oxide ions, $O^{2-}(l)$. The aluminium ions are attracted to the cathode where they pick up electrons to become metal atoms.

$$4Al^{3+}(l) + 12e^- \rightarrow 4Al(l)$$

These collect at the bottom of the tank to be run off.

The oxide ions give up their electrons at the anodes and are released as oxygen gas.

$$6O^{2-}(l) \rightarrow 3O_2(g) + 12e^-$$

Some of the oxygen attacks the graphite anodes, gradually eating them away. Consequently, they are replaced periodically.

A bauxite mine

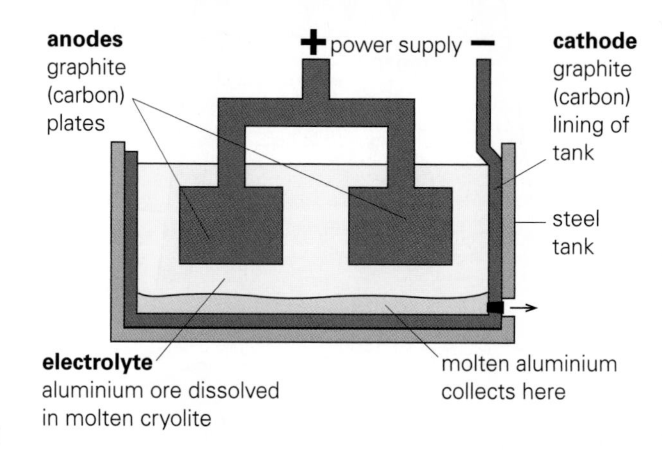

Recycling aluminium

Aluminium is light, easily shaped, resistant to corrosion, and a good conductor of heat and electricity, and it makes very strong alloys. Consequently, it has very many uses. Familiar ones are cooking foil, drinks cans, electricity cables, and aircraft parts.

Because of its widespread use, it makes good sense to recycle aluminium whenever possible. It is also very much cheaper to collect, say, old cans and melt them down for re-use than it is to mine, treat, and electrolyze dwindling supplies of bauxite.

Aluminium cans ready for recycling

Splitting water and anodizing

If a current is passed through water containing some sulphuric acid, hydrogen bubbles form at the cathode and oxygen at the anode. Ions from the sulphuric acid form the conducting path, but it is the hydrogen ions and hydroxide ions from the water which are discharged.

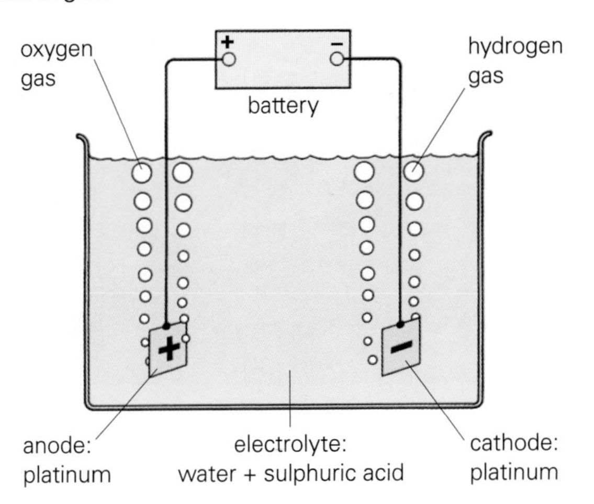

water → hydrogen ions + hydroxide ions

$H_2O(l) \rightarrow H^+(aq) + OH^-(aq)$

At the cathode:

$4H^+(aq) + 4e^- \rightarrow 2H_2(g)$

At the anode:

$4OH^-(aq) \rightarrow O_2(g) + 2H_2O(l) + 4e^-$

If an aluminium article is used as the anode, some of the oxygen reacts with the surface, forming a layer of aluminium oxide. This ***anodized*** layer protects the aluminium from further corrosion and gives an attractive finish. It can also be dyed very easily to give a permanent colour.

An anodized aluminium pan

Purification of copper

Like other metals, copper with impurities in it can be ***refined*** by electrolysis:

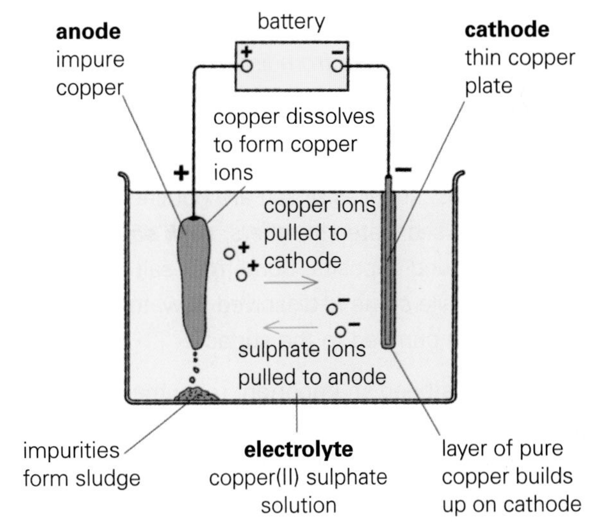

A block of impure copper is made the anode of a cell containing copper(II) sulphate solution as the electrolyte. When a current flows, copper ions are attracted to the cathode (a thin copper plate), where they pick up electrons and are deposited as a layer of pure copper.

$Cu^{2+}(aq) + 2e^- \rightarrow Cu(s)$

At the anode, copper atoms lose electrons and pass into solution as ions to replace those removed at the cathode.

$Cu(s) \rightarrow Cu^{2+}(aq) + 2e^-$

In this way, the impure block dissolves and pure copper collects on the cathode. The impurities either dissolve into solution or fall to the bottom of the cell.

1 What is bauxite? What aluminium compound does it contain?
2 What is meant by **(a)** electrolysis? **(b)** electrolyte?
3 To which terminals of the power supply do you connect the **(a)** anode **(b)** cathode?
4 In the electrolysis of aluminium, what forms the **(a)** anode? **(b)** cathode? **(c)** electrolyte?
5 Why should aluminium cans be recycled.
6 What is anodized aluminium?
7 Explain what happens in the electrolytic refining of copper at the **(a)** anode? **(b)** cathode? **(c)** electrolyte?

3.26 Useful products from rocks

By the end of this spread you should be able to:
- *describe how rock salt is used to make chemicals*
- *list the products made from limestone.*

Mining rock salt

Common salt – sodium chloride

In hot countries, supplies of salt are obtained by the evaporation of sea water. In others, ***rock salt*** is mined from underground deposits. Some rock salt is brought up as solid, while some is dissolved in water and the solution (brine) pumped to the surface.

Salt has always been an important resource as an essential part of our diet and as a food preservative. Today, it is the raw material for many important chemicals and processes.

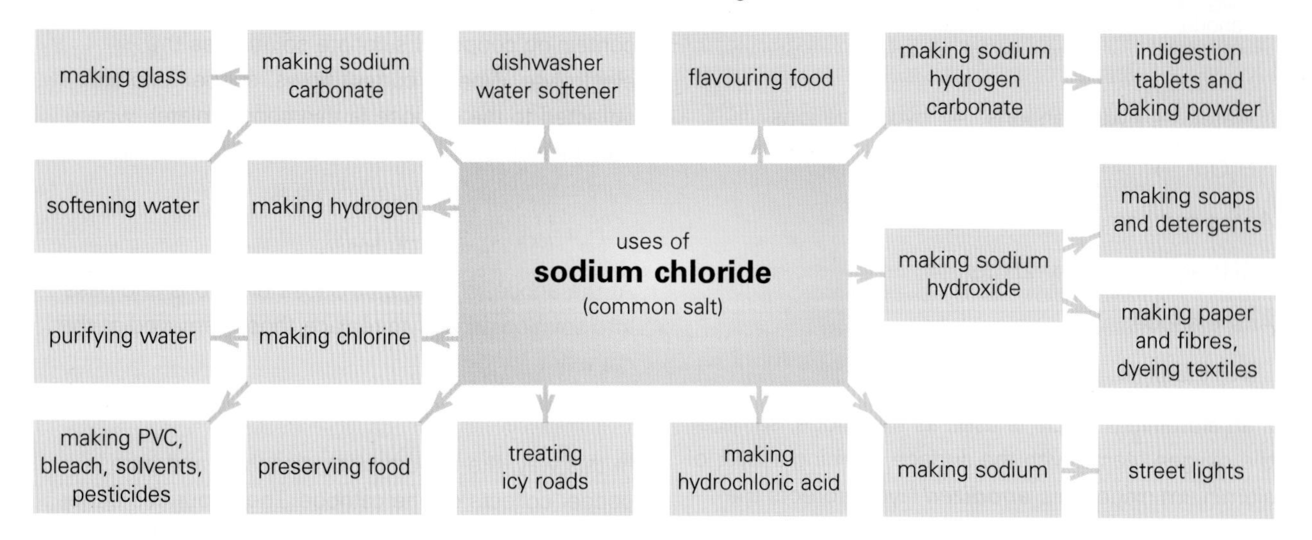

Soda – sodium carbonate

This is made from salt and limestone by a series of reactions involving ammonia. The overall effect is:

salt + limestone *give* sodium carbonate + calcium chloride

$2NaCl$ and $CaCO_3$ *give* Na_2CO_3 and $CaCl_2$

Sodium carbonate is used to treat sewage, to soften water, to make detergents, in processing metals and paper, and, principally, to make glass. This is made by melting together sodium carbonate, limestone, and sand. Very often recycled glass is added to the mixture. This saves raw materials and also lowers the melting temperature, thus saving energy.

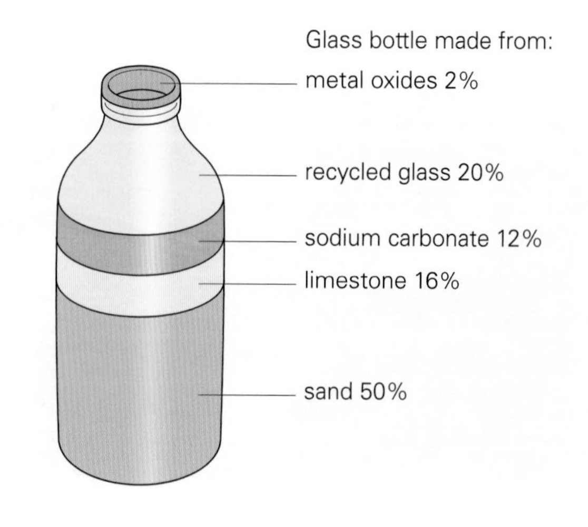

Electrolysis of sodium chloride solution

Sodium chloride solution is commonly known as brine. It contains sodium ions, Na^+*(aq)*, and chloride ions, Cl^-*(aq)*, from the salt plus some hydrogen ions, H^+*(aq)*, and hydroxide ions, OH^-*(aq)*, present in the water.

In the ***membrane cell***, shown below, the electrodes are separated by a porous barrier which lets only water and sodium ions through.

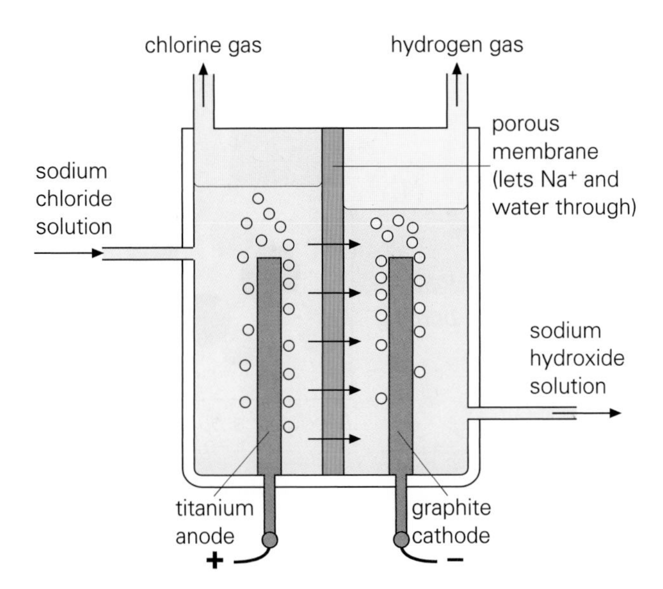

Anode reaction Two chloride ions give up their electrons and are released as chlorine gas:

$2Cl^-(aq) \rightarrow Cl_2(aq) + 2e^-$

Cathode reaction Hydrogen ions take on electrons from the graphite cathode much more easily than do sodium ions. Hydrogen gas is given off:

$2H^+(aq) + 2e^- \rightarrow H_2(g)$

The solution in the cathode compartment contains sodium ions and hydroxide ions. These join to form solid sodium hydroxide when the water is evaporated.

This is a very profitable process because the brine is cheap and all three products, chlorine, hydrogen, and sodium hydroxide, have many uses.

Limestone

Limestone is one of the most common rocks in Britain. It is mainly calcium carbonate and has different forms of varying hardness. Oolitic limestone is the cream-coloured kind often used for public buildings.

Chippings are used in blast furnaces and also for foundations and road building.

Cement is made by roasting limestone, clay, and sand in a kiln, and then grinding the product with gypsum.

Mixed with sand, chippings, and water, it sets hard to form ***concrete***.

Quicklime – calcium oxide

Heated to 1100 °C, calcium carbonate loses carbon dioxide and changes to a white solid, calcium oxide.

calcium carbonate → calcium oxide + carbon dioxide

$CaCO_3(s) \rightarrow CaO(s) + CO_2(g)$

Large quantities of pure quicklime are needed for making steel and other metals.

Slaked lime – calcium hydroxide

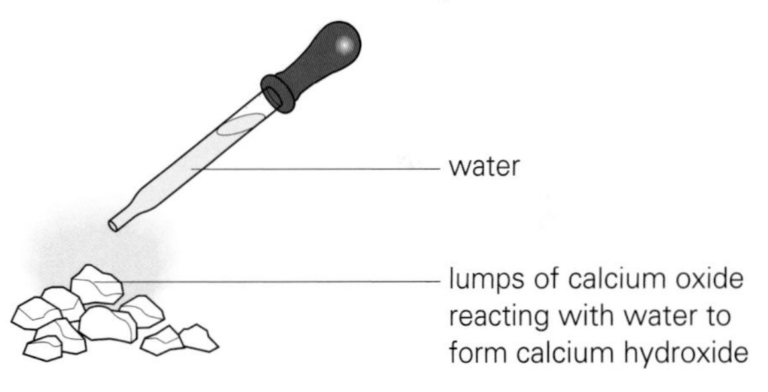

Calcium oxide reacts vigorously with water, giving off heat and crumbling to powdery calcium hydroxide.

calcium oxide + water → calcium hydroxide

$CaO(s) + H_2O(l) \rightarrow Ca(OH)_2(s)$

Slaked lime, like quicklime, is strongly alkaline. It is used by farmers to neutralize acid soils, to lighten clay soils, and to kill pests. Lime is used to neutralize acid wastes and for softening water.

1 How is rock salt obtained?
2 Give *three* direct uses of common salt.
3 Give *four* major uses of sodium carbonate.
4 Give *two* reasons for recycling glass bottles.
5 What are the three products of the electrolysis of brine? Give a major use of each.
6 How are **(a)** quicklime **(b)** slaked lime **(c)** cement made from limestone?

3.27 Petroleum products

By the end of this spread you should be able to:
- *recognize alkanes as the simplest hydrocarbons*
- *describe how crude oil is refined.*

Organic compounds

The molecules of all living (organic) things are made up of carbon compounds. Carbon atoms always form four covalent bonds when they combine. They can link with themselves to form chains or rings. Other elements can then easily join on to these to form compounds.

The simplest organic compounds are ***hydrocarbons***, which have carbon chains with only hydrogen atoms attached.

Hydrocarbons with single covalent bonds in the carbon chain are called ***alkanes***. These all have very similar chemical properties and form a chemical family (a homologous series). Below, their atoms and bonds are shown using ***displayed formulae***.

Butane gas burning

Model of a butane molecule

<table>
<tr><th>Name of alkane</th><th>Number of carbon atoms</th><th>Molecular formula</th><th>Displayed formula</th><th>Boiling point</th></tr>
<tr><td>methane</td><td>1</td><td>CH_4</td><td>H
|
H—C—H
|
H
carbon forms 4 covalent bonds</td><td>−164 °C</td></tr>
<tr><td>ethane</td><td>2</td><td>C_2H_6</td><td>H H
| |
H—C—C—H
| |
H H</td><td>−87 °C</td></tr>
<tr><td>propane</td><td>3</td><td>C_3H_8</td><td>H H H
| | |
H—C—C—C—H
| | |
H H H</td><td>−42 °C</td></tr>
<tr><td>butane</td><td>4</td><td>C_4H_{10}</td><td>H H H H
| | | |
H—C—C—C—C—H
| | | |
H H H H</td><td>−0.5 °C</td></tr>
<tr><td>pentane</td><td>5</td><td>C_5H_{12}</td><td>H H H H H
| | | | |
H—C—C—C—C—C—H
| | | | |
H H H H H</td><td>36 °C</td></tr>
<tr><td>general formula</td><td>n</td><td>C_nH_{2n+2}</td><td>carbon chain increases by
H
|
—C—
|
H</td><td></td></tr>
</table>

boiling points increase (volatility decreases) as carbon chain gets longer

Crude oil – petroleum

Crude oil is a thick, dark, smelly liquid consisting mainly of a mixture of alkanes. It comes from the decayed remains of sea creatures and plants which collected millions of years ago in the mud and sand of the sea bed. Found deep underground, trapped in porous rock like water in a sponge, the crude oil is pumped to the surface and taken to a ***refinery***. Here, the different substances in oil are separated by distillation in a fractionating tower.

Part of an oil refinery

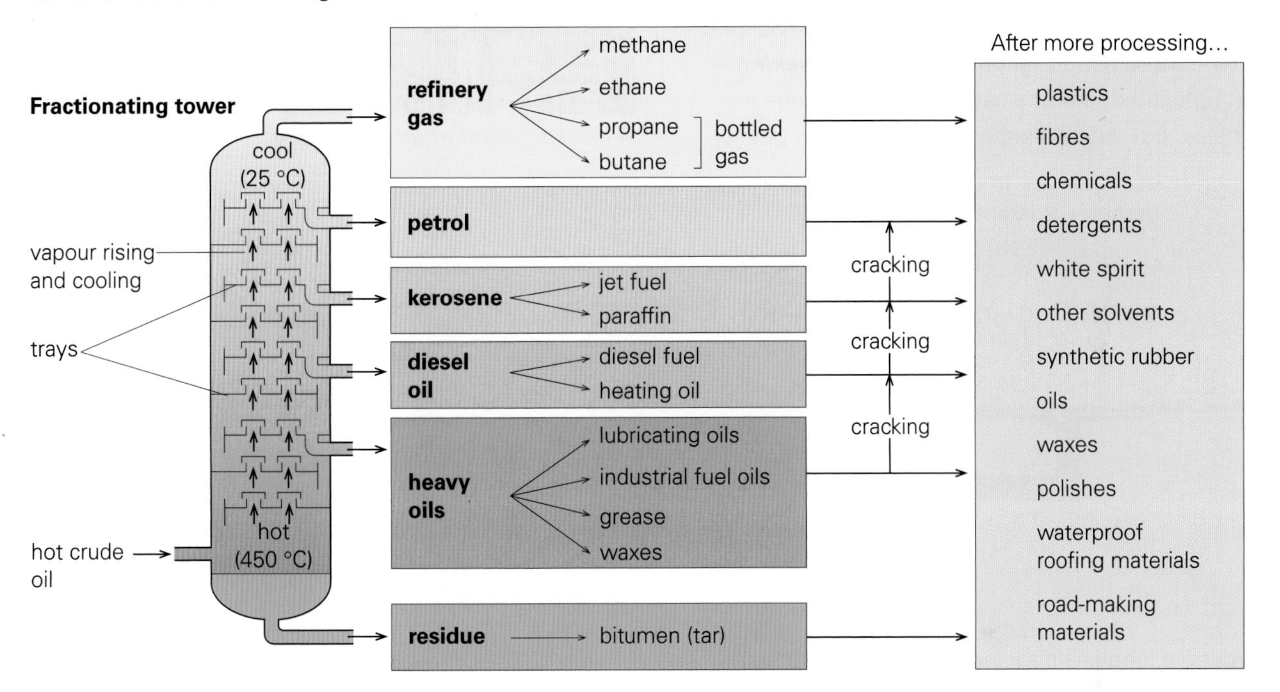

The oil is boiled so that most of it rises up the tower as vapour. As it rises, it cools. The longest-chain hydrocarbons, having high boiling points, condense back to liquid first and are collected at low level. Methane and very short-chain alkanes pass out of the top as gases. Different groups of alkanes with similar chain lengths – ***fractions*** – condense and are collected at intermediate levels.

Each of the fractions is further refined by removing the impurities and distilling again.

Many of the alkanes are used as fuels. They all burn in air, forming carbon dioxide and water and releasing great heat energy (see Spread 3.11). Methane (natural gas) is the gas supplied for domestic cookers and bunsen burners, whilst propane and butane are both used as camping gas.

propane + oxygen → carbon dioxide + water

$C_3H_8(g) + 5O_2(g) \rightarrow 3CO_2(g) + 4H_2O(l)$

The long-chain fractions are more viscous (sticky) and ignite less easily than those with shorter chains.

1 What is an organic compound?
2 What is meant by the terms **(a)** hydrocarbon? **(b)** alkane? **(c)** homologous series?
3 Give the names and formulae of the first four alkanes.
4 What is crude oil and where does it come from?
5 How is petroleum split up into fractions? What is an oil fraction?
6 Write equations for the burning of **(a)** propane **(b)** pentane, C_5H_{12}, in air.

3.28 Addition polymers

By the end of this spread you should be able to:
- *explain the difference between saturated and unsaturated hydrocarbons*
- *describe some common polymers and their uses.*

Cracking

The heavier, long-chain fractions produced by the distillation of crude oil (Spread 3.27) can be changed into the more useful lighter fractions by ***cracking*** – heating with steam or a catalyst. The reactions are complex, but a simplified example is shown below.

A cracking plant at an oil refinery

decane – an alkane in kerosene

```
   H  H  H  H  H  H  H  H  H  H
   |  |  |  |  |  |  |  |  |  |
H—C——C——C——C——C——C——C——C——C——C—H
   |  |  |  |  |  |  |  |  |  |
   H  H  H  H  H  H  H  H  H  H
```

```
   H  H  H  H  H  H  H  H              H     H
   |  |  |  |  |  |  |  |               \   /
H—C——C——C——C——C——C——C——C—H      +       C=C
   |  |  |  |  |  |  |  |               /   \
   H  H  H  H  H  H  H  H              H     H
```

octane – an alkane in petrol   ethene

Cracking a long-chain alkane molecule produces a shorter-chain alkane and also a small hydrocarbon molecule having a double bond between two of its carbon atoms. This is an ***alkene***. Ethene (formed in the reaction above) is the first member of the alkene series. Like all alkenes it is much more reactive than the corresponding alkane. This is because it is ***unsaturated*** – extra atoms can add on to make a bigger molecule when the double bond breaks down to a single bond again.

A few drops of bromine water added to a jar of ethene gas lose their brown colour instantly as the bromine molecules add on across the double bond.

ethene + bromine ⟶ 1,2-dibromoethane

```
H     H                          H  H
 \   /                           |  |
  C=C      +    Br—Br   ⟶     H—C——C—H
 /   \                           |  |
H     H                          Br Br
```

This reaction is used as a test to identify alkenes.

Ethane does not react with bromine water because it is already ***saturated***. This is the term used when the molecules of a hydrocarbon have only ***single covalent*** bonds linking their atoms.

Some of the different alkenes produced by the cracking of oil fractions are shown below. They are used in the manufacture of plastics.

Carbon atoms	Name	Formula
2	ethene	H, H \ / C=C / \ H, H
3	propene	CH_3, H \ / C=C / \ H, H
4	butene	CH_3—CH_2, H \ / C=C / \ H, H

Addition polymers

If there are no other substances available when their double bonds break open, alkene molecules will join up with each other. This is called ***addition polymerization***.

For example, when *ethene* gas is put under pressure and heated, thousands of its molecules link together to form a huge molecule of solid polythene. The diagram below shows just part of the linking

Some of the things made of polythene

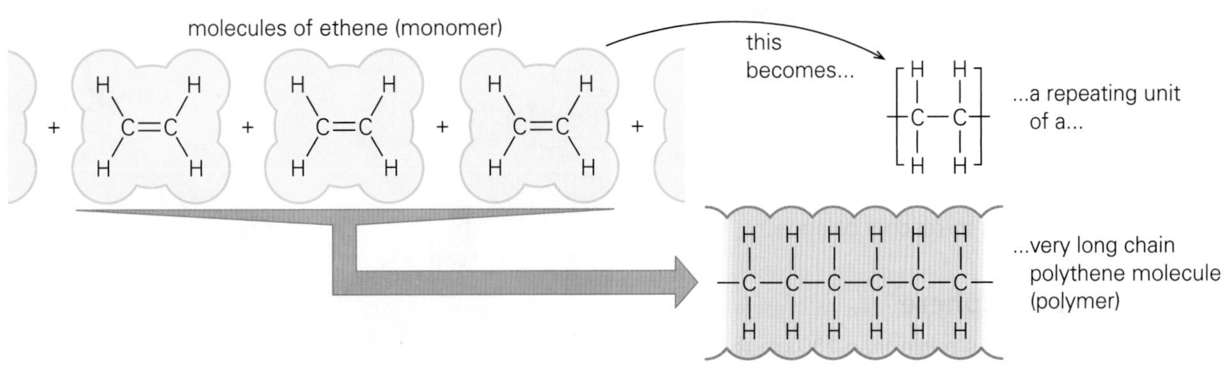

Many other substances having small molecules containing a carbon-to-carbon double bond will polymerize in a similar way. The small molecules are called ***monomers*** and the long chains they form are ***polymers*** ('poly' means many). Because polymers are very easily moulded into different shapes, they are called ***plastics***.

Some of the more common plastics and their uses are shown below.

1 What is meant by *cracking* of oil fractions?
2 What is the difference between saturated and unsaturated hydrocarbons?
3 What is the test used to identify alkenes?
4 What is meant by **(a)** addition polymerization? **(b)** monomer? **(c)** polymer?
5 Briefly describe how polythene is made.
6 Give the repeating unit and a use of **(a)** polypropene **(b)** polystyrene.

Monomer		Polymer	Unit	Uses
propene	CH_3, H, C=C, H, H	polypropene	[CH_3 H / –C–C– / H H]	carpets, crates, ropes
chloroethene (vinyl chloride)	Cl, H, C=C, H, H	PVC	[Cl H / –C–C– / H H]	pipes, gutters, flooring, electrical insulation
cyanoethene	CN, H, C=C, H, H	acrylic	[CN H / –C–C– / H H]	wool substitute
phenylethene (styrene)	C_6H_5, H, C=C, H, H	polystyrene	[C_6H_5 H / –C–C– / H H]	insulation, car parts, cups, packaging, toys

3.29 Sea and air

By the end of this spread, you should be able to:
- *explain why the Earth's seas and oceans are salty*
- *describe how the Earth's atmosphere has evolved.*

Salty seas

The Earth's seas and oceans contain dissolved salts (sodium chloride is the most abundant). Salts get into the sea because rain (which is weakly acid) slowly dissolves rocks, and the dissolved materials are carried into the sea by rivers. However, salts are removed from sea water when the shells of sea creatures form or chemical reactions deposit sediment on the sea bed. It has taken millions of years for the input and output of salts to reach a balance.

Some seas are saltier than others. In the Dead Sea, salt makes the water so dense that you cannot sink!

Gases in the atmosphere

The Earth's atmosphere is more than 100 km thick, but most of the air lies within 10 km of the Earth's surface. Air is a mixture of gases which can be separated and extracted. The chart below shows what they are and how they are used.

The amount of water vapour in the air varies. When the vapour cools, it condenses and forms billions of tiny water droplets which we see as clouds or mist.

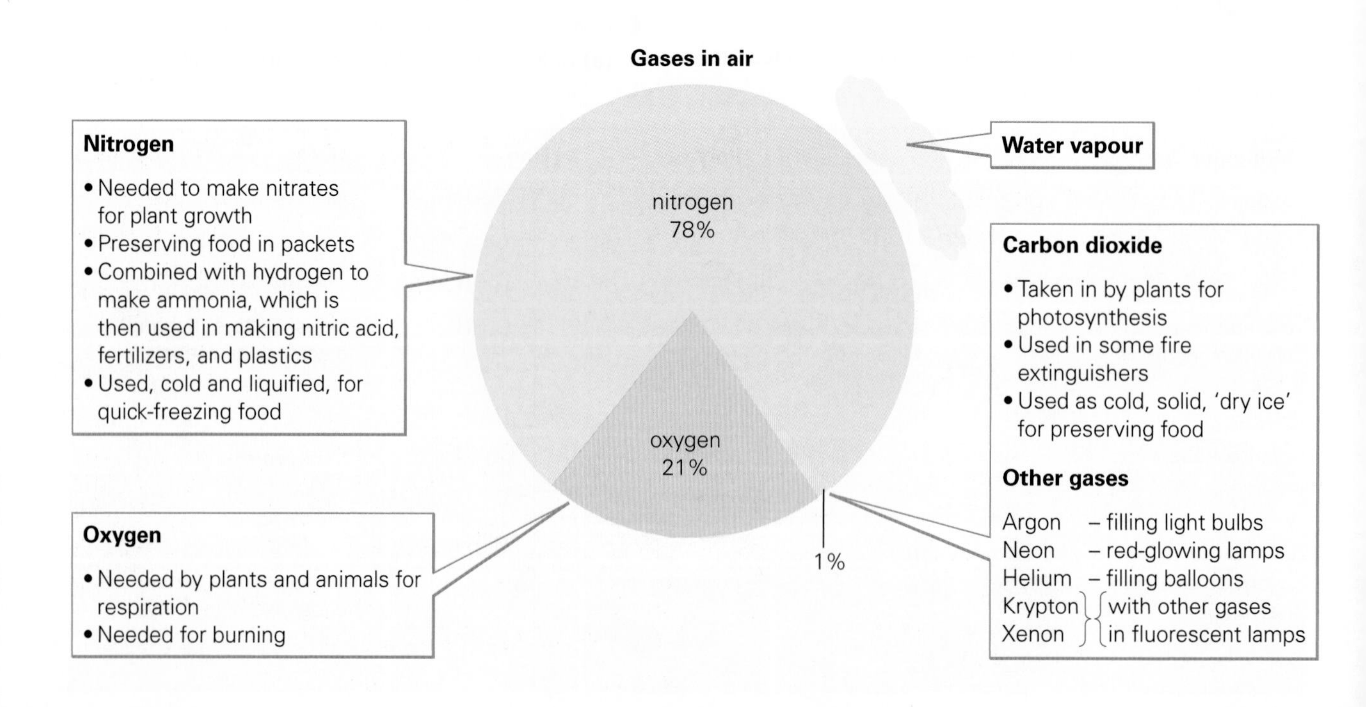

Origins of the atmosphere

The Earth was formed about 4500 million years ago, relatively soon after the Sun (see Spread 4.35). When its surface first solidified, intense volcanic activity created an atmosphere of mainly carbon dioxide gas. Today's atmosphere is largely due to the activities of living organisms. For example, the oxygen has come from plants (see Spread 2.02) and the nitrogen mostly from microbes (see Spread 2.26).

Ozone layer

Some of the oxygen molecules in the atmophere change into ***ozone*** – a form having three oxygen atoms in each molecule (O_3). The ozone has collected in a layer which filters out harmful ultraviolet radiation from the Sun. It is thought that certain gases (chlorofluorocarbons) once used in aerosols and fridges entered the atmosphere and have damaged the ozone layer (see also Spread 2.24).

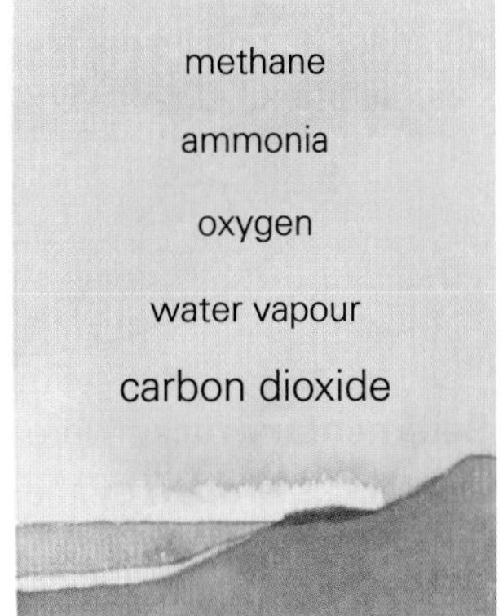

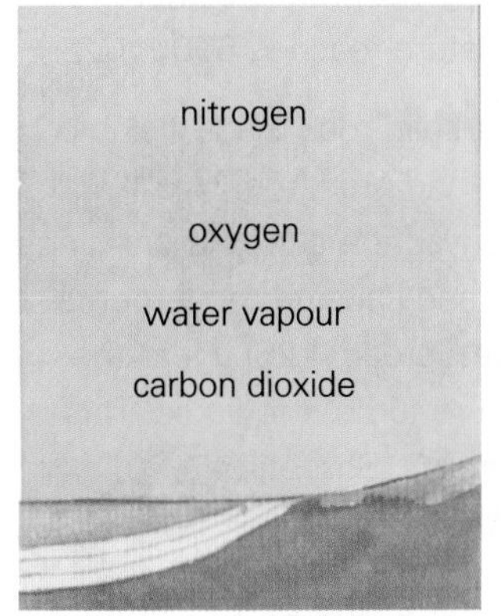

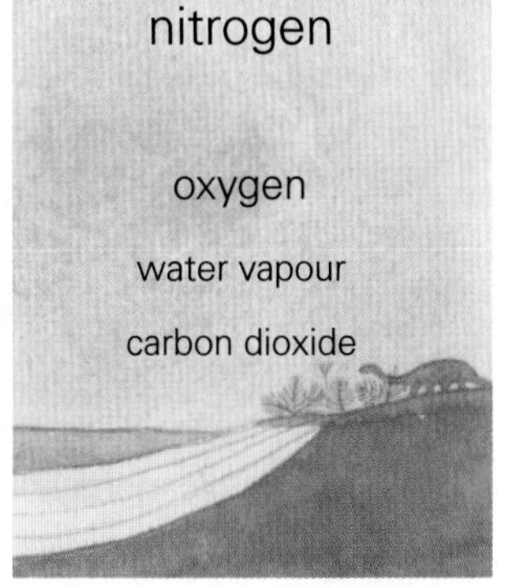

4000 million years ago, the Earth's atmosphere was mainly gases from volcanoes. As the Earth cooled, water vapour condensed to form oceans.

By 500 million years ago, early plants were giving out oxygen, some of which reacted with the methane and ammonia. In time, most of the carbon from carbon dioxide became locked up in sedimentary rocks formed from the skeletons and shells of sea creatures. And microbes were producing nitrogen.

By 200 million years ago, the Earth's atmosphere was similar to that of today. It is now mainly nitrogen (78%) and oxygen (21%).

Keeping the balance

The proportion of carbon dioxide in the air is very small but vital. It is produced by respiration and the burning of fuels, and removed by photosynthesis (see Spread 2.02) and by dissolving in the oceans. If the balance is upset, leaving too much carbon dioxide in the air, wholesale climate changes may result from the ***greenhouse effect*** (see Spread 2.24).

1. Describe a process which **(a)** puts salt into the sea **(b)** removes salt from the sea.
2. What is the ozone layer?
3. Compared with the early atmosphere, why does the Earth's atmosphere now have **(a)** less water vapour? **(b)** less carbon dioxide **(c)** more oxygen?
4. How do the oceans help control carbon dioxide levels in the atmosphere?

A possible result of too much carbon dioxide in the atmosphere – climate change.

3.30 Changes in rocks

By the end of this spread you should be able to:
- *describe the main types of rock found on Earth*
- *describe how rocks are formed and changed.*

The Earth's structure

The core is mostly molten iron and nickel, though the inner core is kept solid by the great pressure there. Deep in the core, the temperature reaches 5000 °C.

The mantle is a thick layer of solid rock made flexible (rather like Plasticine) by heat and pressure. This enables it to move slowly, convection currents bringing very hot rocks to the surface and causing movements of the plates of the crust (see Spread 4.26). Near the surface, any release of pressure turns it to liquid. This hot, molten rock is called ***magma***. Sometimes, it comes out of volcanoes as ***lava***.

The crust is the thin, outer layer of the Earth. The continents are the thickest part (up tp 90 km). They are mainly made of ***granite***. They are like huge rafts which 'float' on the denser mantle underneath. Under the oceans, the crust is thinner (as little as 6 km). It is mainly ***basalt***. In some places, the continental rafts push against each other. Here, the crust buckles and folds, forming mountain ranges.

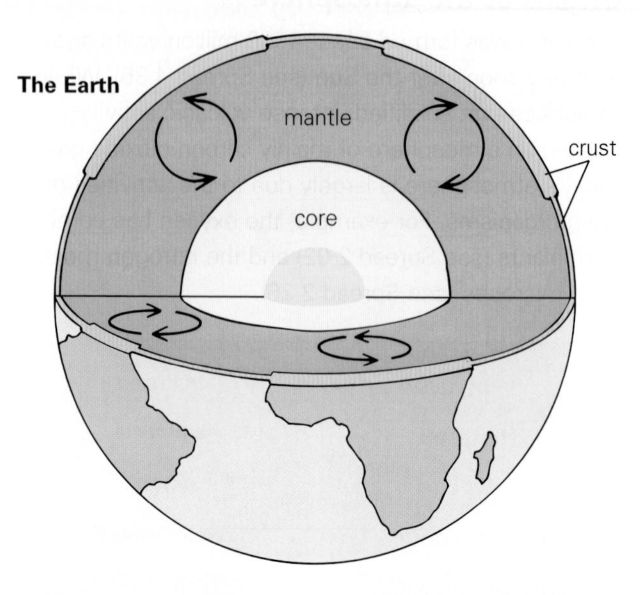

The crust's three types of rock

Igneous rocks, such as granite and basalt, are made of tiny interlocking crystals. They were formed when molten magma cooled and solidified. ***Extrusive*** rocks, such as basalt, have *small* crystals because the magma cooled *quickly* at the *surface*.

If the magma cooled *slowly*, deep underground, the crystals in the ***intrusive*** rocks so formed had time to grow, and are *large*. Granite is an example.

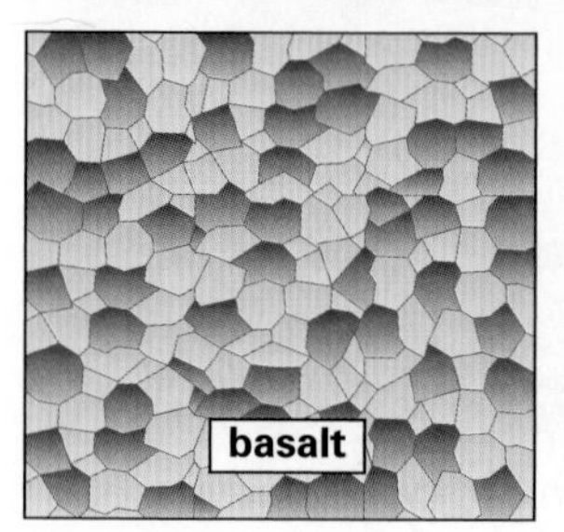

This rock cooled more quickly...

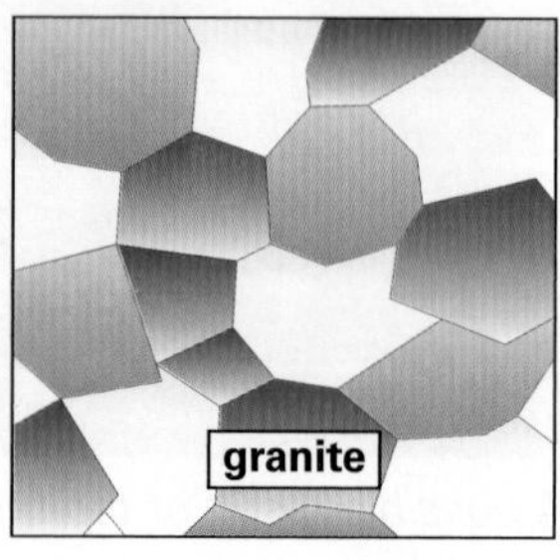

...than this

Sedimentary rocks were formed from fragments of older rocks ***eroded*** by the actions of sun, wind, rain, or ice. The fragments were carried by rivers or winds and deposited as layers of sediment, often at river mouths. The sediments were compressed as more and more material collected above them and cemented by minerals from the evaporating water, hardening in much the same way as concrete sets. You can see the layers, called ***strata***, in sedimentary rocks such as ***sandstone*** and ***shale*** (mudstone).

Strata – rock layers – are visible in a cliff face

Fragments of shells and bones from sea creatures which lived hundreds of millions of years ago also formed sediments which have become ***limestone***. Outlines of shells can be seen as ***fossils*** and used to date the strata in which they are found. The fossilized imprints of plants and animals are found in many sedimentary rocks.

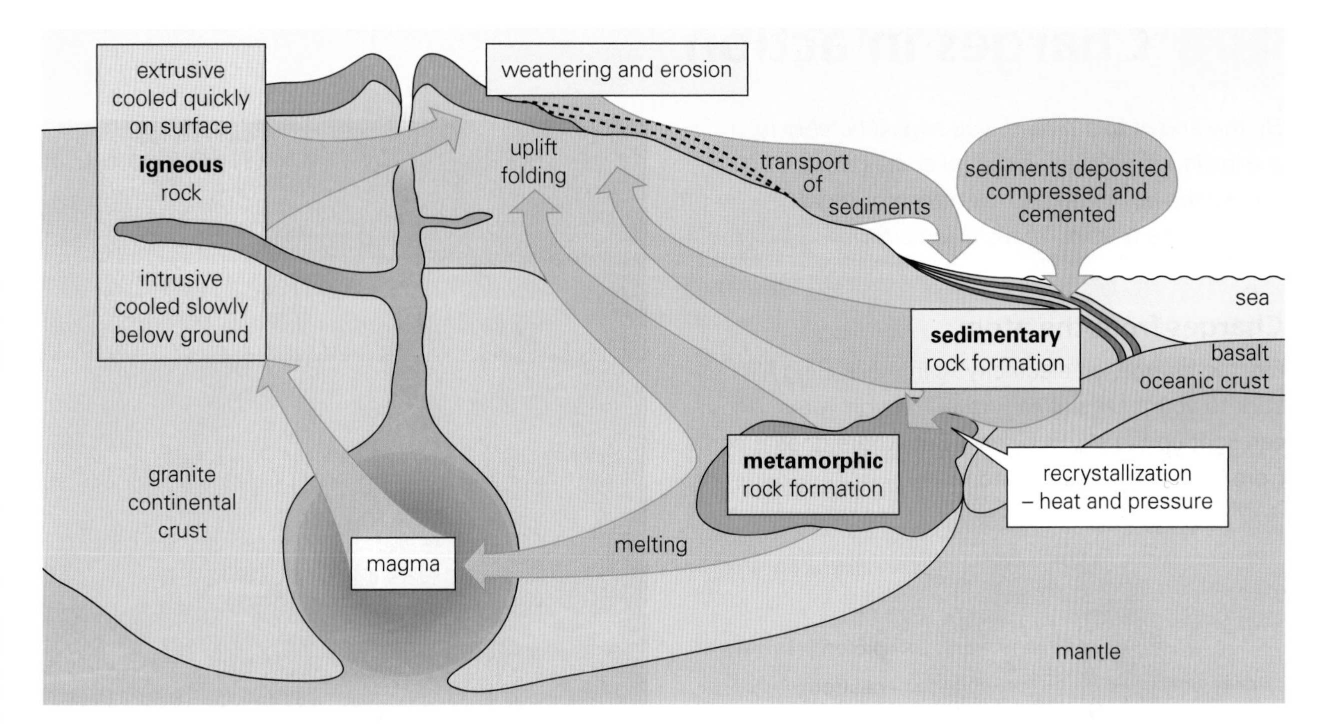

Metamorphic rocks come from sedimentary rocks which were buried deep underground by Earth movements. Baked by great heat and squeezed by intense pressure, the rocks ***metamorphosed*** – changed form. The layers (and fossils) were destroyed or greatly distorted, and the rocks became much harder like ***marble*** – made from limestone.

Pressure causes many metamorphic rocks to have minerals arranged in parallel sheets with a crumpled, layered look. They sometimes split easily into hard, flat sheets like ***slate*** (metamorphosed shale).

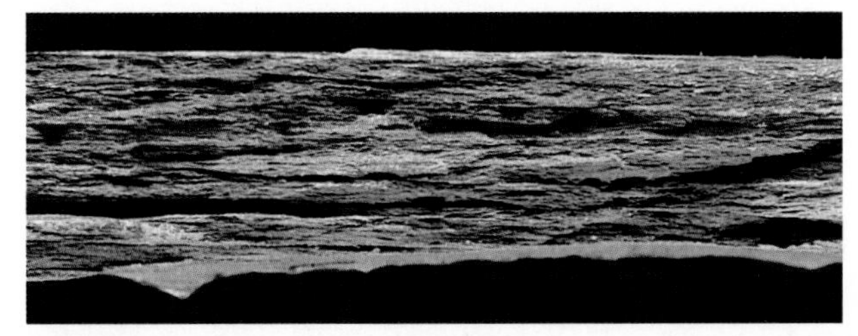

This hard slate was once shale (mudstone)

The rock cycle

Materials from rocks have been used over and over again in the ***rock cycle*** (above). The processes took millions of years and are still happening.

Where rocks are exposed, as in a cliff face, it is often possible to tell under what conditions they were formed. Below, the sandstone formed some distance from an ancient shoreline, and the pebbles in the conglomerate were laid down much later at the shore when the sea had retreated.

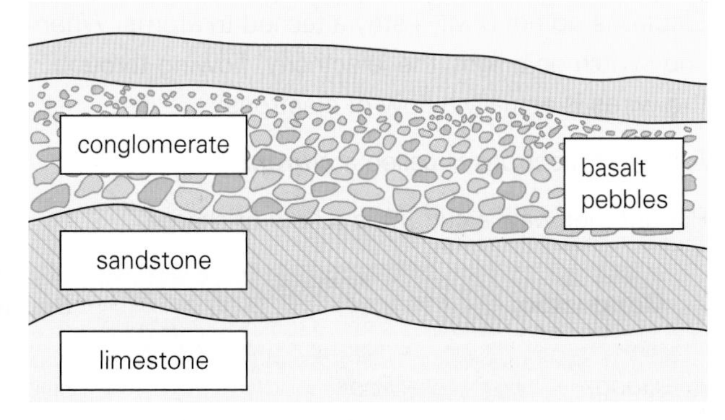

1 Briefly describe the three zones of the Earth.
2 What is *magma* and how is it formed?
3 What are **(a)** *extrusive* **(b)** *intrusive igneous* rocks? How can you tell which is which?
4 How were *sedimentary* rocks formed?
5 Where might the pebbles in the conglomerate shown above have come from?
6 What are *fossils* and what can they tell you about how the rocks containing them were formed?
7 What are *metamorphic* rocks ?

4.01 Charges in action

By the end of this spread, you should be able to:
- *explain where electric charge comes from*
- *describe how charges may attract or repel*
- *give some uses of 'static electricity'.*

Charges from the atom

Electric ***charge***, or 'electricity', can make cling film stick to your hands. It can travel through wires. And it can light up the sky in a flash. But where does it come from? The answer is the atom:

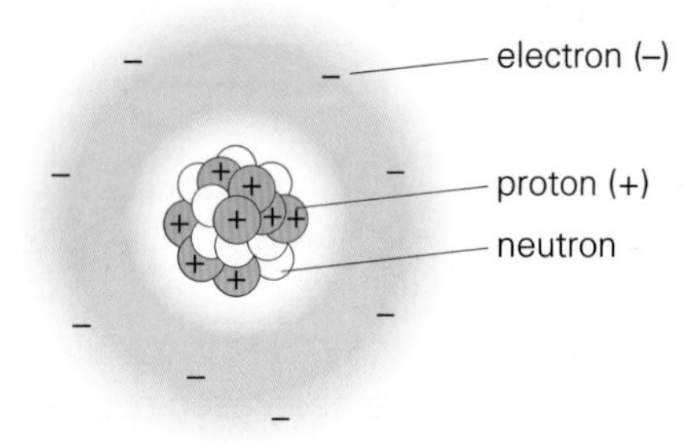

An atom has a tiny ***nucleus***, made up of smaller particles *(**protons** and **neutrons**)*. Around this are even smaller particles called ***electrons***.

In an atom, there are two types of charge. Electrons have a ***negative** (–)* charge, protons have a ***positive** (+)* charge. Atoms normally have the same number of electrons as protons, so, overall, they are uncharged.

Electrons do not always stay attached to atoms. When you switch on a light, the 'electricity' flowing through the wires is actually a flow of electrons:

A flow of electrons is called a ***current***.

Put another way: a current is a flow of charge.

From conductors to insulators

Conductors are materials which let electrons flow through. In a conductor such as copper, some electrons are so loosely attached to their atoms that they are free to flow between them.

Air and water can conduct, but only if they contain ***ions*** (see 3.07). Ions are charged atoms (or groups of atoms). In gases and liquids, they are free to move, so they can transfer charge from one place to another.

Insulators are materials which do not let electrons flow through. Their electrons are held tightly to atoms and are not free to move.

Semiconductors are 'in-between' materials. They can be treated chemically to control how much they conduct, and are used in microchips.

Conductors		Semiconductors	Insulators
Good	*Poor*	silicon	plastics
metals,	human body	germanium	e.g.
especially	water		PVC
silver	air		polystyrene
copper			Perspex
aluminium			glass
carbon			rubber

Static electricity

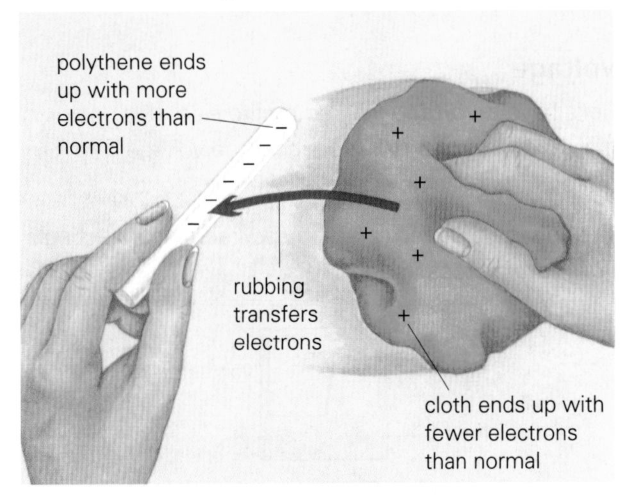

Insulators can become charged when rubbed. People say that there is 'static electricity' on them.

If you rub a polythene rod with a cloth, it pulls electrons from the cloth. The polythene ends up negatively charged, and the cloth positively charged.

If you rub Perspex with a cloth, the effect is opposite: the cloth pulls electrons from the Perspex.

The rubbing action doesn't make electric charge. It just separates charges already there. It works with insulators because, once the charges are separated, they tend to stay where they are.

Forces between charges If charged rods are held close, there are forces between them:

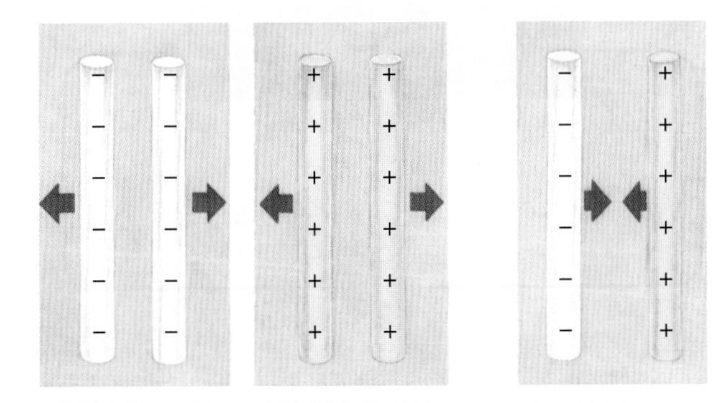

Like charges repel. Unlike charges attract.

A charge will also attract something uncharged. That is why dust is attracted to the charged screen of a TV. Being uncharged, the dust has equal amounts of + and –, so it feels attraction and repulsion. But the attracted charges are pulled slightly closer than the repelled ones, so the force on them is stronger.

Earthing

If enough charge builds up on something, electrons may be pulled through the air and cause sparks – as with lightning. To prevent charge building up, objects can be ***earthed***: they can be connected to the ground by a conductor so that unwanted charge flows away.

An aircraft and its tanker must be earthed to prevent sparks when the fuel 'rubs' against the pipe.

Using static electricity

Electrostatic precipitators are fitted to the chimneys of some power stations and factories to reduce pollution. They use electric attraction to remove bits of ash from the gases going up the chimney.

Photocopiers use electric attraction to pull powdered ink onto a charged drum (or plate). The drum is light-sensitive, and an image of the original document is projected onto it so that the powder only sticks in some places. The powder is transferred to a sheet of paper when the drum is rolled over it, and the paper heated so that the ink melts and sticks to it.

Inkjet printers Some use electric attraction to direct charged ink droplets to the right place on the paper.

1 How is a conductor different from an insulator?
2 What will happen if a negatively charged rod is brought close to a small plastic ball if the ball is **(a)** positively charged? **(b)** negatively charged?
3 If you rub Perspex with a cloth, electrons are transferred from the Perspex to the cloth. What type of charge does this leave **(a)** on the Perspex? **(b)** on the cloth?
4 Give an example of how a build-up of electric charge can be dangerous.
5 Give *two* examples of how 'static electricity' can be useful.

4.02 Circuits and cells

By the end of this spread, you should be able to:
- *explain how current and voltage are measured*
- *describe the features of series and parallel circuits.*

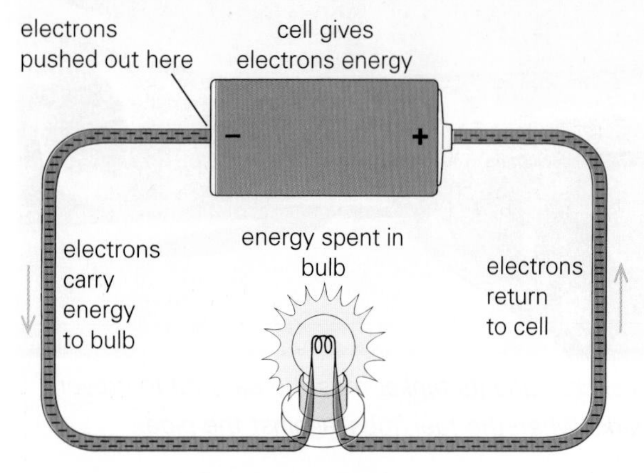

An electric ***cell*** will produce a current (flow of electrons) if there is a conducting material between its terminals. A chemical reaction inside the cell pushes electrons out of the negative (–) terminal and round to the positive (+) terminal. The conducting path through the bulb, wires, and cell is called a ***circuit***. The electrons transfer energy from the cell to the bulb where it is given off as heat and light.

Current

Current is measured in ***amperes*** (***A***). The higher the current, the greater the flow of electrons.

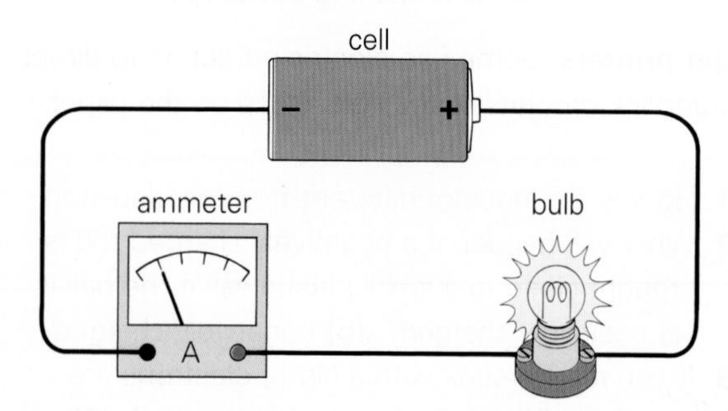

Current is measured with an ***ammeter***, as above. The ammeter can be connected anywhere in the circuit, because the current is the same all the way round. Also, putting in the ammeter doesn't affect the current.

Small currents are sometimes measured in ***milliamperes (mA)***. 1000 mA = 1A

Voltage

Cells have a ***voltage*** marked on the side. It is measured in ***volts (V)***. The higher the voltage, the more energy each electron is given to spend.

The voltage of a cell can be measured by connecting a ***voltmeter*** across its terminals:

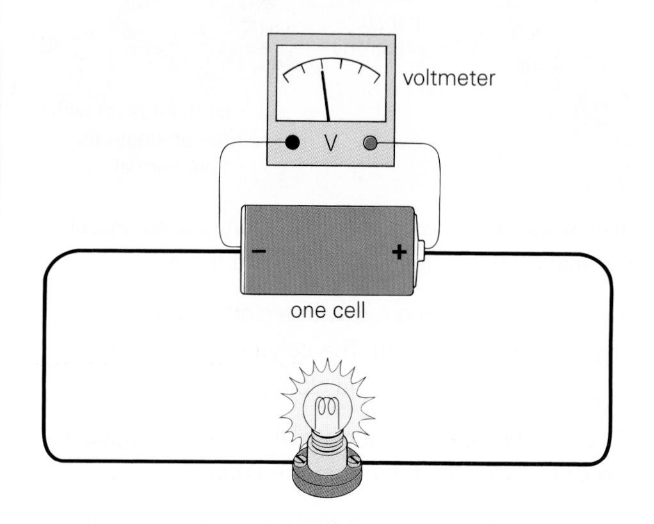

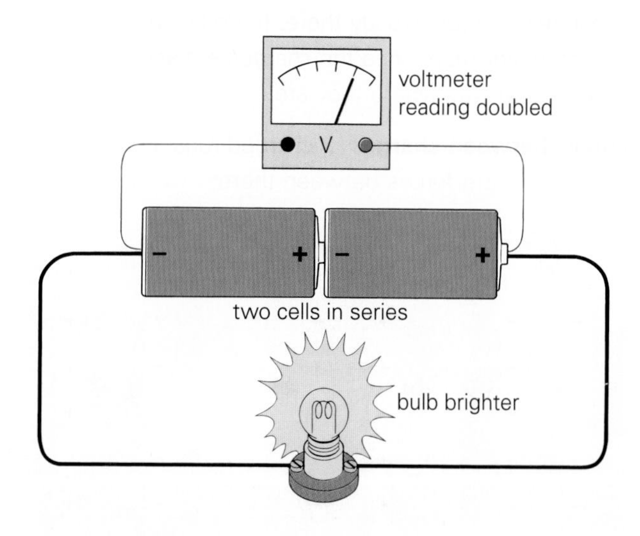

If two cells are connected in ***series*** (in a line), the total voltage is doubled. Also, the bulb is brighter because more current is pushed through it:

- The more cells in series, the higher the voltage.
- The higher the voltage, the higher the current.

A collection of cells is called a ***battery***, although the word is sometimes used for a single cell as well.

Series and parallel

Here are two different ways of adding an extra bulb to the previous circuit:

Bulbs in series The bulbs glow dimly. It is more difficult for the electrons to pass through two bulbs than one, so there is less current than before.

Adding more bulbs makes them even dimmer. And if one bulb is removed, the circuit is broken. So all the bulbs go out.

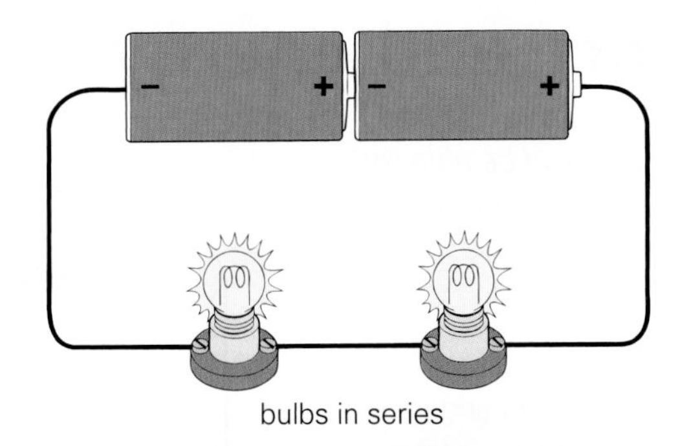

bulbs in series

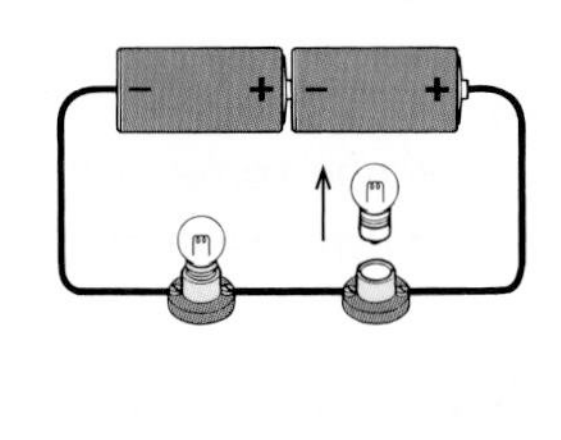

Bulbs in parallel The bulbs glow brightly, because each is getting the full voltage from the cells. However, together, two bright bulbs take more current than a single one, so the cells will not last as long.

If one bulb is removed, there is still a complete circuit through the other bulb, so it keeps glowing brightly.

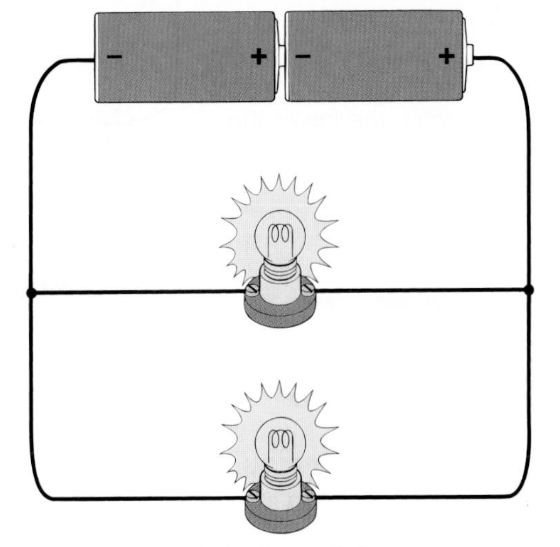

bulbs in parallel

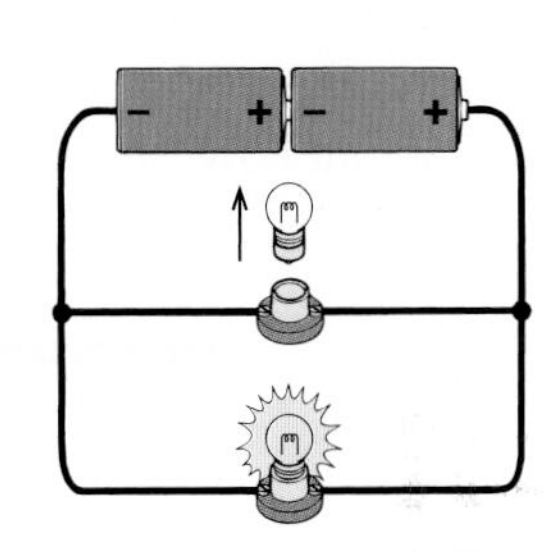

Circuit symbols

It can take a long time to draw pictures of circuits! That is why scientists and electricians prefer to use ***symbols***.

On the right, you can see the circuit with bulbs in parallel, drawn using symbols.

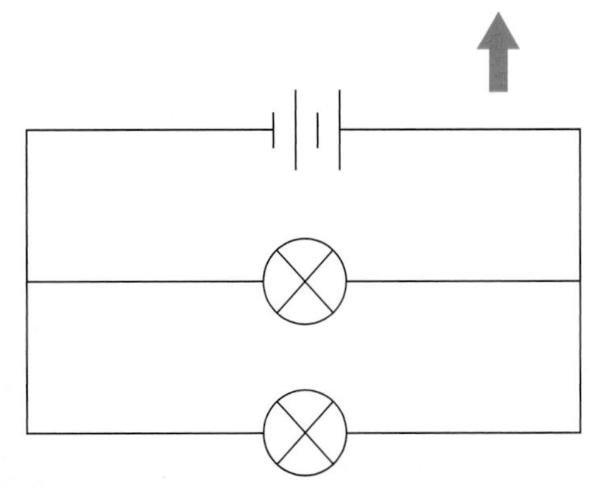

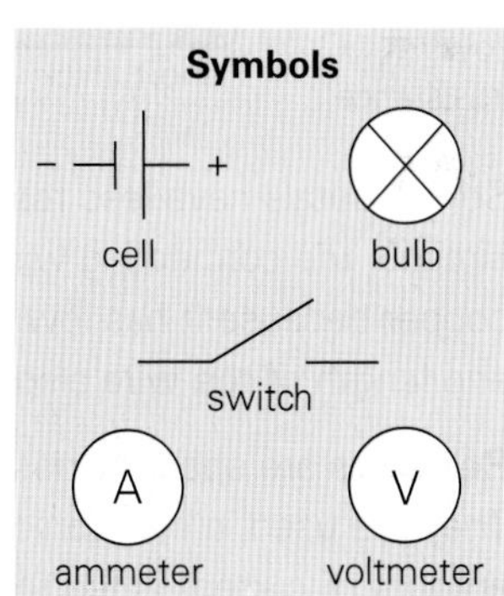

1 In the circuit on the right, what type of meter is X? What reading would you expect to see on it?

2 What type of meter is Y?

3 Redraw the circuit so that it has two cells instead of one, and Y is across both. What difference would you expect to see in
(a) the brightness of the bulb? **(b)** the reading on X?
(c) the reading on Y?

4 If an extra bulb is added to your new circuit in series, how will this affect **(a)** the brightness? **(b)** the current?

5 What are the advantages of connecting the extra bulb in parallel?

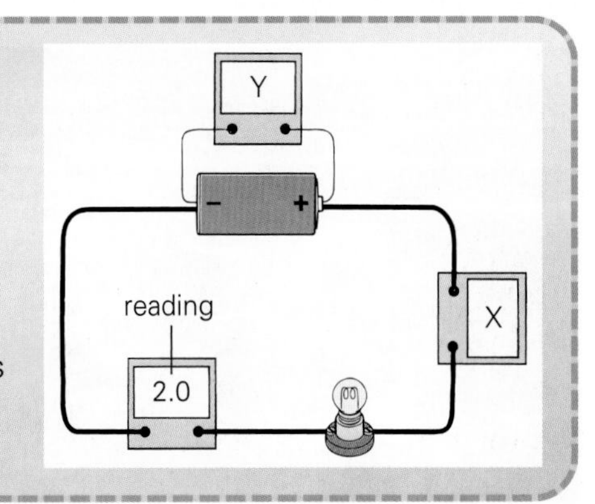

4.03 Resistance and energy

By the end of this spread, you should be able to:
- *describe some of the effects of resistance*
- *calculate the cost of running appliances.*

Resistance and resistors

Bulbs do not conduct as well as connecting wire. Scientifically speaking, they have more ***resistance***. Energy must be spent overcoming this resistance. A bulb gives off this energy as heat and light.

The more resistance there is in a circuit, the lower the current. Resistance can be measured and calculated. To find out how, see the next spread, 4.04.

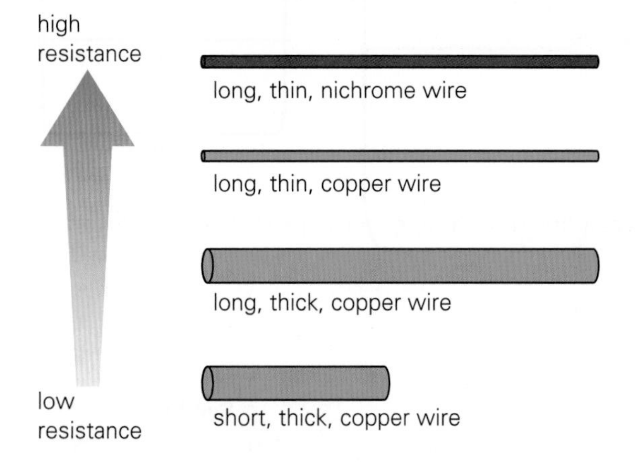

Some metals have less resistance then others. In circuits, the connecting wires are usually made of copper because it has low resistance. The thickness and length of the wire also affect the resistance.

Resistors are specially designed to provide resistance. They are used in electronics circuits so that the right amount of current is fed to different components (parts) to make them work properly.

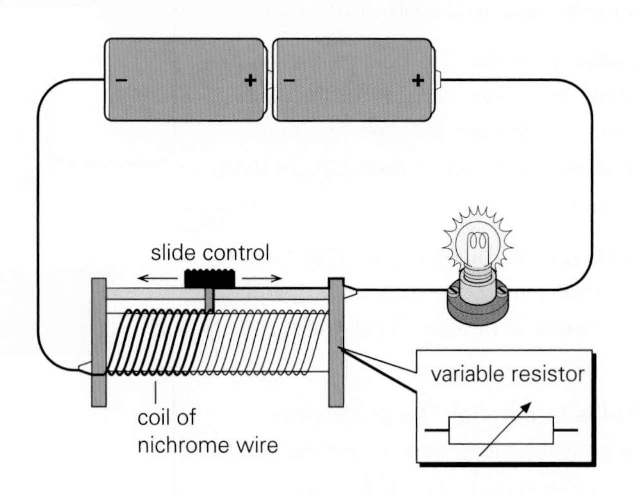

Above, a ***variable resistor*** is being used to control the brightness of a bulb. The variable resistor contains a long coil of thin nichrome wire. Sliding the control to the right puts more resistance into the circuit, so the bulb gets dimmer.

Heat from resistance

Whenever current flows through a resistance, heat is given off. This idea is used in the filament of a light bulb. It is also used in the ***heating elements*** in ***appliances*** such as kettles, irons, and toasters. The elements normally contain lengths of nichrome wire.

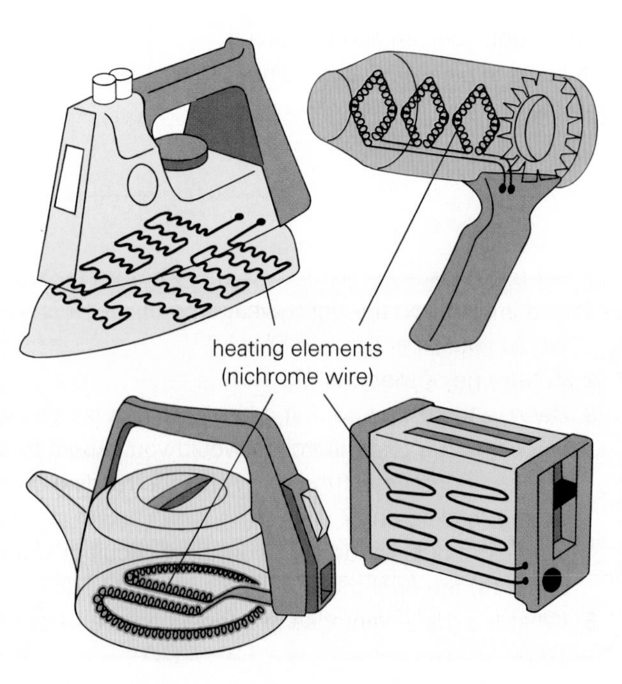

Paying for energy

Most appliances get their energy from the ***mains***. This energy has to be paid for.

Appliances usually have a ***power*** marked on them, in ***watts (W)*** or in ***kilowatts (kW)***. There are some examples on the right. The higher the power, the quicker the appliance takes energy from the mains. (For more on energy and power, see 4.17 and 4.22).

The energy supplied depends on the power of the appliance and on how long it is switched on for. Electricity companies measure energy in ***kilowatt hours (kW h)***, also called ***Units***. For example:

If 1 kW is switched on for 1 hour, then 1 kW h of energy is supplied.

If 2 kW is switched on for 3 hour, then 6 kW h of energy is supplied.

These results can be worked out with an equation:

energy in kW h	=	power in kW	×	time in hours

Electricity companies charge for each kW h (Unit). On the right, you can see how to calculate the cost of running an appliance.

If the power is given in watts, you must change this into kilowatts before using the equation.

1 kW = 1000 W. So, for example, 100 W = 0.1 kW.

Reading the meter

Every house has an 'electricity meter'. Really, it is an energy meter. Below, you can see how the meter reading in one house changed over a 24-hour period. To work out how many Units (kW h) were supplied, you take one number from the other.

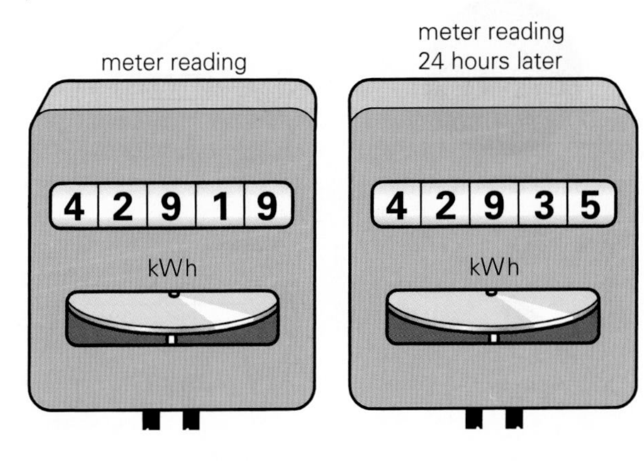

Typical powers

	in kW		in W
kettle	2.4	electric mower	900
fan heater	2	electric drill	500
hairdrier	2	food mixer	500
hotplate	1.5	TV set	120
iron	1	table lamp	60
toaster	1	stereo system	60

1 kW = 1000 W

Using a 3 kW heater for 4 hours

energy = power × time
= 3 kW × 4 h
= 12 kW h

The electricity company charges 10p per kW h (Unit). So

cost = 12 × 10p
= £1.20

1 Copper connecting wire is used in circuits.
(a) Why is thick copper wire better than thin?
(b) Why is nichrome wire not used instead?
(c) What is nichrome wire used for? Why?

2 In the circuit at the top of the opposite page, what will happen to the bulb if the variable resistor control is moved to the left? Why?

The table on this page gives the powers of some appliances. Assume that energy costs 10p per Unit:

3 How much energy (in kW h) is needed to run the toaster for 2 hours? What will it cost?

4 What is the cost of running **(a)** the fan heater for 4 hours? **(b)** the food mixer for 2 hours?

5 Use the meter readings on the left to work out the cost of the energy supplied.

4.04 Resistance, voltage, and current

By the end of this spread, you should be able to:
- *calculate resistance*
- *describe how current varies with voltage for resistors and other components.*

Ohm's law and resistance

The circuit on the right can be used to find how the current through a conductor depends on the voltage across it. The conductor in this case is a coil of nichrome wire, kept at a steady temperature. The table shows some typical results. Note that:

- If voltage doubles, current doubles, and so on.
- Voltage divided by current always has the same value (5 in this example).

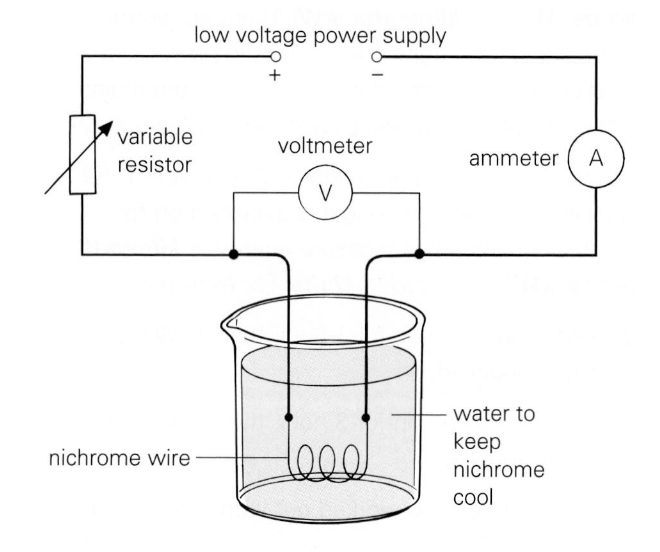

Voltage	Current	Resistance = $\frac{\text{voltage}}{\text{current}}$
1.0 V	0.2 A	5.0 Ω
2.0 V	0.4 A	5.0 Ω
3.0 V	0.6 A	5.0 Ω
4.0 V	0.8 A	5.0 Ω
5.0 V	1.0 A	5.0 Ω
6.0 V	1.2 A	5.0 Ω

Mathematically, these mean that ***the current is proportional to the voltage*** (provided the temperature does not change).

This is called ***Ohm's law***. All metals obey Ohm's law.

The value of voltage/current for a conductor is called its ***resistance***. It is measured in ***ohms*** (Ω).

$$\text{resistance} = \frac{\text{voltage}}{\text{current}}$$

resistance in Ω
voltage in V
current in A

So, the nichrome wire has a resistance of 5.0 Ω.

The resistance of a metal increases with temperature, but only slightly unless the temperature rise is large.

Resistance components

Some of the components (parts) used in circuits are designed to have a resistance which can vary:

A thermistor has a high resistance when cold but a low resistance when hot. It can be used in electronic circuits which detect temperature change – for example, those in fire alarms or thermometers.

A light-dependent resistor has a high resistance in the dark but a low resistance in the light. It can be used in circuits which switch on lights automatically.

A diode has a very high resistance in one direction but a low resistance in the other. In effect, it only allows current to flow through it one way. It is used in radios, power adaptors, and other electronic items.

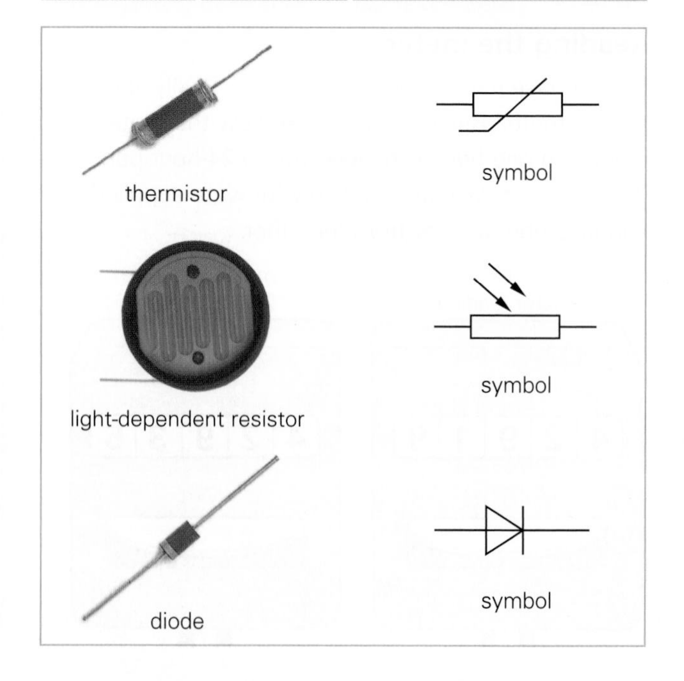

Current–voltage graphs

With readings taken using a circuit like the one on the left, current–voltage graphs can be produced for different components. Here are some examples:

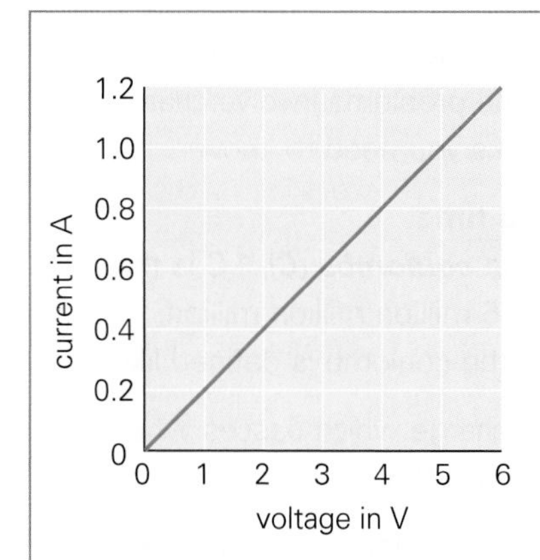

Metal resistor at a steady temperature (e.g. nichrome wire on opposite page) The graph is a straight line through the origin. Voltage/current is the same at all points. In other words, the resistance doesn't change.

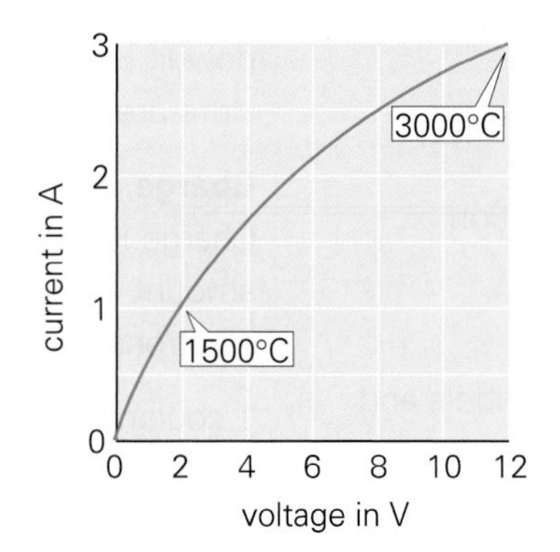

Tungsten filament (in a bulb) As the current increases, the temperature rises and the resistance goes up. Voltage/current is not the same at all points. The current is not proportional to the voltage.

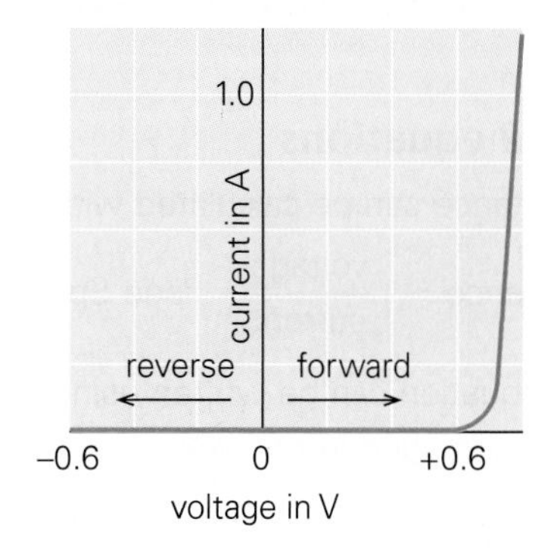

Diode The current is not proportional to the voltage. And if the voltage is reversed (by connecting the diode the other way round), the current is almost zero. In effect, the diode 'blocks' current in the reverse direction.

Adding resistances

If you connect more and more resistors in series, the total resistance is increased:

If resistances are connected in series:

total resistance = sum of separate resistances

For example, on the right, resistances of 3 Ω and 6 Ω in series have a total resistance of 9 Ω.

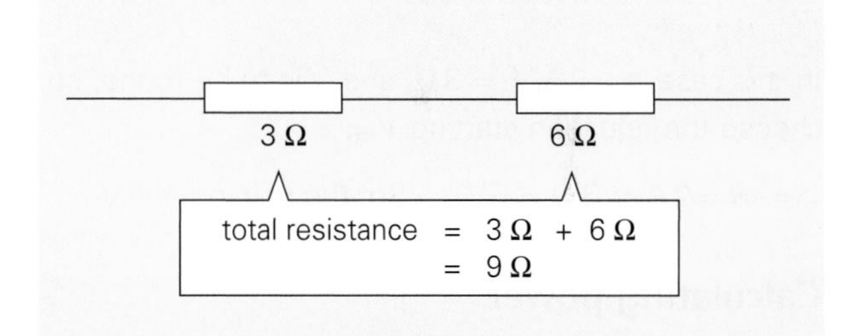

1

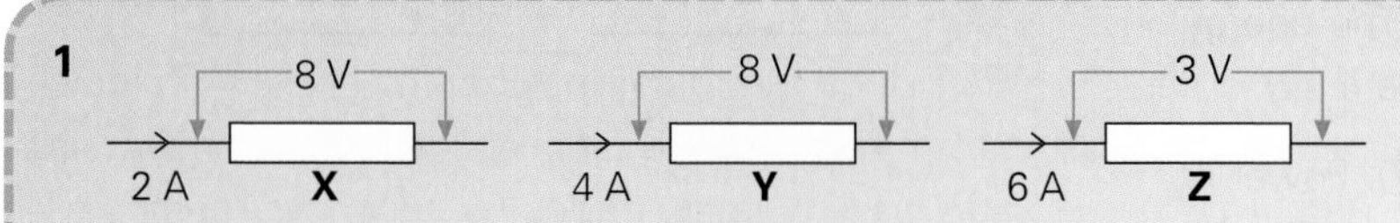

The resistors above are metal wires kept at a steady temperature.
(a) What is the resistance of each?
(b) If the voltage is doubled, what is the current through each?

2 Name a component whose resistance is
(a) high when cold but low when hot
(b) high in one direction but low in the other.

3 Look at the middle graph above. Calculate the resistance of the filament at **(a)** 1500 °C **(b)** 3000 °C.

4 Look at the graph on the right. Decide which resistor has the higher resistance, and explain your answer.

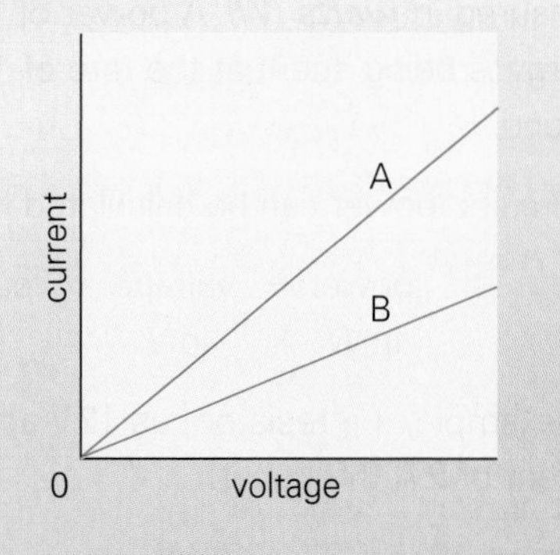

4.05 Circuit equations and rules

By the end of this spread, you should be able to:
- *use the equations linking voltage, current, resistance, energy, power, and charge*
- *give the basic rules for series and parallel circuits.*

V, I, R equations

Resistance can be calculated with this equation:

$$\text{resistance} = \frac{\text{voltage}}{\text{current}}$$ (see Spread 4.04)

The equation can be written using letter symbols and also rearranged in two ways:

$R = \frac{V}{I}$ $\quad I = \frac{V}{R}$ $\quad V = IR$

R = resistance in Ω
V = voltage in V
I = current in A

Here is an example of how to use one of these:

A current of 2 A flows through a 3 Ω resistor. What is the voltage across the resistor?

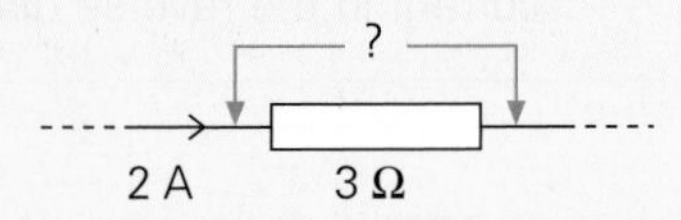

In this case, I = 2 A, R = 3 Ω, and V is to be found, so choose the equation starting V = ...

$V = IR$ = 2 A × 3 Ω = 6 V   So, the voltage is 6 V.

Calculating power

When a battery pushes a current through a resistor, energy is spent. This is given off as heat. The energy spent per second is called the ***power*** (see 4.22).

Energy is measured in joules (J) (see 4.17). Power is measured in ***watts (W)***. A power of 1 watt means that energy is being spent at the rate of 1 joule per second.

In circuits, power can be calculated like this:

power	=	voltage	×	current
in W		in V		in A

For example, if a resistor has 12 V across it and a current of 2 A through it:

power = 12 V × 2 A = 24 W.

Further links

Most circuit problems can be solved by using the links between voltage, current, resistance, and power. However, some problems involve charge. Here are the extra links you need to know:

Charge, current, and time

Charge is measured in ***coulombs (C)***. 1 C is the amount of charge on 6 million million million electrons! However, the coulomb is defined like this:

1 coulomb (C) is the charge which passes when a current of 1 ampere (A) flows for 1 second (s).

So, if a current of 3 A flows for 2 s, a charge of 6 C passes. There is an equation for working this out:

charge	=	current	×	time
in C		in A		in s

Energy, charge, and voltage

There is a link between energy in joules (J), charge in coulombs (C), and voltage in volts (V):

If the battery voltage is 12 V, then 12 J of energy is given to each coulomb of charge pushed out.

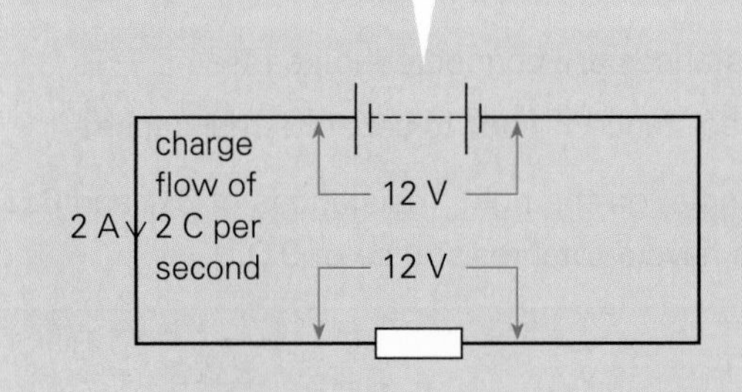

If the voltage across a resistor is 12 V, then 12 J of energy is spent in pushing each coulomb of charge through the resistor.

In the above example, energy is being transferred from the battery to the resistor, and this equation applies:

energy transferred	=	charge	×	voltage
in J		in C		in V

Basic circuit rules

The basic rules for series and parallel circuits are illustrated by the examples below.

Note that for any resistor, $V = IR$ is always true (and of course the other versions of the equation as well).

In the diagrams, the current arrows run from the + terminal of the battery round to the –. This is called the ***conventional*** direction. It is the direction you would expect positive charge to move. Electrons have negative charge, so they flow the other way.

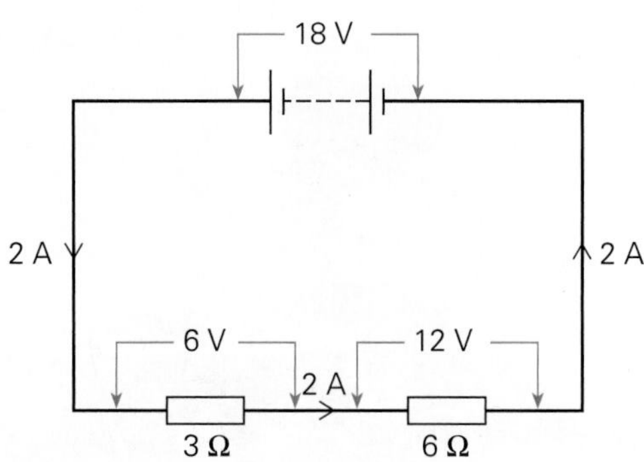
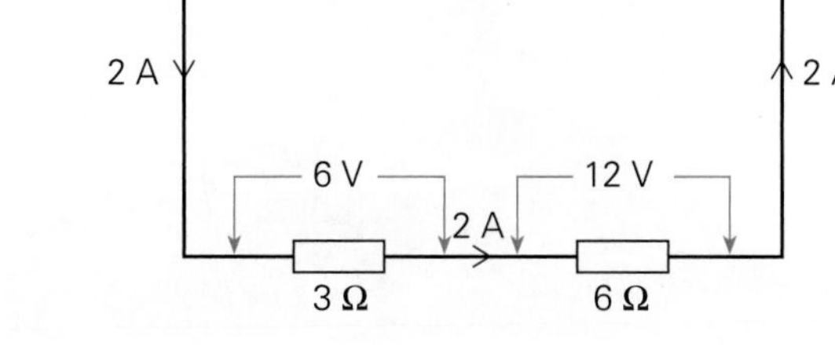

When a battery is connected across resistors (or other components) in ***series***:

- each resistor gets the full current from the battery.
- the voltages across the resistors add up to equal the battery voltage (because the resistors share the battery voltage).

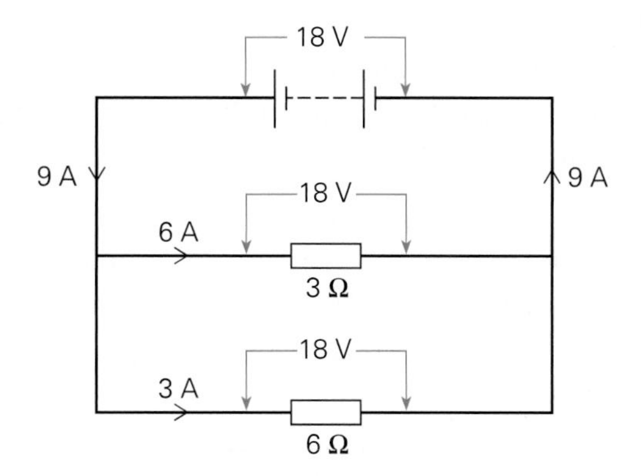

When a battery is connected across resistors (or other components) in ***parallel***:

- each resistor gets the full battery voltage.
- the currents through the resistors add up to equal the current from the battery (because the resistors share the current from the battery).

Cell arrangements

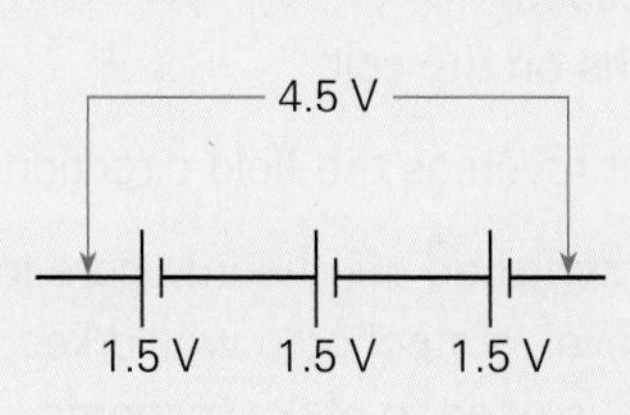

Cells in series: you add up their voltages to find the total voltage.

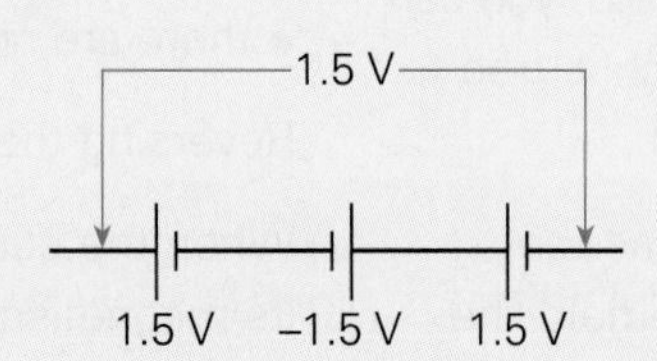

Here, one cell is the wrong way round, so it cancels out one of the others.

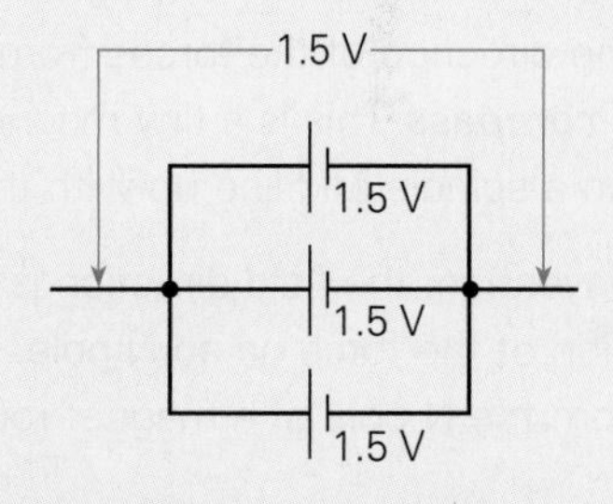

The voltage across parallel cells is the same as from a single cell.

1 What is the current through a 4 Ω resistor if the voltage across it is **(a)** 8 V? **(b)** 12 V? **(c)** 2 V?

2 When a 6 V battery is connected across a resistor, the current is 3 A. Calculate
(a) the resistance **(b)** the battery's power output.

3 In question 2:
(a) How much charge does the battery deliver in one second?
(b) How much energy does the battery deliver in one second?

4 A 24 V battery is connected across resistors of 6 Ω and 12 Ω in parallel. Draw a diagram of this arrangement, then calculate
(a) the current through the 6 Ω resistor
(b) the current through the 12 Ω resistor
(c) the current from the battery.

4.06 Magnets and electromagnets

By the end of this spread, you should be able to:
- *describe the effects of magnets and electromagnets*
- *explain some of the uses of electromagnets.*

Magnets

A few metals are ***magnetic***. They are attracted to magnets and can be magnetized. The main magnetic metals are iron and steel (but not stainless steel).

The force from a magnet seems to come from two points near the ends. These are called the ***north pole (N)*** and the ***south pole (S)*** of the magnet. When the ends of magnets are brought close together, you can feel a force:

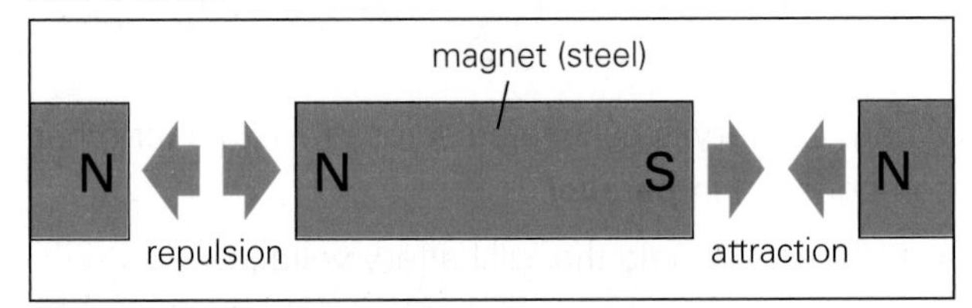

Like poles repel. Unlike poles attract.

Magnetic fields

A magnet will push or pull on other magnets and attract unmagnetized bits of iron and steel. Scientists say that the magnet has a ***magnetic field*** around it. To find the direction of the forces from this field, you can use a ***compass***. This is a tiny magnet which is free to turn on a spindle and line up with the field.

By convention, the field direction is taken as the direction of the force on an N pole. So the field lines run from the N pole of a magnet round to the S pole.

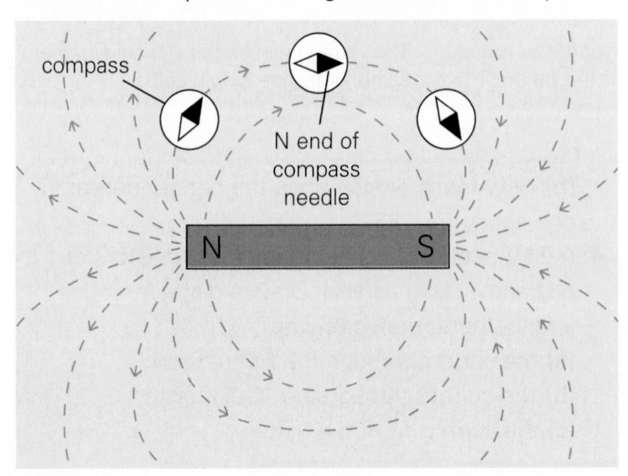

Electromagnets

If a current flows through a wire, it produces a magnetic field (see the next page). The effect is used in the ***electromagnet***, shown below. This is made up of a ***coil*** of copper wire wound round an iron ***core***.

The electromagnet produces a field rather like the one around a bar magnet. The iron ***core*** makes the field much stronger. The field is even stronger if

- the current is increased
- there are more turns on the coil.

Reversing the current reverses the field direction.

When the current is switched off, the iron core loses its magnetism. However, a steel core would keep its magnetism. This idea is used to make magnets.

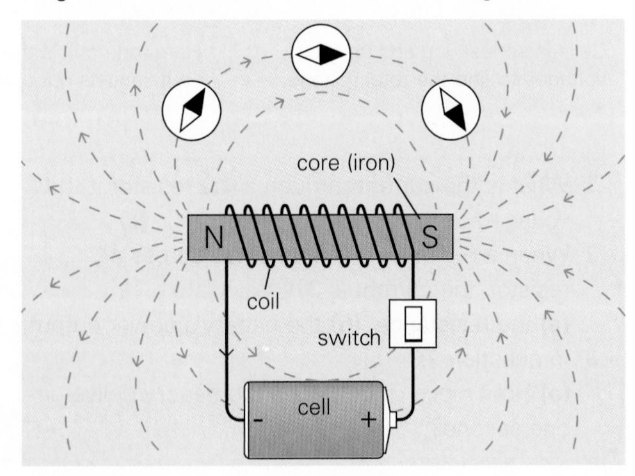

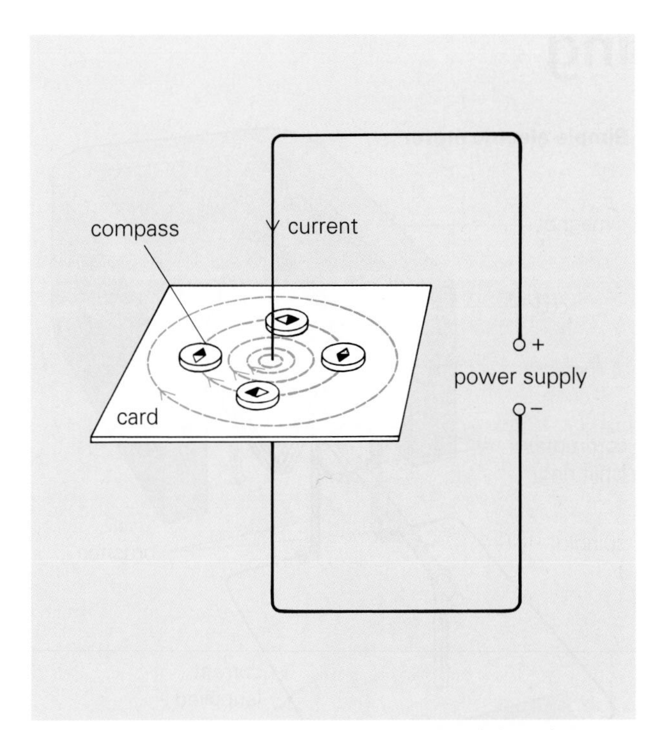

Magnetic field around a current in a straight wire.

Using electromagnets

Using an electromagnet, the pulling force can be 'switched' on and off. The photograph on the opposite page shows one use. Here are two more:

Magnetic relay This is a switch operated by an electromagnet. With a relay, it is possible to use a tiny switch with thin wires to turn on the current in a much more powerful circuit – for example, a mains circuit with a big electric motor in it.

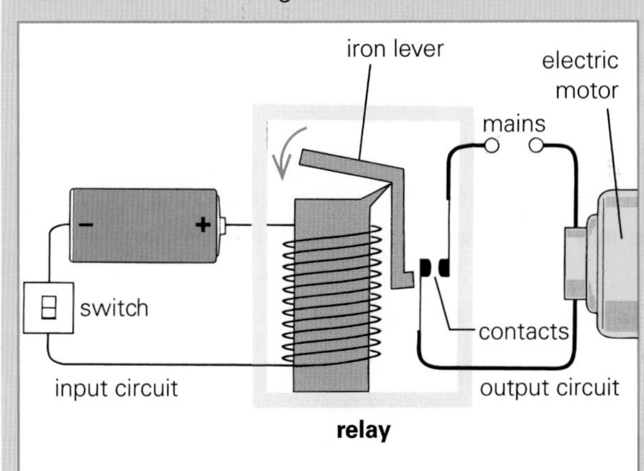

How it works If you switch on the current in the input circuit, the electromagnet pulls on an iron lever. This closes two contacts in the output circuit.

Circuit breaker This is an automatic safety switch. It cuts off the current in a circuit if this gets too high. It can be used instead of a fuse (see 4.08).

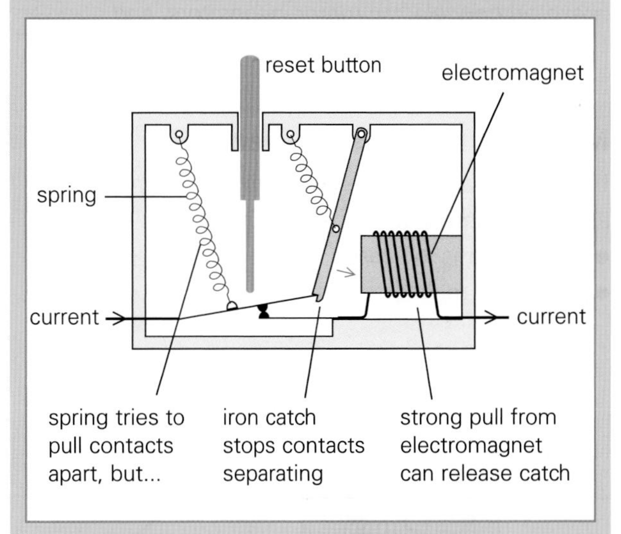

How it works A spring is trying to pull two contacts apart, but an iron catch stops them from opening. If the current gets too high, the pull of the electromagnet becomes strong enough to release the catch, so the contacts open and break the circuit.

Loudspeaker This has a paper or plastic cone attached to a coil which is in the field of a magnet. The amplifier in your radio or stereo makes current flow backwards, forwards...and so on, through the coil. this makes the coil move in and out, so the cone vibrates and gives out sound waves (see 4.24).

1 How can you show that there is a magnetic field around a magnet?
2 What changes would you make to an electromagnet to give it a stronger pull?
3 Describe how you could magnetize a steel nail.
4 Describe the magnetic field pattern around the current through a long straight wire.
5 Explain why, with a relay, it only takes a small current to switch on an electric motor, even though the motor takes a big current.
6 **(a)** What is a circuit breaker used for?
(b) In the circuit breaker shown above, what is the purpose of the electromagnet?

4.07 Turning and changing

By the end of this spread, you should be able to:
- *explain how electric motor, generators, and transformers work.*

Magnetic force on a current

If a wire is carrying a current across a magnetic field as below, there is a force on it. The force is larger if
- the current is increased
- a stronger magnet is used.

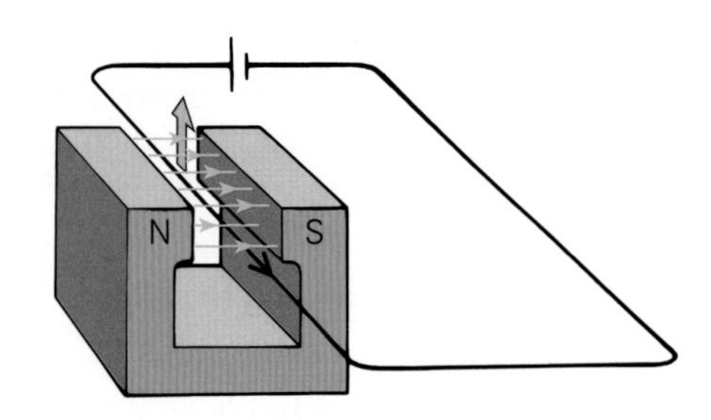

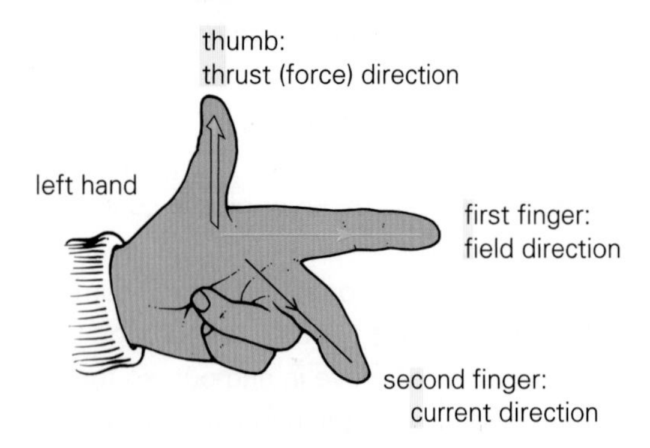

The direction of the force is given by ***Fleming's left hand rule*** as shown above. When using the rule:

- The current direction is the conventional direction (see Spread 4.05).
- The field direction is from the N to S pole.

Electric motors

In an electric motor, the force on a current in a magnetic field is used to produce a turning effect.

The motor in the diagram has a coil which can spin between the poles of a magnet. The cell supplies the coil with current through two contacts called ***brushes***.

Simple electric motor

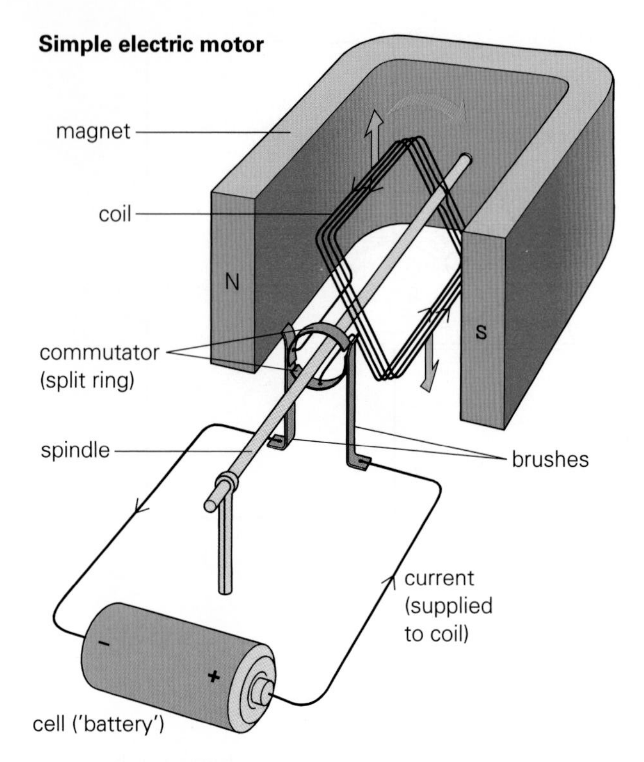

The magnetic field exerts an upward force on one side of the coil and a downward force on the other (because the current flows in opposite directions along the two sides). Together, the forces have a turning effect.

The ***commutator*** reverses the current through the coil every half-turn. This is necessary to keep the coil turning. For example, once the upward force has pushed one side of the coil right up, it must become a downward force to pull it down another half-turn, then an upward force to push it up...and so on.

Practical motors often use electromagnets rather than ordinary magnets. Also, for smoother running, they usually have several coils set at different angles.

Generators, AC, and DC

If you take a simple electric motor, remove the battery, replace it with a meter, and spin the coil, the motor produces a current. It has become a ***generator***.

Whenever wires cut through a magnetic field, or are in a changing magnetic field, a voltage is produced. The effect is called ***electromagnetic induction***. The ***induced*** voltage can make a current flow.

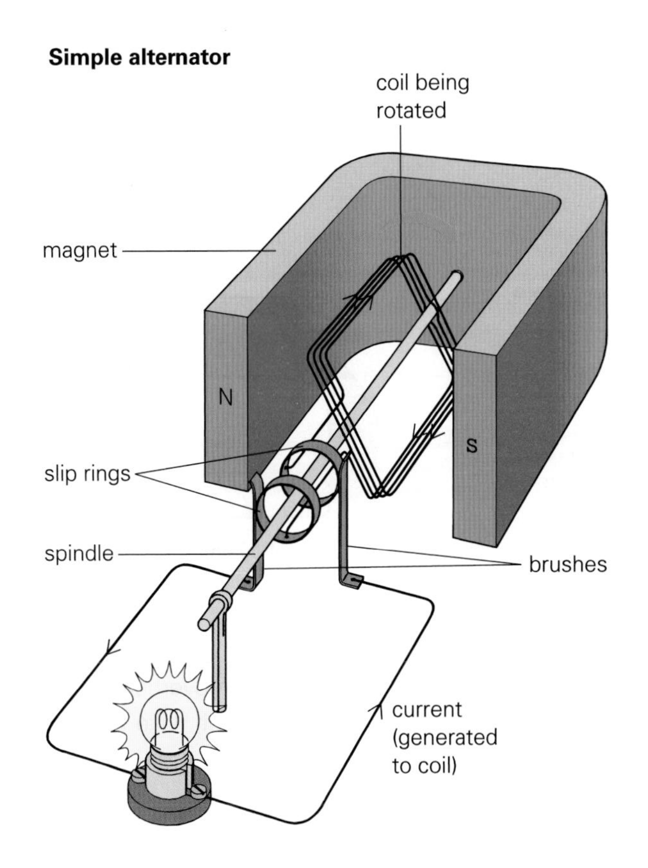

Most modern generators are ***alternators***. The current they produce alternates in direction: it flows backwards, forwards, backwards, forwards...and so on as the generator turns. Current like this is called ***alternating current (AC)***. For example, the electricity which comes from the mains is AC. It flows backwards and forwards 50 times every second.

The current from a battery always flows in the same direction. It is called ***direct current (DC)***.

1. In an electric motor, what job is done by **(a)** the brushes? **(b)** the commutator?
2. Give *three* ways in which you could make an electric motor produce a stronger turning effect.
3. What is the difference between alternating current and direct current? Which type comes from **(a)** a battery? **(b)** the mains? **(c)** an alternator?
4. In a simple alternator: **(a)** How is the current generated? **(b)** What are the slip rings for?
5. In the experiment on the right, what happens when the electromagnet is switched on, then off?

In the simple alternator on the left, the brushes rub against two metal ***slip rings***. As the coil faces first one way and then the other, the current generated in the coil flows backwards and then forwards. This makes alternating current flow through the bulb.

Mains electricity comes from huge alternators in power stations. The alternators have fixed coils round the outside with rotating electromagnets in the middle.

More on electromagnetic induction

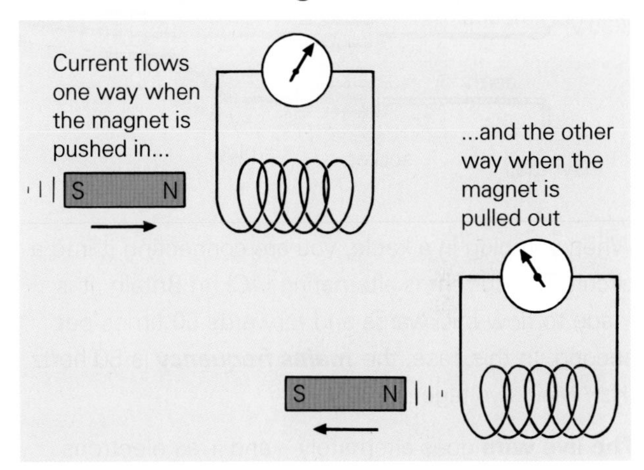

The experiment above shows the principle of electromagnetic induction. If the magnet is moved in or out of the coil, a small voltage is generated.

The voltage is increased if:
- a stronger magnet is used
- the magnet is moved faster
- the coil has more turns.

Moving the magnet in and out generates AC. Switching an electromagnet on and off, as can be done in the experiment below, has the same effect. A similar idea is used in the ***transformer*** (see Spread 4.09).

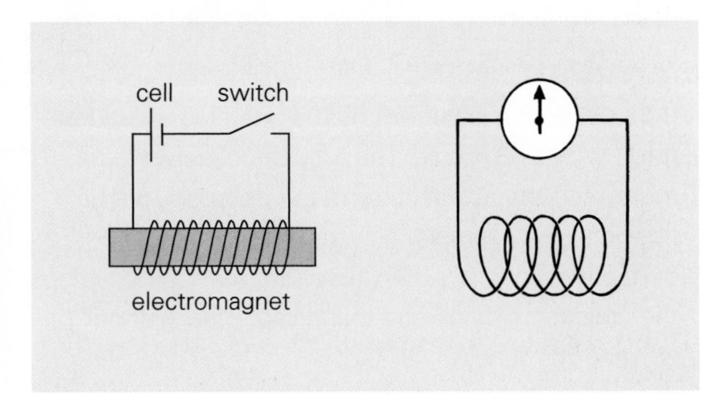

4.08 Mains electricity

By the end of this spread, you should be able to:
- *describe the features of the mains electricity supply, including those which make it safe.*

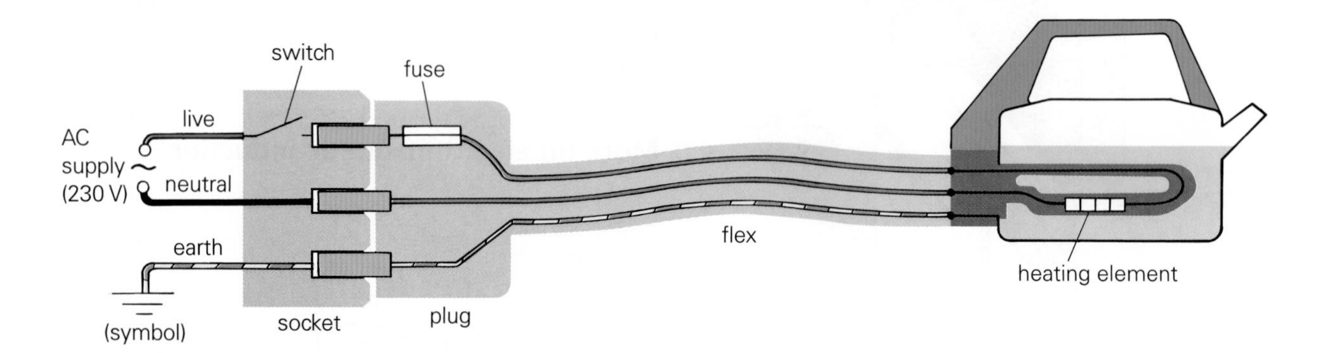

When you plug in a kettle, you are connecting it into a circuit. The current is alternating (AC). In Britain, it is made to flow backwards and forwards 50 times per second. In this case, the ***mains frequency*** is 50 hertz (Hz). Mains voltage is 230 V.

The live wire goes alternately – and + as electrons are pushed and pulled round the circuit.

The neutral wire is kept at zero voltage by the electricity company.

The switch is fitted in the live wire. This is to make sure that none of the wire in the flex is live when the switch is off.

The fuse is a piece of thin wire which overheats and melts if too much current flows through. Like the switch, it is placed in the live wire.

The earth wire is a safety wire. It connects the metal body of the kettle to a conductor which is kept at zero voltage. If, say, the live wire comes loose and touches the metal body, a current flows to earth, blows the fuse, and breaks the circuit. This means that the kettle is then safe to touch.

Appliances with an insulating plastic case do not usually have an earth wire. As their flex is also insulated, they are ***double insulated***.

Three-pin plug

In the UK, appliances such as kettles are connected to the mains using a three-pin plug. For safety, this must be wired up correctly. It is also important that a fuse of the correct value is fitted.

Plugs are normally fitted with 3 A or 13 A fuses. The value tells you the current needed to 'blow' the fuse.

If a kettle takes a current of 10 A, its plug should be fitted with a 13 A fuse. The fuse should always be more than the actual current, but as close to it as possible. For example, a TV taking 0.5 A will work perfectly well with a 13 A fuse. But if a fault develops, the circuits could overheat and catch fire without blowing the fuse.

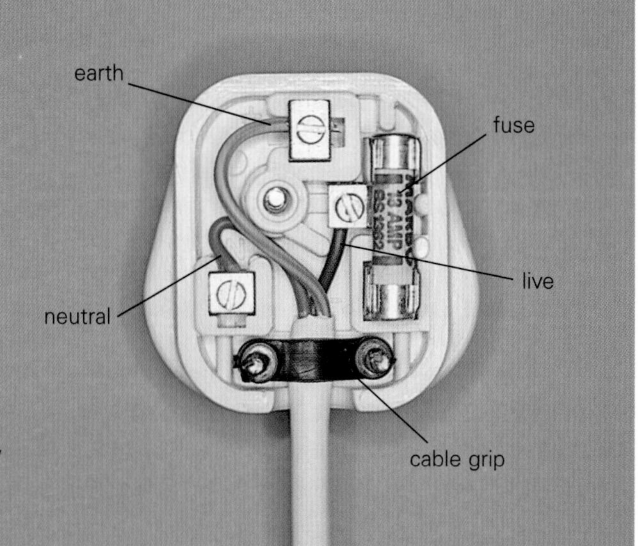

RCCB

This 'power breaker' plug is a ***residual current circuit breaker (RCCB)***. It compares the currents in the live and neutral wires. They should be the same. If they are different, then current must be flowing to earth – perhaps through someone touching a faulty wire. The RCCB senses the difference and switches off the power before any harm can be done.

Two-way switches

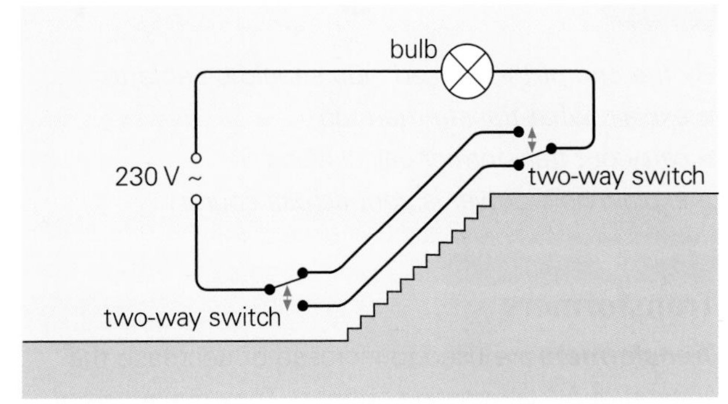

In most houses, you can turn the landing light on or off using upstairs or downstairs switches. These have two contacts instead of one. They are ***two-way switches***. In the diagram, if both switches are up or down, then a current flows through the bulb. But if one is up and the other is down, the circuit is broken. Each switch reverses the effect of the other.

Circuits round the house

In a house, the electricity company's cable branches into several parallel circuits. These carry power to the lights, cooker, immersion heater, and mains sockets. Each circuit passes through a fuse or circuit breaker (see Spread 4.06) in the ***consumer unit*** ('fuse box').

More about AC and DC

An oscilloscope (CRO) can be used to demonstrate the difference between AC and DC (the one-way direct current which flows from a battery). The oscilloscope plots a graph on its screen very rapidly, over and over again. The line shows how the voltage across a component (such as a bulb) varies with time.

1. In mains circuits: **(a)** In which wire is the switch fitted, and why? **(b)** Why do metal appliances need an earth wire? **(c)** What colour is the live wire's insulation in a three-pin plug?
2. An electric heater has a power of 460 W. **(a)** What current does it take from the 230 V mains? (power = voltage x current) **(b)** What fuse should its plug have, 3 A or 13 A? **(c)** Why should you not fit a 13 A fuse to a 60 W lamp?
3. Redraw screen A below to show what the line might be like if the AC voltage were reduced.
4. Redraw screen C below to show what the line might be like if the DC voltage were reduced.

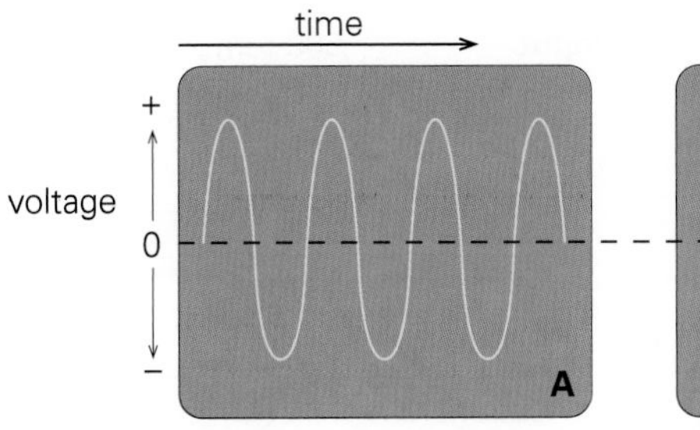

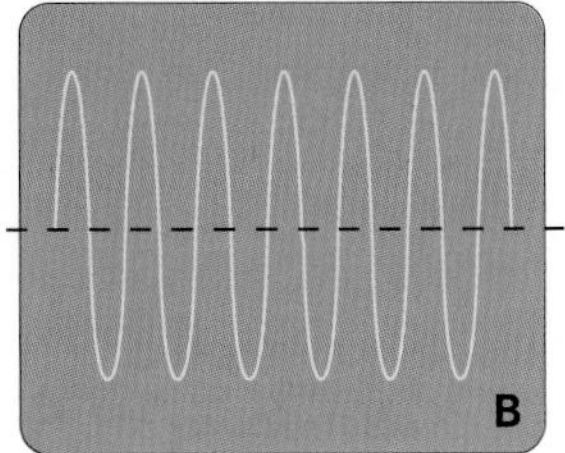

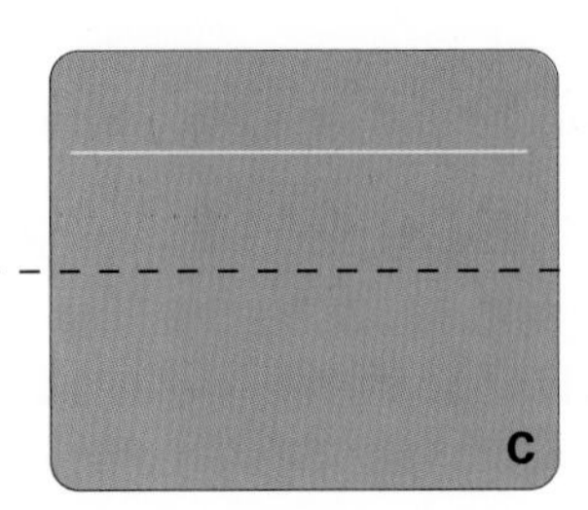

Experiments on the measurement of mains AC must only be carried out by a suitably qualified person.

AC The voltage keeps changing direction. Each peak represents the maximum voltage in the forward direction.

Higher-frequency AC The voltage changes direction more times per second. So the peaks are closer together.

DC The voltage is constant, and always in the same direction. So the graph line is level.

4.09 Passing on the power

By the end of this spread, you should be able to:
- *explain what transformers do*
- *carry out transformer calculations*
- *explain how power is sent across country.*

Transformers

Transformers are used to increase or decrease the voltage of AC (alternating current). They make use of electromagnetic induction (see Spread 4.07).

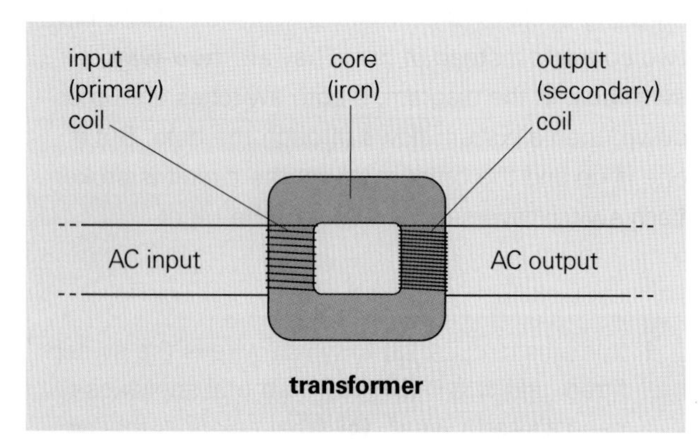

In a transformer, the ***primary coil*** is as an electromagnet. Connected to an AC supply, it produces an alternating magnetic field. This generates AC in the ***secondary coil***. For the right output voltage, the coils have to be carefully chosen. For example, if the number of secondary turns is twice the number of primary turns, then the output voltage is twice the input voltage, as shown by this equation:

$$\frac{V_2}{V_1} = \frac{n_2}{n_1}$$

V_2 = output voltage
V_1 = input voltage
n_2 = output turns
n_1 = input turns

Applying this to the transformer on the right, we get

$$\frac{24}{12} = \frac{1000}{500}$$

This transformer increases voltage. It is a ***step-up*** transformer. Some transformers are ***step-down***.

In computers, TVs, and stereos, small transformers reduce mains voltage to the 9 V or so needed by the electronic circuits. (Extra components convert the transformer's AC output into the DC required by the circuits.)

Step-up transformer	Step-down transformer
Symbol:	Symbol:
More turns on output coil than turns on input coil	Fewer turns on output coil than turns on input coil
Output voltage more than input voltage	Output voltage less than input voltage

Transformer power

Power is the energy supplied per second. It is measured in ***watts (W)*** (see Spreads 4.05 and 4.22). If a transformer wastes no energy, its power output and input are the same. Power = voltage × current. So:

$$V_2 \times I_2 = V_1 \times I_1$$

(power output) = (power input)

V_2 = output voltage
V_1 = input voltage
I_2 = output current
I_1 = input current

You can see an example of this in the diagram below. Note that as the voltage goes up, the current must go down so that the power stays the same.

Real transformers waste some power. Their copper coils have resistance. Also, the changing magnetic field generates currents in the iron core.

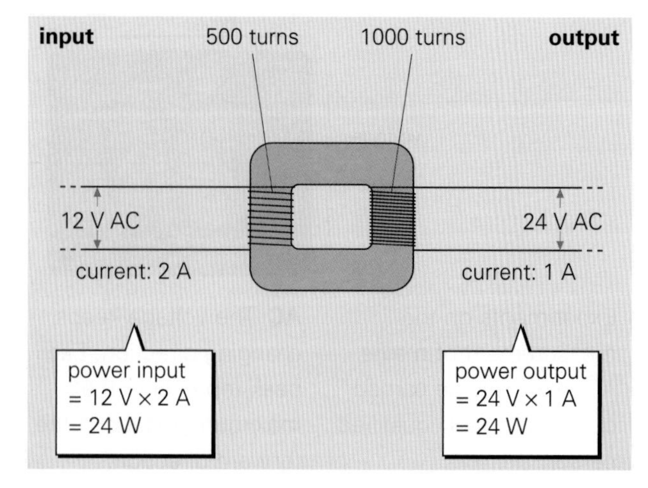

Power across country

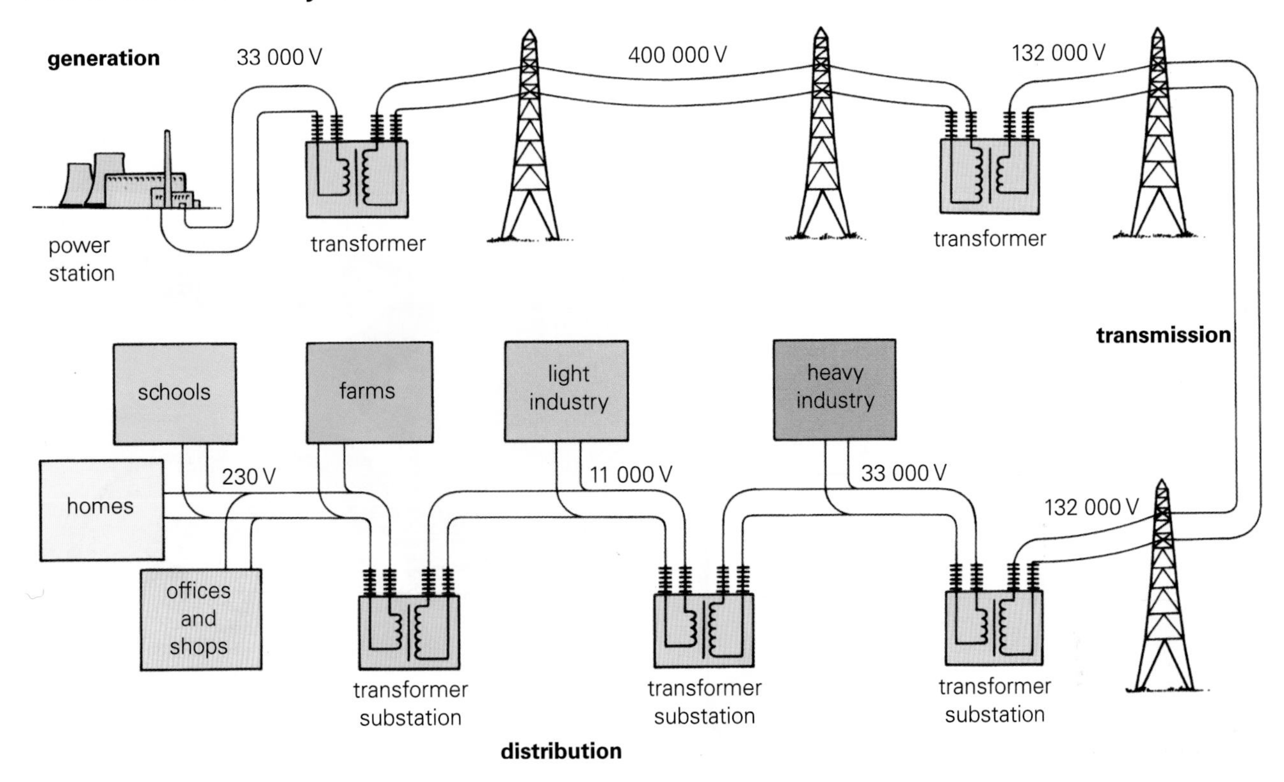

Power stations pass their power to a national distribution network called the ***Grid***. If one region needs more electricity, it can come from power stations in other regions. Power is normally sent across country by overhead cables. Transformers step up the voltage before the power is passed to the cables. At the far end, more transformers step down the voltage before the power reaches consumers.

On the left, you can see why the voltage is stepped up before transmission. Increasing the voltage reduces the current, so that thinner, lighter, and cheaper cables can be used to carry the same amount of power.

In areas of outstanding natural beauty like the one on the right, the transmission cables are put underground. But this is very expensive.

Overhead cables are not allowed here

1 Give *two* uses of transformers.

2 Why will transformers work on AC but not DC?

3 A transformer has 2000 turns on its input coil and 200 turns on its output coil. It is being used to light a bulb and is taking a current of 0.2 A from the 230 V mains. No power is being wasted:
(a) Is the transformer step-up or step-down?
Calculate the **(b)** output voltage **(c)** input power **(d)** output power **(e)** output current.

4 Why is mains power transmitted across country at high voltages?

5 100 kW of power is to be sent along a transmission line. What is the current in the cable if the power is sent at **(a)** 1 kV? **(b)** 100 kV?

4.10 Moving and stopping

By the end of this spread, you should be able to:
- *calculate speed*
- *describe how friction is sometimes a nuisance and sometimes useful*
- *explain how speed affects road safety.*

Speed

Here is a simple method of measuring speed. You could use it to work out the speed of a cyclist:

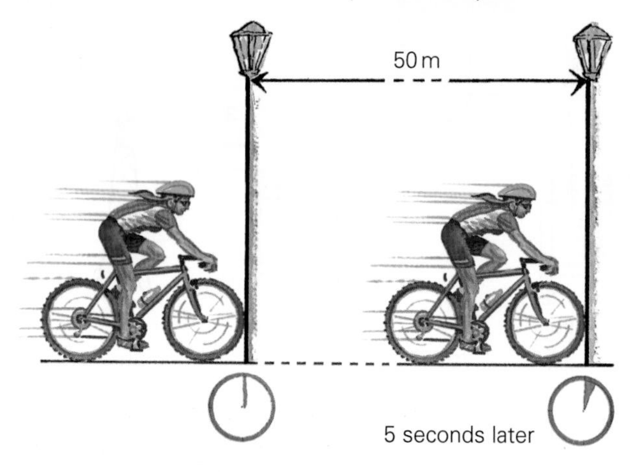

Measure the distance between two points on a road, say two lamp posts. Measure the time taken to travel between these points. Then use this equation:

$$\text{speed} = \frac{\text{distance travelled}}{\text{time taken}}$$

distance in metres (m)
time in seconds (s)
speed in m/s

If the cyclist travels 50 metres in 5 seconds, her speed is 50/5, which is 10 metres per second. This is written 10 m/s for short.

This calculation really gives her average speed, as her actual speed may vary during the 5 seconds. To find an actual speed, you need to know the distance travelled in the shortest time you can measure.

Friction

Friction is the force that tries to stop materials sliding past each other. There is friction between your hands when you rub them together. And there is friction between your shoes and the ground when you walk. ***Air resistance*** is also a type of friction. It slows you down when you ride a bike.

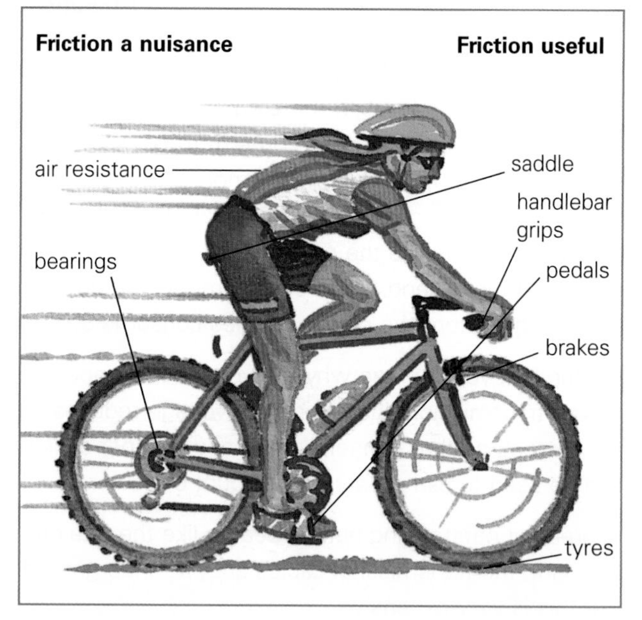

Using friction Friction can be useful. Without friction between the tyres and the ground, you would not be able to ride a bike. It would be like trying to ride on ice. You could not speed up, turn, or stop.

Brakes rely on friction. Cycles are slowed by rubber blocks pressed against the wheel rims. Cars are slowed by fibre pads pressed against discs attached to the wheels.

Problems with friction Friction can also be a nuisance. Moving things are slowed by friction. Friction also produces heat. In machinery, grease and oil reduce friction so that moving parts do not overheat and seize up. Ball bearings and roller bearings also reduce friction. Their rolling action means that a wheel does not rub against its shaft.

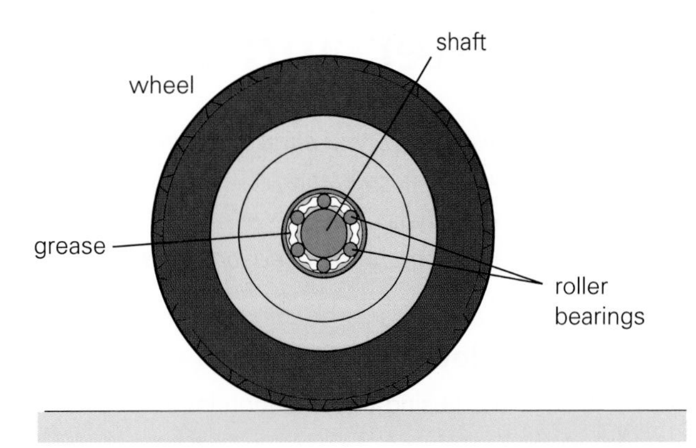

Reducing friction in a wheel

Speed and safety

In an emergency, the driver of a car may have to react quickly and apply the brakes.

The car's stopping distance depends on two things:

- The ***thinking distance***. This is how far the car travels before the brakes are applied, while the driver is still reacting.
- The ***braking distance***. This is how far the car then travels, after the brakes have been applied.

It takes an average driver about 0.6 seconds to react, and press the brake pedal. This is the driver's ***reaction time***. During this time, the car does not slow down. And the higher its speed, the further it travels.

This is how to work out the thinking distance for a car travelling at 20 m/s (45 mph). The driver's reaction time is 0.6 seconds:

$$\text{speed} = \frac{\text{distance}}{\text{time}}$$

So, $\text{distance} = \text{speed} \times \text{time}$

$= 20 \times 0.6 = 12$ metres

So, the thinking distance is 12 metres.

1 Siân cycles 100 metres in 10 seconds. What is her average speed?

2 Look at the photograph on the opposite page. What features can you see for reducing friction?

3 For a car, where is friction **(a)** useful? **(b)** a nuisance? Give *two* examples of each.

4 In the chart below, what is the thinking distance **(a)** at 25 m/s? **(b)** at 30 m/s? Why does the thinking distance go up, even though the driver's reaction time stays the same?

5 Alcohol slows people's reactions. If a driver has a reaction time of 2 seconds, **(a)** what will his thinking distance be at 56 mph (25 m/s)? **(b)** what will his stopping distance be? (You will need information from the chart to answer this.)

The chart below shows the stopping distances for cars at different speeds. The figures are for a dry road. If the road is wet or icy, or the driver's reactions are slow, the stopping distances will be even greater.

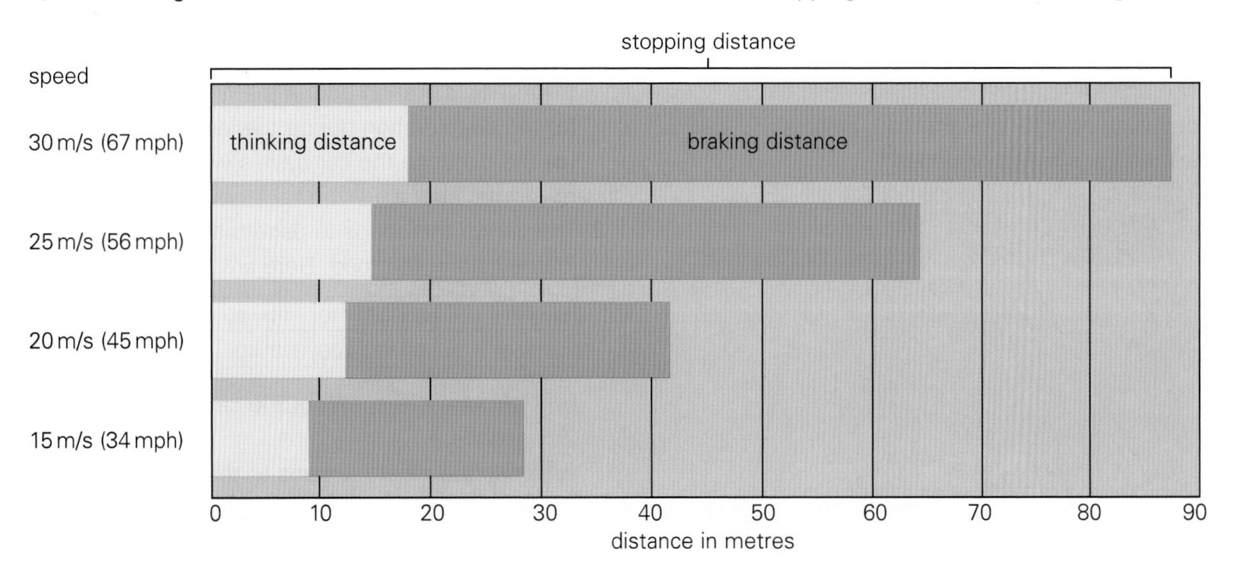

4.11 Faster and slower

By the end of this spread, you should be able to:
- *calculate speed and acceleration*
- *explain what velocity is*
- *interpret distance–time and speed–time graphs.*

Speed and velocity

Speed is calculated using this equation (see 4.10):

$$\text{speed} = \frac{\text{distance travelled}}{\text{time taken}}$$

distance in metres (m)
time in seconds (s)
speed in m/s

Velocity means speed in a particular direction. One way of showing the direction is to use a + or –.

For example:

+10 m/s (velocity of 10 m/s to the right)

–10 m/s (velocity of 10 m/s to the left)

Acceleration

A car is going faster and faster. If its velocity goes up by 4 m/s every second, it has an ***acceleration*** of 4 m/s^2. Acceleration can be calculated like this:

$$\text{acceleration} = \frac{\text{change in velocity}}{\text{time taken}}$$

velocity in m/s
time in s
acceleration in m/s^2

When a car slows down, it loses velocity. If it loses 4 m/s every second, it has a ***deceleration*** (also called a ***retardation***) of 4 m/s^2. Mathematically, this is an acceleration of –4 m/s^2.

Motion graphs

On the right, a car is travelling along a straight road, away from a post. The car's distance from the post is measured every second. The graphs show four different examples of what the car's motion might be.

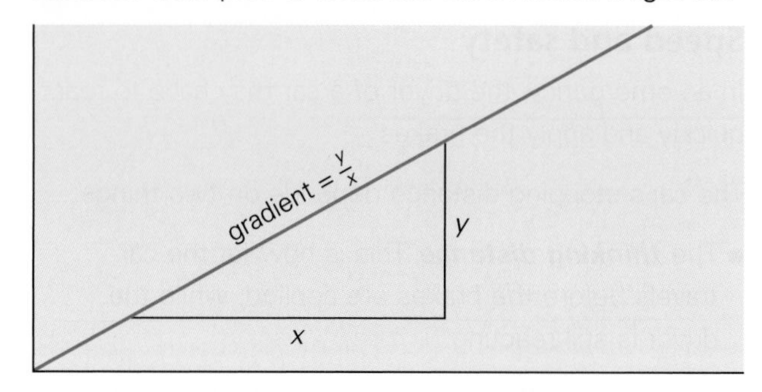

You calculate the ***gradient*** of a graph line like this:

On a straight-line graph, the gradient has the same value wherever you measure *y* and *x*.

On a distance–time graph the gradient tells you how much extra distance is travelled every second. So:

The gradient gives the speed.

On a speed–time graph the gradient tells you how much extra speed is gained every second. So:

The gradient gives the acceleration.

Also:

The area under the line gives the distance travelled.

- The area must be calculated using the scale numbers on the axes. It isn't the 'real' area.
- Area of a triangle = $\frac{1}{2}$ × base × height.

1 The speed–time graph below is for a car travelling along a straight road. Find:
(a) the car's maximum speed

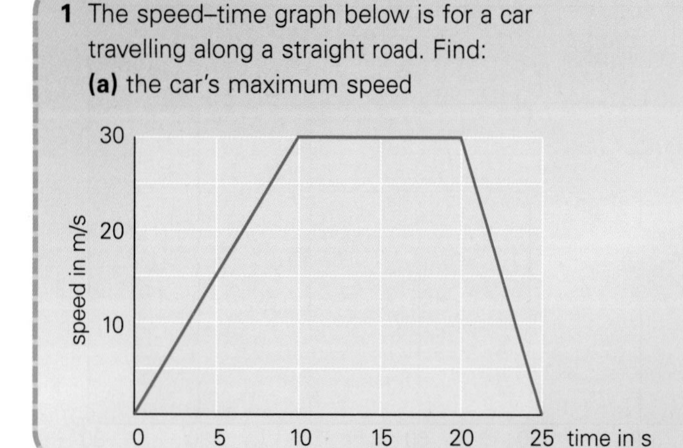

(b) the acceleration during the first 20 s
(c) the deceleration during the last 10 s
(d) the distance the car travels while it is at its maximum speed.

2 The car on the opposite page makes another journey along the road. Here are the readings:

time in s	0	5	10	15	20	25	30	35	40
distance in m	0	100	200	300	400	450	500	500	500

(a) Calculate the speed of the car over the first 20 seconds.
(b) Describe what happens to the speed of the car during the whole 40 seconds.

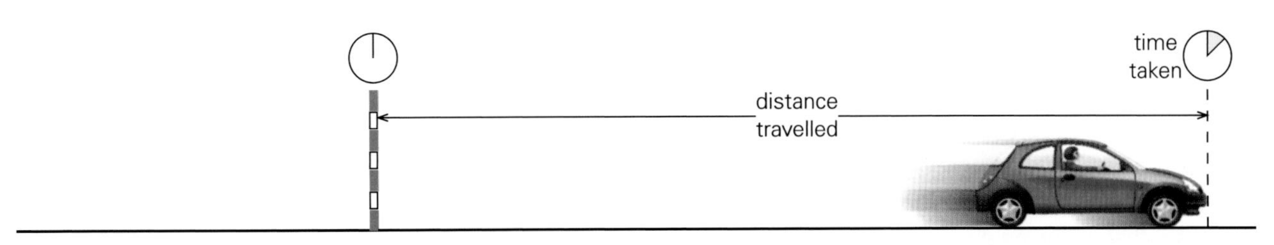

Distance–time graphs

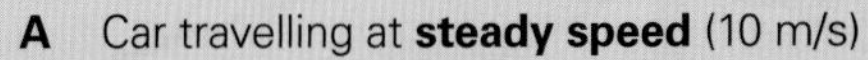

A Car travelling at **steady speed** (10 m/s)

time in s	0	1	2	3	4	5
distance in m	0	10	20	30	40	50

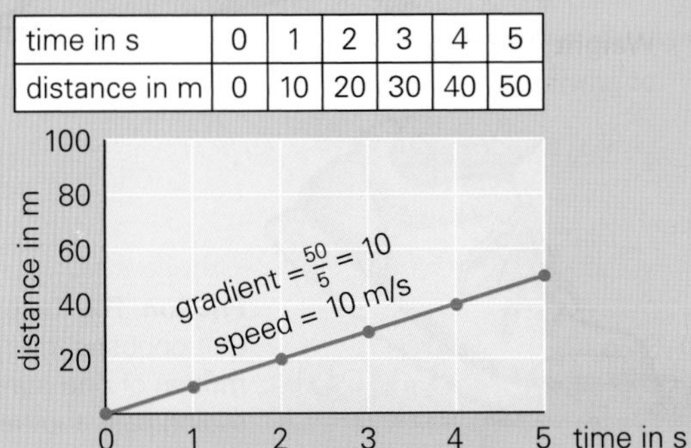

The line rises 10 m on the distance scale for every 1 s on the time scale.

B Car travelling at **higher steady speed** (20 m/s)

time in s	0	1	2	3	4	5
distance in m	0	20	40	60	80	100

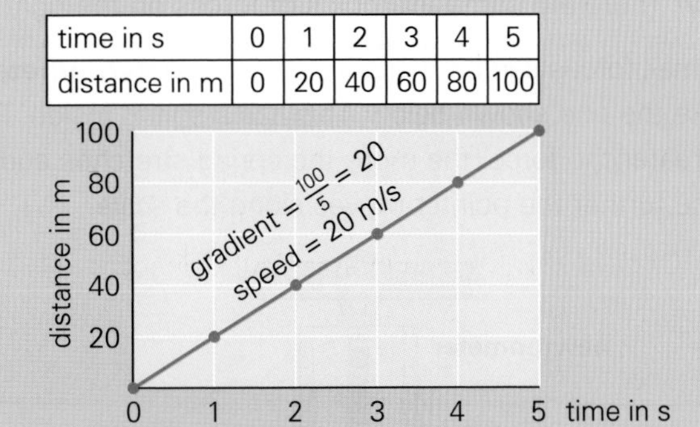

The line is steeper than before. It rises 20 m on the distance scale for every 1 s on the time scale.

C Car **accelerating**

time in s	0	1	2	3	4	5
distance in m	0	10	25	45	70	100

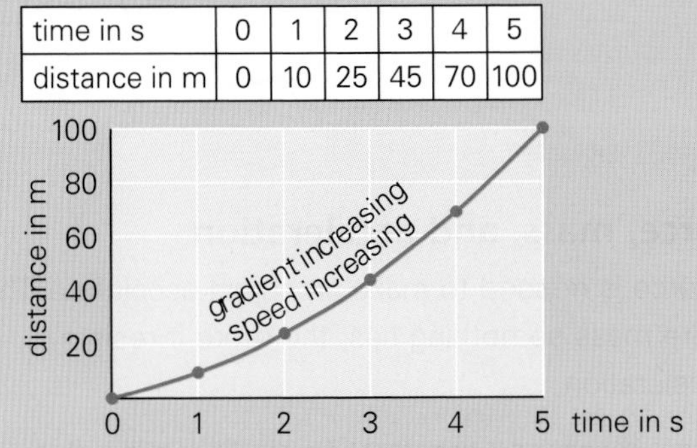

The speed rises. So the car travels further each second than the one before, and the line curves upwards.

D Car **stopped**

time in s	0	1	2	3	4	5
distance in m	50	50	50	50	50	50

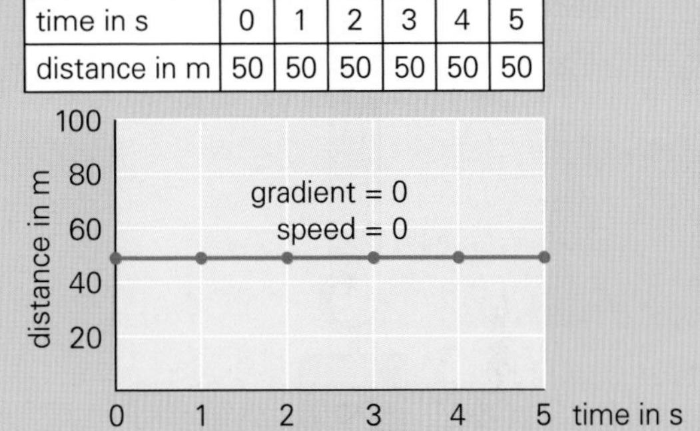

The car is parked 50 m from the post, so this distance stays the same.

Speed–time graphs

E Car travelling at **steady speed** (15 m/s)

time in s	0	1	2	3	4	5
speed in m/s	15	15	15	15	15	15

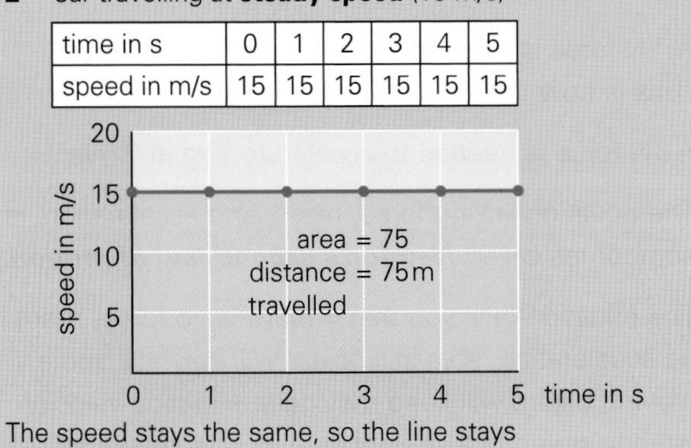

The speed stays the same, so the line stays at same level.

F Car travelling at **steady acceleration** (4 m/s^2)

time in s	0	1	2	3	4	5
speed in m/s	0	4	8	12	16	20

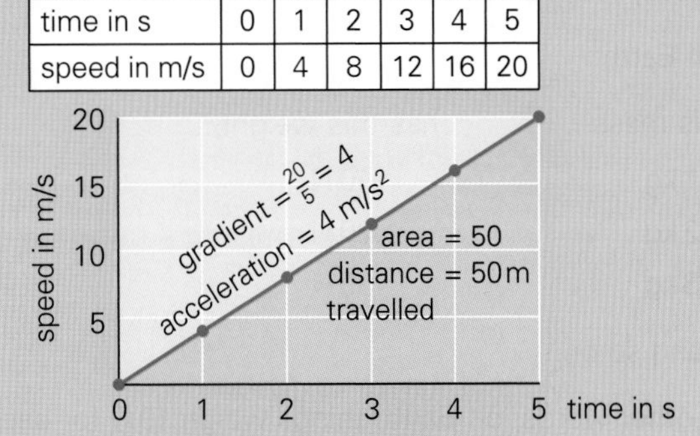

As the car gains speed, the line rises 4 m/s on speed scale for every 1 s on time scale.

4.12 Forces and motion

By the end of this spread, you should be able to:
- *explain what is meant by weight, and g*
- *describe how acceleration and force are linked*
- *explain that all forces exist in pairs.*

A force is a push or pull. It is measured in ***newtons (N)***. There are some examples of forces on the right.

Small forces can be measured using a ***newtonmeter*** like the one shown below. This has a spring inside. The greater the force, the more the spring stretches and the further the pointer moves along the scale.

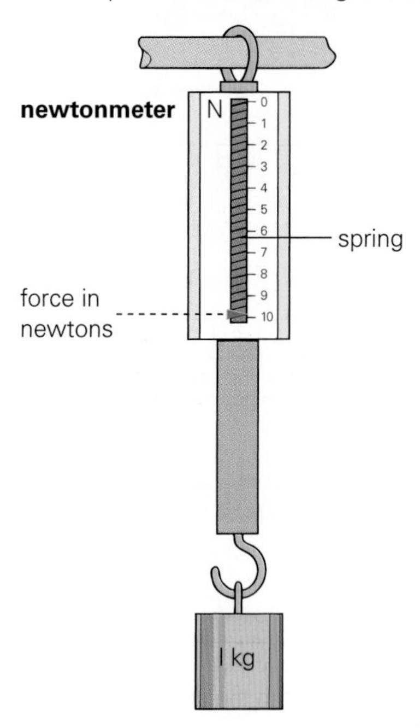

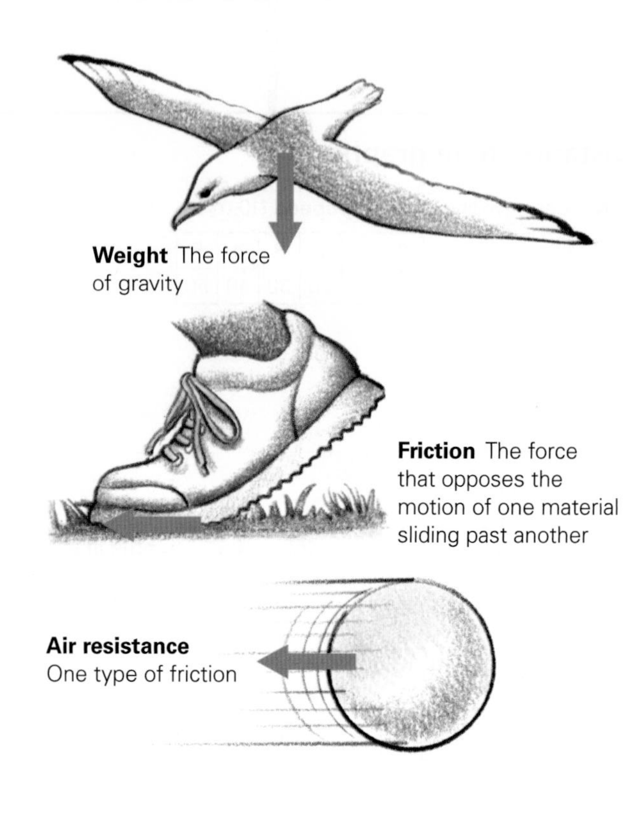

Weight

On Earth, every object has the downward force of gravity on it. This force is called its ***weight***. Like other forces, it is measured in newtons (N).

On Earth:

this mass...	has this weight...
1 kg	10 N
2 kg	20 N
3 kg	30 N

...and so on.

In other words, on Earth, things weigh 10 N for each kilogram of mass. Scientifically speaking, the Earth's ***gravitational field strength*** is 10 N/kg.

Force, mass, and acceleration

A force is needed to make an object accelerate. The more mass something has, the more it resists acceleration.

Force, mass, and acceleration are linked like this:

$$\underset{\text{in N}}{\text{force}} = \underset{\text{in kg}}{\text{mass}} \times \underset{\text{in m/s}^2}{\text{acceleration}}$$

For example:

A 1 N force is needed to accelerate 1 kg at 1 m/s^2. (This is how the newton is defined.)

A 6 N force is needed to accelerate 2 kg at 3 m/s^2.

The equation linking force, mass, and acceleration, is sometimes called ***Newton's second law of motion***.

The equation tells you that if there is no force, there is no acceleration, so a still object will stay still and a moving object will keep moving at a steady velocity (steady speed in a straight line). This is known as ***Newton's first law of motion***.

Acceleration of free fall *(g)*

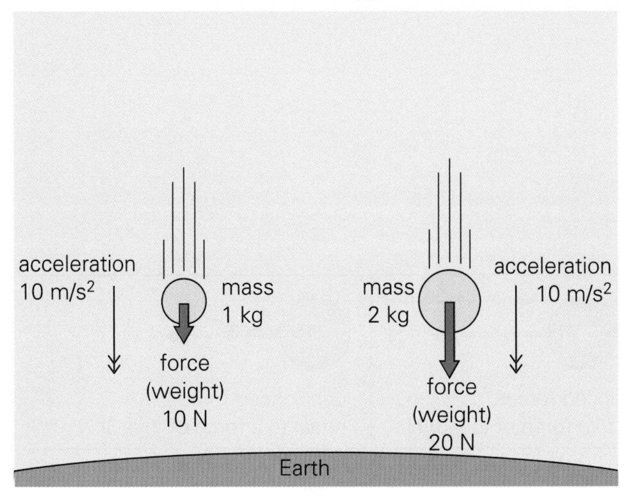

With no air resistance, things falling near the Earth have an acceleration of 10 m/s². This is the ***acceleration of free fall*, *g***. It is the same for all masses, light or heavy.

Air resistance tends to affect light things more than heavy ones. But without it, a falling feather would speed up just as quickly as a falling rock.

The equation force = mass x acceleration shows why the acceleration of free fall is the same for all masses:

This force...	acting on this mass...	produces this acceleration...
10 N	1 kg	10 m/s²
20 N	2 kg	10 m/s²
30 N	3 kg	10 m/s²

force = mass × acceleration

g has *two* meanings. You can think of it as:

- an acceleration (10 m/s²), or...
- a force per kg (10 N/kg).

Action and reaction

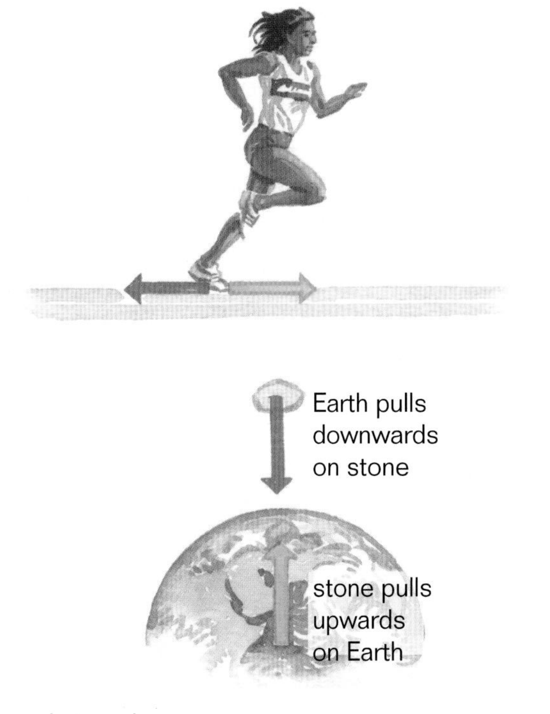

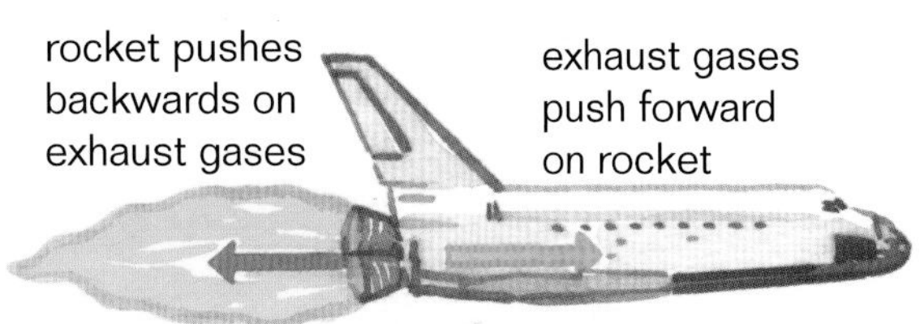

Forces are pushes or pulls between two things. So forces always exist in pairs. One force acts on one thing. Its equal but opposite partner acts on the other. This idea is known as ***Newton's third law of motion***. You can see three examples above.

The forces in each pair are known as the ***action*** and the ***reaction***. But it does not matter which you call which. One cannot exist without the other.

1 A 5 kg stone is dropped near the ground. What is **(a)** its weight in N? **(b)** its acceleration in m/s²?

2 In question 1, why would a 10 kg stone have the same acceleration as the 5 kg stone?

3 What force is needed to give a mass of 3 kg an acceleration of 5 m/s²?

4 If a force of 2 N acts on a mass of 4 kg, what acceleration is produced?

5 The middle diagram above shows that a stone exerts an upward force on the Earth. The stone accelerates downwards. Why can't you detect the Earth accelerating upwards?

6 The Moon's gravity is weaker than the Earth's. On the Moon, a 1 kg rock only weighs 1.6 N.
(a) How much would a 5 kg rock weigh?
(b) What is its acceleration of free fall?

4.13 Balanced and unbalanced forces

By the end of this spread, you should be able to:
- *describe how objects will behave when the forces on them are balanced, and unbalanced.*

When forces act together, they can have the same effect as a single force, called the ***resultant***. For example:

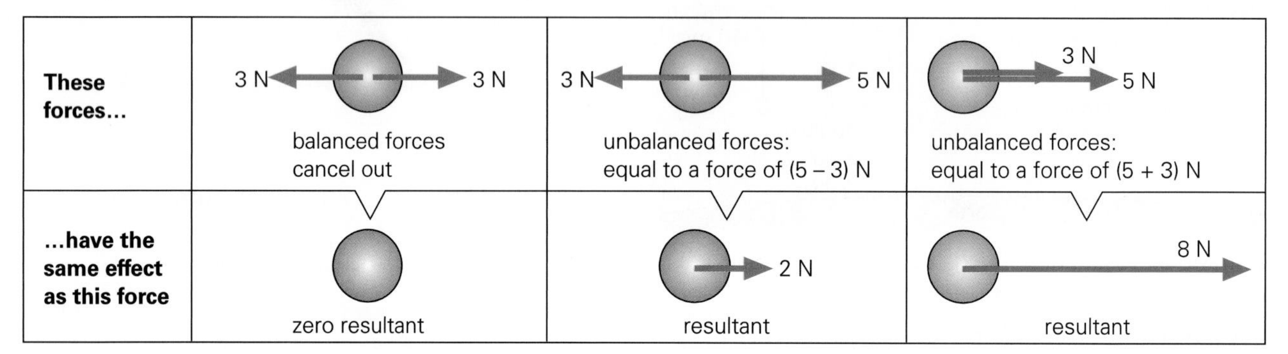

Unbalanced forces

A resultant force always produces an acceleration. You can work out the acceleration using the equation on the right (see Spread 4.12). Here are two examples:

force in N = mass in kg × acceleration in m/s^2

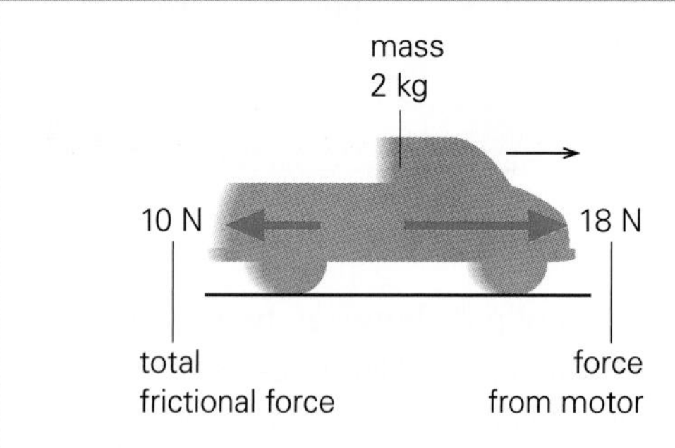

This model car has two opposing forces on it. Together, they are equivalent to a single force of (18 – 10) N to the right. So the resultant force is 8 N to the right.

force = mass × acceleration

So: 8 N = 2 kg × acceleration

Rearranged, this gives:

acceleration = 4 m/s^2 (to the right)

Whenever you use this equation:
force = mass × acceleration
Remember that 'force' means the resultant force.

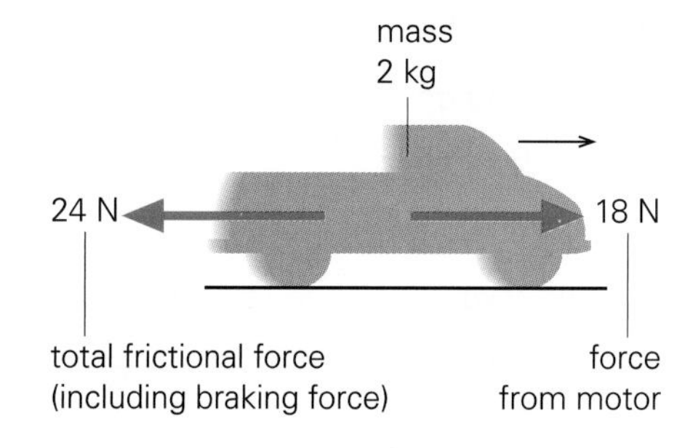

The model car has brakes. When applied, they increase the total frictional force to 24 N. Now, the resultant is (24 – 18) N to the left. So it is 6 N to the left.

force = mass × acceleration

So: 6 N = 2 kg × acceleration

Rearranged, this gives:

acceleration = 3 m/s^2 (to the left)

If the car is travelling to the right, but accelerating to the left, it is losing speed (see Spread 4.11). So it has a deceleration (retardation) of 3 m/s^2.

Balanced forces

In each of the following examples, the forces are balanced, so there is no resultant force:

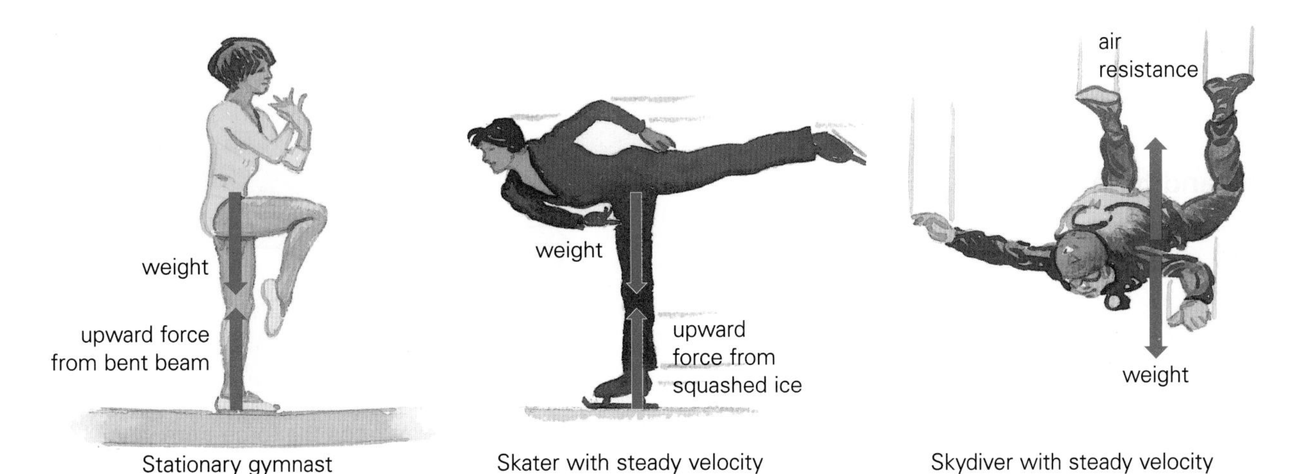

Stationary gymnast

Skater with steady velocity

Skydiver with steady velocity

With no resultant force on it, an object is either at rest or moving at a steady velocity. If the resultant force is zero, the equation force = mass × acceleration tells you that the acceleration is also zero, so the velocity doesn't change.

Terminal velocity

If a skydiver jumps from a hovering helicopter, the air resistance on her increases as her speed rises. Eventually, the air resistance is enough to balance her weight, and she gains no more speed. She is at her ***terminal velocity***. Typically, this is about 60 m/s, though the actual value depends on air conditions, as well as the size, shape, and weight of the skydiver.

When the skydiver opens her parachute, the extra area of material increases the air resistance. She loses speed rapidly until the forces are again in balance.

Skydivers falling at their terminal velocity

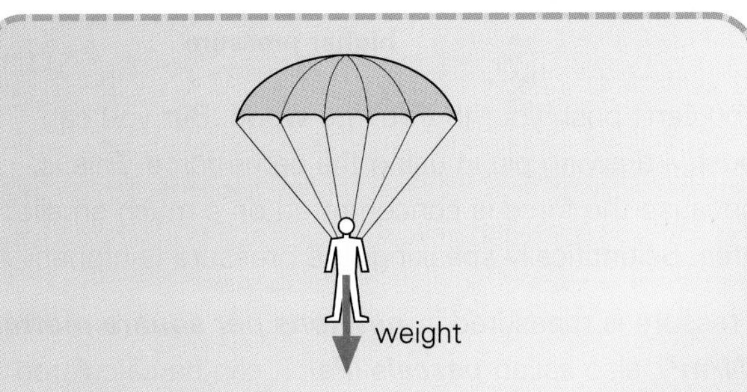

1 The parachutist above is descending at a steady velocity.
(a) What name is given to this velocity?
(b) Copy the diagram. Mark in and label another force acting.
(c) How does this force compare with the weight?
(d) If the parachutist used a larger parachute, how would this affect the steady velocity reached? Explain why.

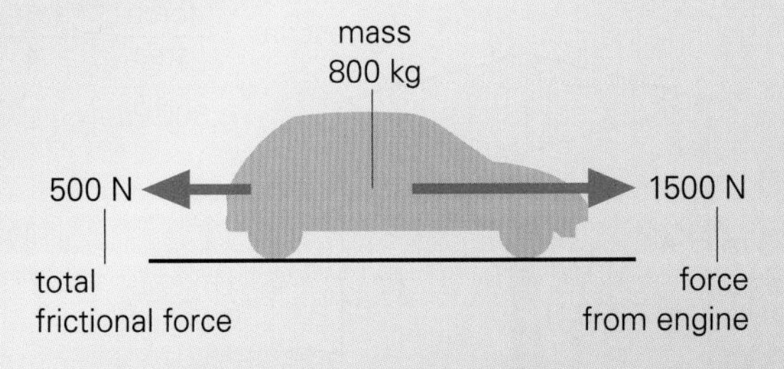

2 **(a)** What is the resultant force on the car above?
(b) What is the car's acceleration?
(c) If the total frictional force rises to 1500 N, what happens to the car?

4.14 Pressure

By the end of this spread, you should be able to:
- *explain what pressure is and how to calculate it*
- *describe some of the effects of pressure*
- *describe how hydraulic machines work.*

Low and high pressure

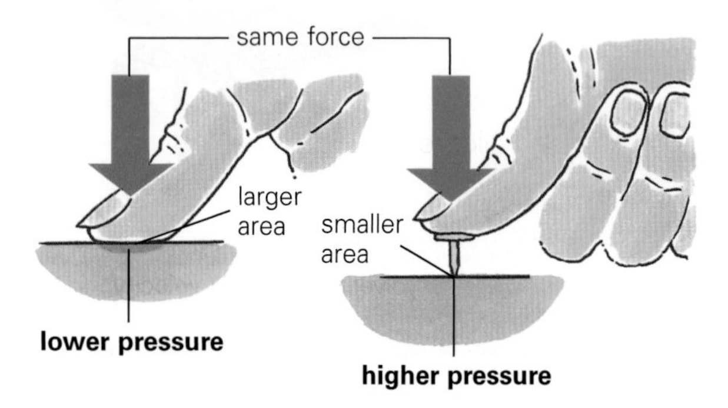

You can't push your thumb into wood. But you can push a drawing pin in using the same force. This is because the force is concentrated on a much smaller area. Scientifically speaking, the pressure is higher.

Pressure is measured in ***newtons per square metre (N/m^2)***, also called ***pascals (Pa)***. It can be calculated with this equation:

$$\text{pressure} = \frac{\text{force}}{\text{area}}$$

force in newtons (N)
area in square metres (m^2)
pressure in pascals (Pa)

For example, the block below weighs 1200 N and rests on a face measuring 2 m × 2 m:

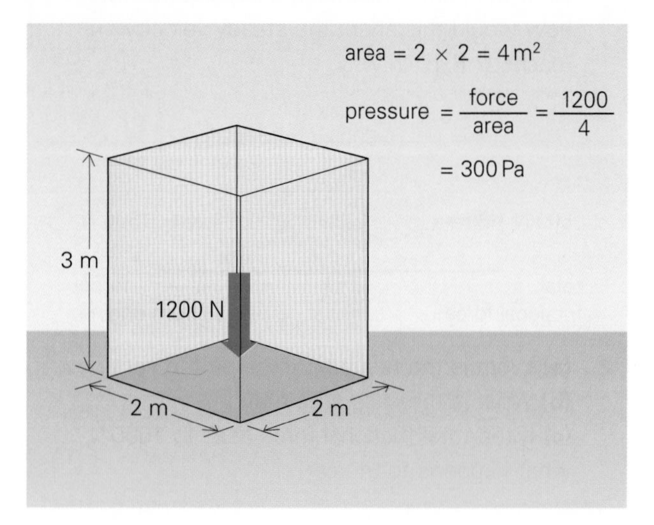

Concentrating a force on a *small area* gives...

high pressure

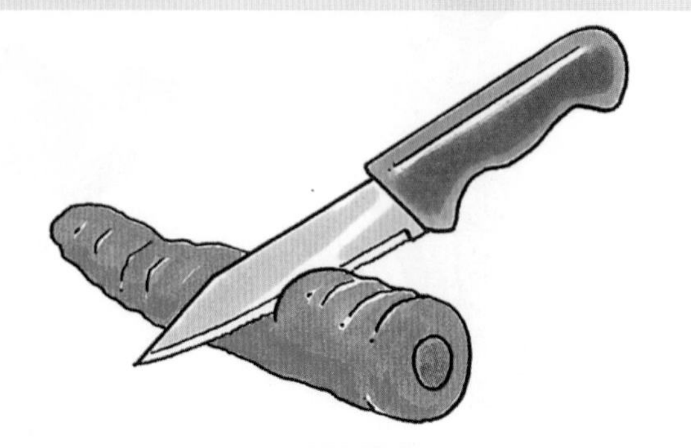

A sharp blade concentrates the force from your hand so that cutting is easy.

Spreading a force over a *large area* gives...

low pressure

This ski spreads the skier's weight so that the foot does not sink into soft snow.

When the nut is tightened, the washer spreads the force, so that the nut does not sink into the wood.

Liquid pressure

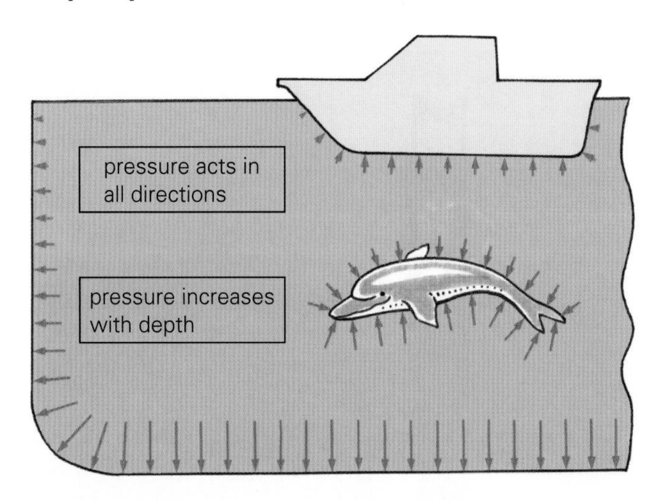

The deeper you go into a liquid, the greater the pressure becomes. This pressure pushes in all directions. It is the pressure from water which keeps a boat afloat. Water pressing on the hull produces an upward push called an ***upthrust***. This is strong enough to support the weight of the boat.

Hydraulic machines

These are machines in which liquids are used to transmit forces. Machines like this rely on two features of liquids:

- Liquids cannot be squashed. They are virtually incompressible.
- If a trapped liquid is put under pressure, the pressure is transmitted throughout the liquid.

The diagram below shows a simple hydraulic jack. When you press on the narrow piston, the pressure is transmitted by the oil to the wide piston. It produces an output force which is larger than the input force.

Follow the sequence of circled numbers 1–4 on the diagram. They show you how to use the link between pressure, force, and area to calculate the output force.

Car brakes work hydraulically. When the brake pedal is pressed, a piston puts pressure on trapped brake fluid. The pressure is transmitted, by pipes, to the wheels. There, the pressure pushes on pistons which move the brake pads.

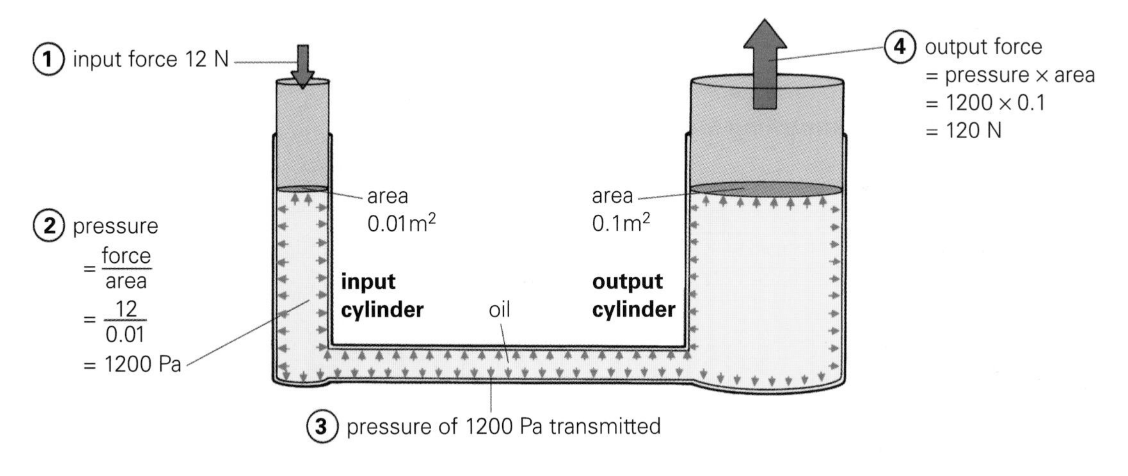

1 Use your ideas about pressure to explain why:
(a) it is easier to walk on soft sand if you have flat shoes rather than shoes with small heels
(b) it is easier to cut through something with a knife if the knife has a very sharp blade.

2 A rectangular block measures 4 m × 3 m × 2 m. It weighs 600 N and rests with one face on level ground. Draw a diagram to show the position of the block when the pressure under it is
(a) as high as possible **(b)** as low as possible.
Calculate the pressure in each case.

3 In the simple hydraulic system below:
(a) What is the pressure of the oil?
(b) What is the output force?
(c) If the diameter of the output cylinder were greater, how would this affect the output force?

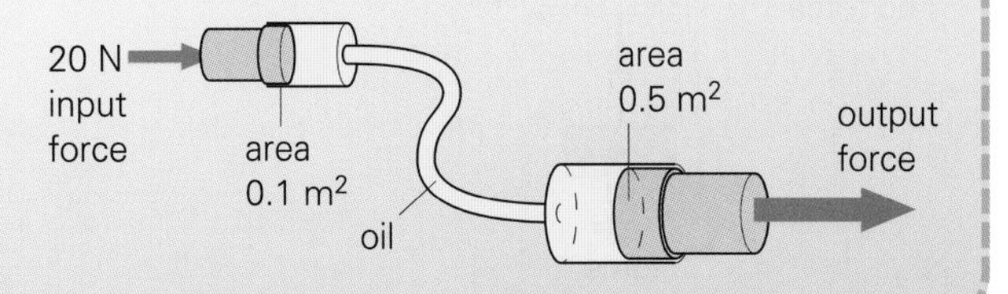

4.15 Stretching and compressing

By the end of this spread, you should be able to:

- *explain what Hooke's law is*
- *describe how the pressure of a gas (at constant temperature) depends on its volume.*

Stretching a wire

In some bridges, the roadway is suspended by steel cables. The cables are stretched slightly by the weight of the roadway, so the designer must make sure that they are strong enough to take the load.

The diagram, table, and graph below show the effect of a stretching force on a long, thin, steel wire. As the force increases, so does the ***extension*** (the length by which the cable stretches).

Up to point X on the graph:

- Each extra 100 N (newtons) of force produces the same extra extension (1 mm in this case).
- If the force doubles, the extension doubles, and so on.

Mathematically, these mean that ***the extension is directly proportional to the stretching force***.

This is called ***Hooke's law***.

Steel and other metals obey Hooke's law. So do coil springs made out of steel. But rubber and many plastics do not. With these materials, the graph would be a curve, not a straight line.

Point E on the graph is called the ***elastic limit***. Up to this point, the cable will return to its original length if the force is removed. Scientists say that the material is ***elastic***. However beyond E, the cable becomes permanently stretched – and at Y, it breaks.

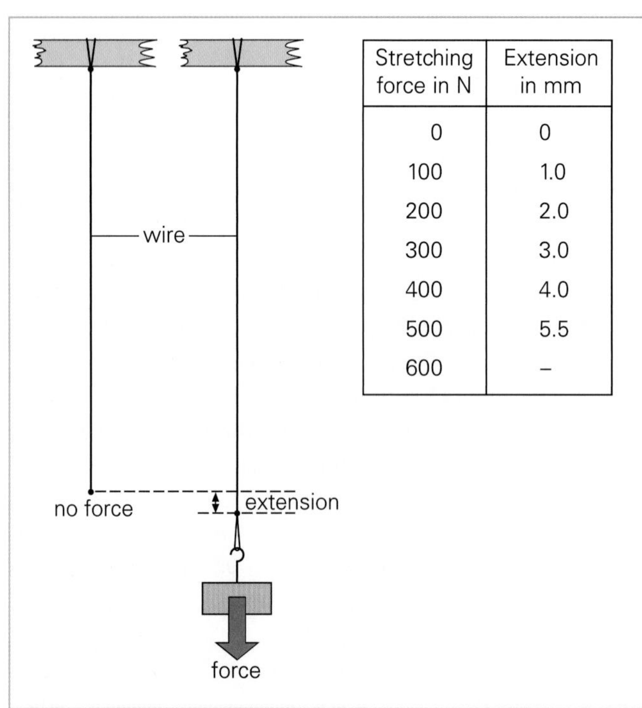

Stretching force in N	Extension in mm
0	0
100	1.0
200	2.0
300	3.0
400	4.0
500	5.5
600	–

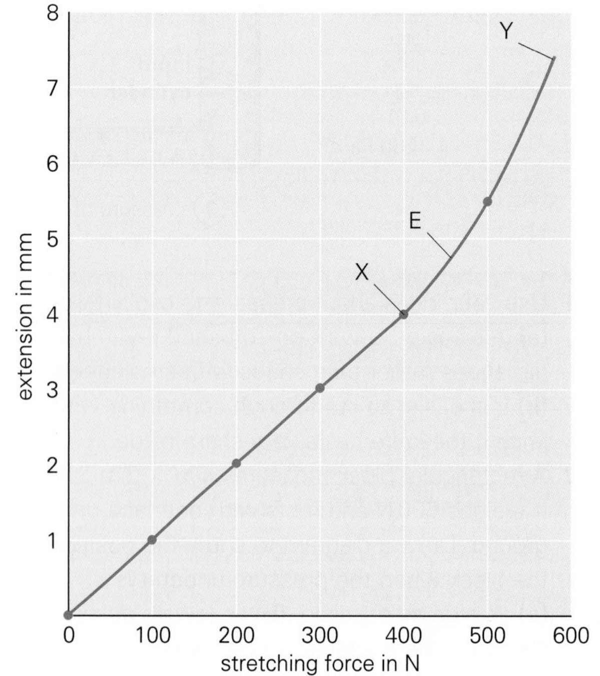

Compressing a gas

If you squash a gas into a smaller volume, its pressure rises. The experiment on the right is designed to find out more about the link between pressure and volume. Some gas (air) is trapped in a cylinder. The gas is compressed by pushing in the piston, and the pressure and volume measured at different stages.

Compressing the gas gas heats it up, which also affects the pressure. So the compression must be carried out very slowly in order that the gas can lose heat and keep a steady temperature.

The table under the diagram give some typical readings. Here are two ways of describing the connection between them:

- If the volume of the gas is halved, then the pressure is doubled...and so on.
- pressure × volume keeps the same value each time (10 000 in this case.)

These results can be summed up as follows:

If a fixed mass of gas is kept at a steady temperature, pressure × volume stays the same.

This is known as ***Boyle's law***. The air around us obeys Boyle's law. So do most other gases.

Explaining Boyle's law A gas is made up of tiny particles (atoms or molecules) which move about at high speed. They bounce of the sides of the container, and it is this that causes the pressure. If the gas is squashed into half the volume, the particles are more concentrated than before, and twice as many hit each square centimetre of the sides every second. So the pressure doubles.

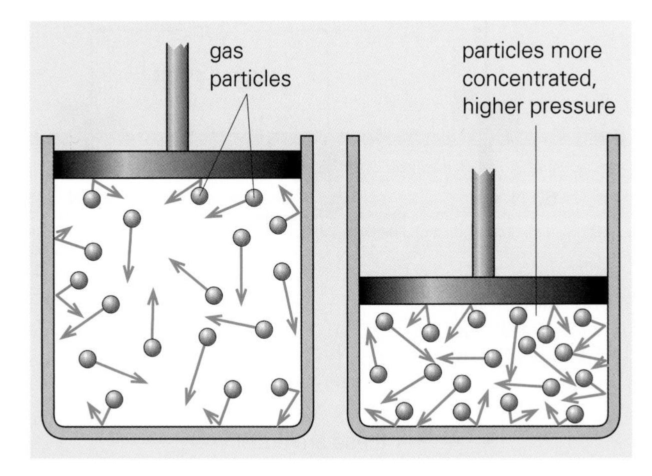

Pressure can be measured in pascals (see Spread 4.14). However, a larger unit, the ***kilopascal (kPa)*** is often more convenient: 1 kPa = 1000 Pa.

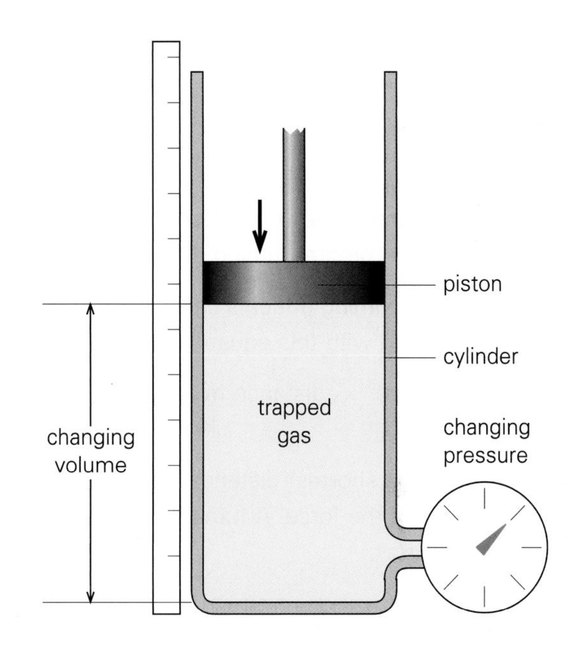

Volume in cm^3	50	40	25	20	10
Pressure in kPa	200	250	400	500	1000

You will need graph paper for questions 1 and 2.

Stretching force in N	0	1	2	3	4	5
Length in mm	40	49	58	67	79	99

1 When a spring was stretched, the readings above were taken.
(a) Make a table and draw a graph of extension against force. **(b)** How can you tell if the spring obeys Hooke's law? **(c)** If the spring obeys this law, up to what point does it do so? **(d)** What force produces an extension of 21 mm?

2 **(a)** Use the readings under the diagram above to plot a graph of pressure against volume.
(b) Describe what the graph shows.
(c) When the gas is at a pressure of 300 kPa, what is its volume?

3 A balloon contains 6 m^3 of helium. As it rises through the atmosphere, the pressure falls from 100 kPa to 50 kPa, but the temperature stays the same. What is the new volume of the balloon?

4.16 Turning forces

By the end of this spread, you should be able to:
- *calculate the turning effect of a force*
- *explain how forces can be magnified.*

Moments

Forces can have a turning effect.

Below, someone is using a spanner to turn a bolt. With a longer spanner, they could use the same force to produce an even greater turning effect.

The strength of a turning effect is called a ***moment***. It can be calculated with this equation:

moment	=	force	×	distance from turning point
in N m		in N		in m

The distance is the shortest distance from the turning point to the line of the force. A turning point is also known as a ***pivot***.

Forces can have a turning effect

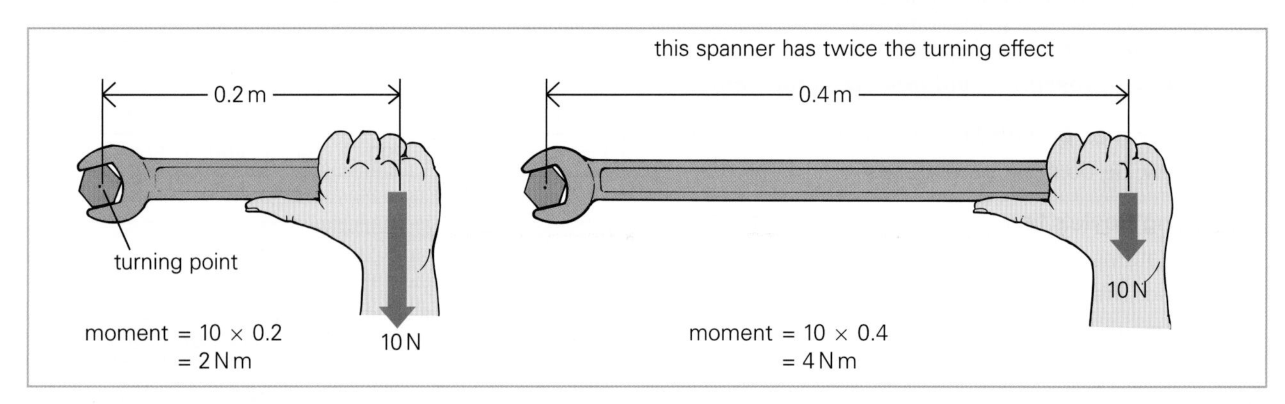

Moments in balance

On the right, a plank has been balanced on a log. Different weights have been placed on both sides of the plank. They have been arranged so that the plank still balances.

One weight has a turning effect to the left. The other has a turning effect to the right. The two turning effects are equal, and cancel each other out. That is why the ruler balances.

In other words, if something balances:

moment turning to the left	=	moment turning to the right

This is an example of the ***law of moments***.

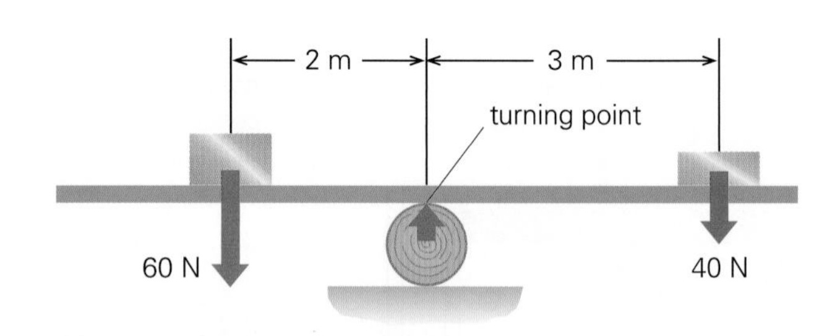

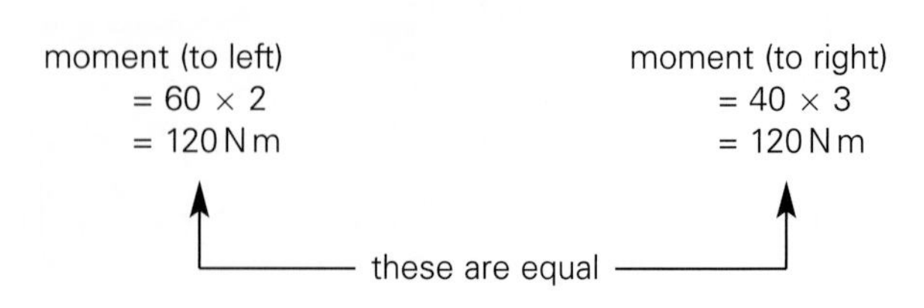

Centre of mass

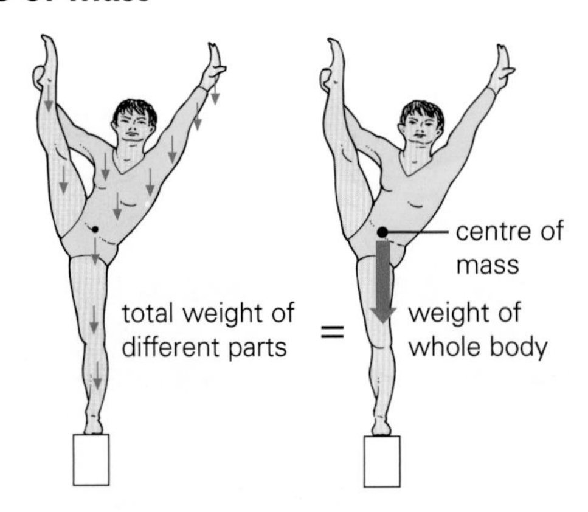

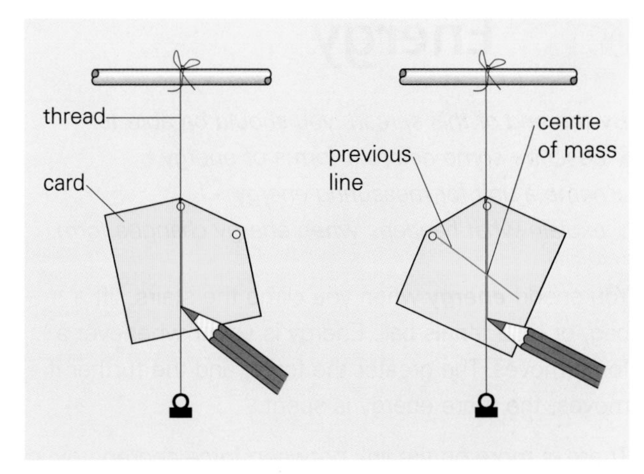

Balancing on a beam is difficult. The secret lies in how you position your weight. All parts of the body have weight. Together, they act like a single force pulling at just one point. This point is called your ***centre of mass*** (or ***centre of gravity***). To balance on a beam, you have to keep your centre of mass over the beam.

If you suspend a piece of card from some thread, it always hangs with its centre of mass in line with the thread. You can use this idea to find the centre of mass. Suspend the card from one corner and draw a vertical line on it. Do the same using another corner. Then see where the two lines cross.

Stability

If something is in a ***stable*** position, it will not topple over.

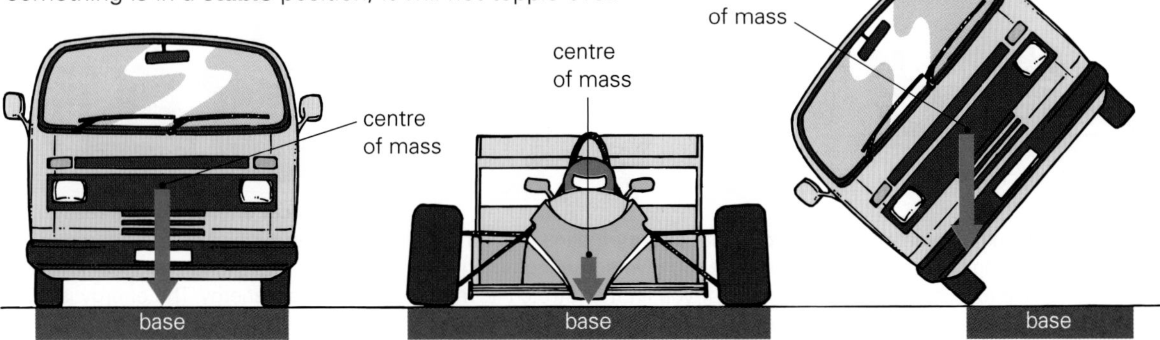

This van is in a stable position. If it starts to tip, its weight will pull it back again. As long as its centre of mass stays over its base, it will not topple.

This racing car is even more stable than the van. It has a lower centre of mass and a wider base. It has to be tipped much further before it starts to topple.

The van is now in an unstable position. Its centre of mass has passed beyond the edge of its base. So its weight will pull it over.

The model crane on the right has a movable counterbalance.

1 Why does the crane need a counterbalance?

2 Why must the counterbalance be movable?

3 What is the moment of the 100 N force (about O)?

4 To balance the crane, what moment must the 400 N force have?

5 How far from O should the counterbalance be placed?

6 What is the maximum load the crane should lift?

7 Give *two* ways of making the design of the crane more stable.

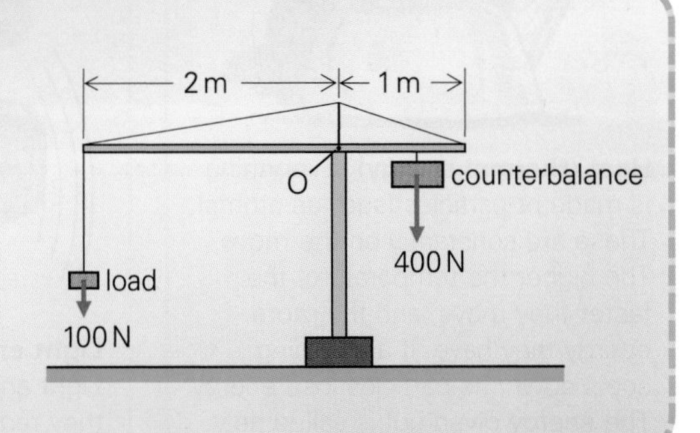

4.17 Energy

By the end of this spread, you should be able to:
- *describe some different forms of energy*
- *name a unit for measuring energy*
- *explain what happens when energy changes form.*

You spend ***energy*** when you climb the stairs, lift a bag, or hit a tennis ball. Energy is spent whenever a force moves. The greater the force, and the further it moves, the more energy is spent.

There is more on the link between force and energy in Spread 4.22.

Forms of energy

Energy can take different forms. Here are some of the names used to describe them:

Kinetic energy means 'movement energy'. A moving thing has energy because it can make forces move when it hits something else.

Potential energy means 'stored energy'. You give someting potential energy if you lift it or stretch it. The energy is released when you let it go.

Chemical energy This is really another type of stored energy. Foods, fuels, and batteries have chemical energy. The energy is released by chemical reactions.

Heat (thermal energy) Everything is made of particles (such as atoms). These are constantly on the move. The higher the temperature, the faster they move, and the more energy they have. If a hot thing cools down, its particles lose energy. The energy given out is called heat.

Light energy and **sound energy** Light and sound carry energy as they radiate from their source.

Electrical energy This is the energy carried by an electric current.

Nuclear energy is stored in the nucleus of the atom. It is released by nuclear reactions.

Energy is measured in ***joules (J)***.

On the right, you can see some examples of different amounts of energy. Large amounts are sometimes measured in ***kilojoules (kJ)***. 1 kJ = 1000 J.

Typical energy values	
Potential energy:	
stretched rubber band	1 J
you, on top of a step-ladder	500 J
Kinetic energy:	
kicked football	50 J
small car at 70 mph	500 000 J
Heat (thermal energy):	
hot cup of tea	150 000 J
Chemical energy:	
torch battery	10 000 J
chocolate biscuit	300 000 J
litre of petrol	35 000 000 J

Energy chains

Just like money, energy doesn't vanish when you spend it. It just goes somewhere else! Below is an example of how energy can change from one form to another. It is sometimes called an ***energy chain***:

In every energy chain, the total amount of energy stays the same. Scientists express this idea in the ***law of conservation of energy***:

Energy can change into different forms, but it cannot be made or destroyed.

In any chain, some energy is always wasted as heat. For example, you give off heat when you exercise, which is why you sweat! However, the total amount of energy (including the heat) stays the same.

1 Give an example of something which has **(a)** kinetic energy **(b)** chemical energy **(c)** potential energy.
2 A fire gives out 10 kJ of energy. What is this in joules?
3 What type of energy is supplied to a car engine? What happens to this energy?
4 Describe the energy changes that take place when you apply the brakes on a moving cycle.
5 Describe the energy changes which take place when you throw a ball up into the air.
6 Scientists say that energy can 'never be destroyed'. Explain what they mean.

Energy changers

Here are some examples of energy changers in action:

Energy input		Energy changer		Energy output
electrical energy	→	heating element	→	heat (thermal energy)
sound energy	→	microphone	→	electrical energy
electrical energy	→	loudspeaker	→	sound energy
kinetic energy	→	brakes	→	heat (thermal energy)

4.18 Energy on the move

By the end of this spread, you should be able to:
- *explain that heat is different from temperature*
- *describe how energy can be transferred by conduction and convection.*

Heat and temperature

Everything is made of tiny particles. These are constantly on the move (see Spread 3.02). If something hot cools down, its particles lose energy. The energy given out is called ***heat***.

Heat is not the same as temperature. The hot materials in the two photographs below are at the same temperature, but the amounts of energy they hold are quite different.

The sparks from this sparkler are at a temperature of 1600 °C. But they hold so little energy that they do not burn you when they touch your skin.

This molten (melted) iron is also at 1600 °C. It holds lots of energy and would be very dangerous to touch.

For heat to flow from one place to another, there must be a temperature difference between them. Here are two of the ways in which the energy can be transferred (moved):

Conduction

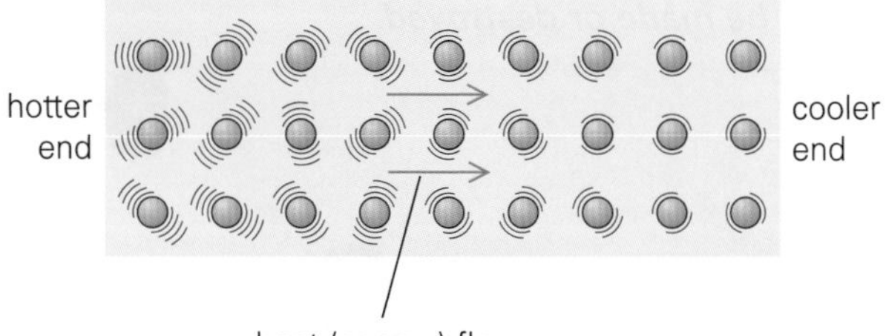

If one end of a bar is heated, its particles vibrate faster. In time, their extra movement is passed on to particles right along the bar. Scientists say that energy is being transferred by ***conduction***.

Metals are the best ***conductors*** of heat. Their atoms have some loosely attached electrons (see Spread 4.01). These are free to move through the metal and carry energy rapidly from one end to another.

Most non-metals are poor conductors. Poor conductors of heat are called ***insulators***.

Good conductors	Insulators (poor conductors)
metals	glass
especially	water
silver	plastic
copper	wood
aluminium	materials with air trapped in them — wool, fibrewool, plastic foam, fur, feathers
	air

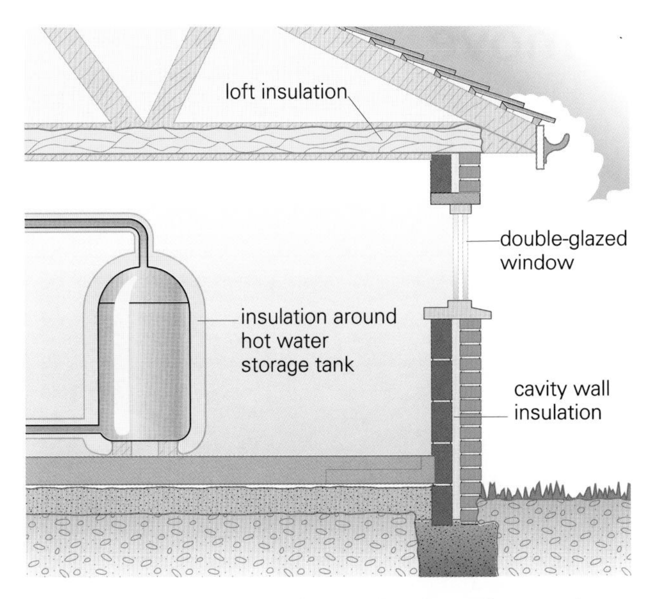

Air is a poor conductor of heat. Feathers, fur, wool, and plastic foam are all good insulators because they contain tiny pockets of trapped air. Most insulating materials used in the home contain trapped air, including double-glazing, lagging, and loft insulation.

Convection

If air is free to circulate, it can quickly transfer energy from one place to another:

When air is heated, it expands (takes up more space) and becomes less dense. It floats upwards as cooler, denser air sinks and moves in to take its place. The result is a circulating flow called a ***convection current***. Convection can occur in other gases as well as air. And it can occur in liquids, such as water.

Most rooms are heated by convection:

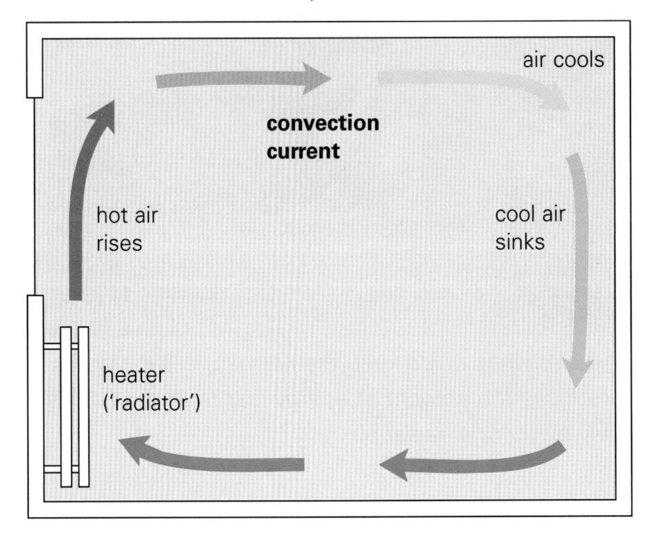

Convection has a part to play in the weather. You can see an example above. During the day, the land warms up more quickly than the sea. This sets up a convection current so that breezes blow in from the sea. Where warm, damp air rises and cools, clouds may form.

1 *conduction* *convection*
Which of the above processes is the reason for each of the following?
(a) The handle of a metal teaspoon gets hot if the spoon is left standing in a hot drink.
(b) In a room, the air near the ceiling is usually warmer than the air near the floor.
(c) If a canned drink is taken out of a 'fridge, the drink inside the can warms up.

2 Explain each of the following:
(a) In most saucepans, the base is made of metal, but the handle is plastic.
(b) Feathers, fur, and wool are good insulators.

3 Below, heat is flowing through a metal bar. The bar is made of tiny particles (atoms) which are vibrating. Which end of the bar has
(a) the higher temperature?
(b) the faster particles?

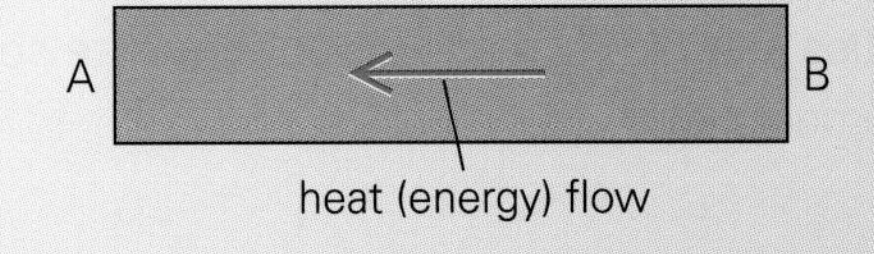

4.19 More energy on the move

By the end of this spread, you should be able to:
- *describe how energy can be transferred by radiation and evaporation.*

Energy can be transferred by conduction and convection (see Spread 4.18). Here are two more ways in which it can be transferred:

Radiation

On Earth, we are heated by the Sun, as shown on the right. The Sun's energy travels to us as rays of ***electromagnetic radiation*** (see Spread 4.28). This includes ***light*** rays (which we can see) and ***infrared*** rays (which are invisible). If we absorb any of this radiation, it heats us up, so it is sometimes called ***thermal radiation***. Often, people just call it 'radiation', although there are other types of radiation as well.

All warm or hot surfaces give off thermal radiation. The hotter something is, the more energy it radiates.

	dull black	shiny black	white	silvery
Giving off radiation	best	- - - - - - -	- - - - - - -	worst
Reflecting radiation	worst	- - - - - - -	- - - - - - -	best
Absorbing radiation	best	- - - - - - -	- - - - - - -	worst

Black surfaces are the best at giving off radiation. They are also the best at absorbing it.

Silvery or white surfaces are good at reflecting radiation – which means that they are poor at absorbing it. In hot, sunny countries, buildings are often painted white so that they absorb as little of the Sun's radiation as possible.

Silvery or white surfaces are also poor at giving off radiation. Kettles are usually made silvery or white so that they lose heat slowly.

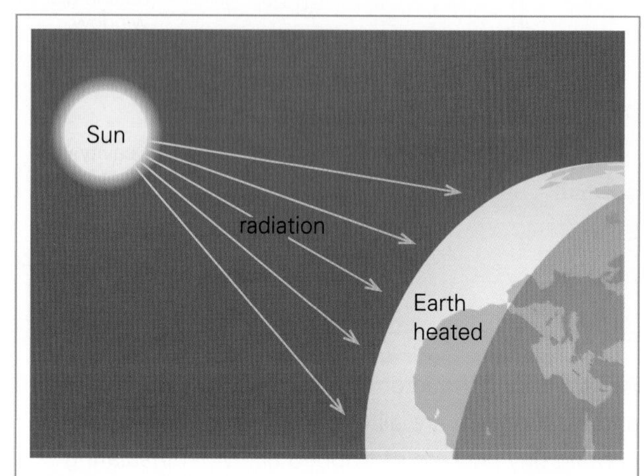

There is empty space between the Earth and the Sun. Energy can travel through space in the form of electromagnetic radiation. Moving away from the Sun, its heating effect is reduced because the radiation becomes more spread out.

The Sun's energy cannot reach us by conduction or convection because those processes depend on the movements of the tiny particles in a solid, liquid, or gas.

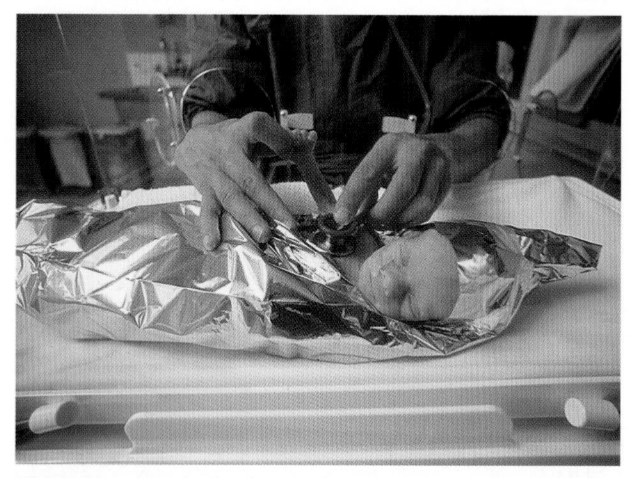

This shiny bag helps keep the premature baby warm. It reduces the amount of energy lost from the baby's body by thermal radiation.

The vacuum flask

A vacuum flask can keep drinks hot for hours. On the right, you can see the features which a flask has to reduce the amount of energy lost by conduction, convection, and radiation.

The flask can also keep chilled drinks cold because it is just as difficult for energy to flow in as out.

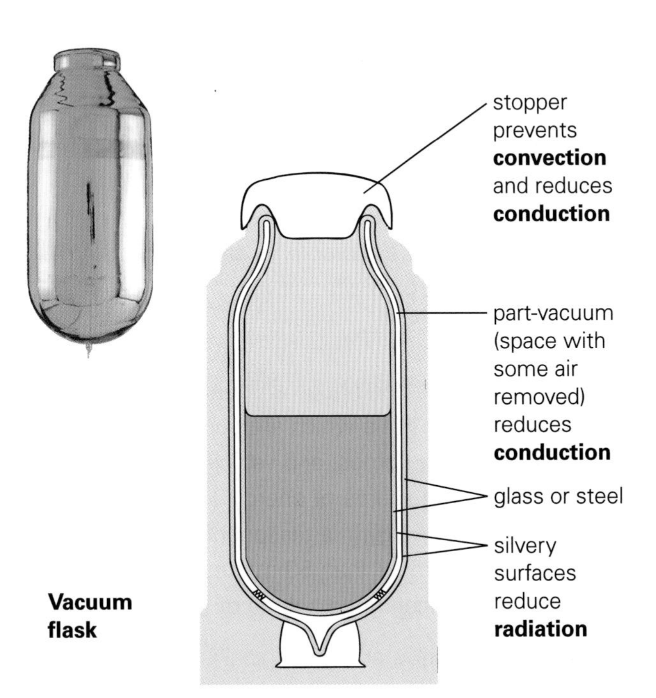

Vacuum flask

Evaporation

Wet hands dry out in a few minutes. This is because the water ***evaporates*** (changes into vapour). If air is blowing from a drier, as below, the water evaporates much more quickly.

As your hands dry, they feel colder. This is because energy is needed to turn liquid water into vapour. The vapour takes the energy from your hands, so they cool down. Overall, energy is transferred from your hands to the air.

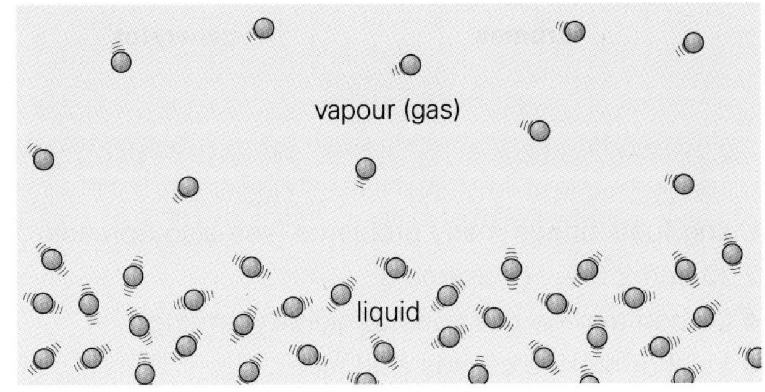

As particles escape from a liquid to form a vapour, they take energy with them, so the liquid cools down.

To answer these questions, you may also need information from the previous spread, 4.18.

1 Explain why
(a) houses in hot countries are often painted white
(b) it is better for a kettle to be white or silvery on the outside rather than black.

2

If the cars above are left standing in a car park on a sunny day, which one would you expect to be hotter inside? Give a reason for your answer.

3 What features does a vacuum flask have to stop energy losses by **(a)** conduction? **(b)** radiation?

4 Why is a vacuum flask good at keeping drinks cold as well as hot?

5 *conduction convection radiation evaporation*
Which of the above processes
(a) is the method by which the Sun's energy reaches the Earth?
(b) depend on the movements of tiny particles in a solid, liquid, or gas?

4.20 Supplying the energy

By the end of this spread and the next (4.21), you should be able to:
- *explain how the world gets its energy*
- *describe some of the problems caused by using fuels, and give some alternatives*
- *explain what efficiency means.*

Industrial societies need huge amounts of energy. Most comes from fuels which are burnt in power stations, factories, homes, and vehicles. Fuels are a very concentrated source of energy. For example, there is enough energy in a teaspoonful of petrol to move a large car more than 50 metres. Our fuel, food, is also a very concentrated source of energy.

Most of the world's energy originally came from the Sun. To find out how, see the next spread, 4.21.

Turbine

Power stations

Mains electricity comes from generators in power stations. In many power stations, the generators are turned by ***turbines***, blown round by the force of high-pressure steam (see below). For other ways of turning generators, see the next spread.

Ready with the power

If the demand for electricity suddenly rises, hydroelectric power stations (see Spread 4.21) are very quick at getting their generators up to speed. So are small, gas-burning power stations. Other types are slower, and nuclear power stations slowest of all.

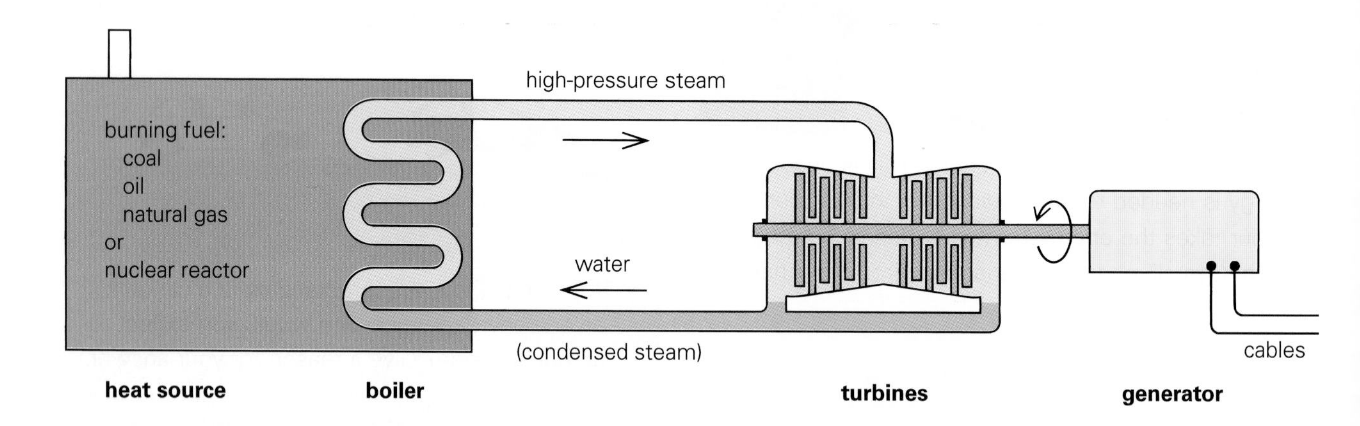

Burning fuels

Fuels use up oxygen when they burn. With most fuels (including food), this reaction takes place:

fuel + oxygen $\rightarrow$ carbon dioxide + water

There may be other products as well. For example: burning coal produces some sulphur dioxide.

Using fuels brings many problems (see also Spreads 2.23 and 2.24). For example:
- Carbon dioxide gas adds to global warming.
- Sulphur dioxide causes acid rain.
- Transporting fuels can cause pollution.
- Supplies of most fuels will eventually run out.

Renewable or non-renewable?

Coal, oil, and natural gas are called ***fossil fuels*** (see the next spread).They took many millions of years to form. Once used up, they cannot be replaced. They are ***non-renewable***.

Some fuels are ***renewable***. For example, if wood is burnt, it can be replaced by growing more trees.

You can see some examples of renewable and non-renewable energy sources below. For more details on each one, see the next spread.

Non-renewable energy sources	Renewable energy sources
fossil fuels: oil natural gas coal nuclear fuel: uranium-235	hydroelectric energy tidal energy wave energy wind energy solar energy geothermal energy biofuels (fuels from plant and animal matter)

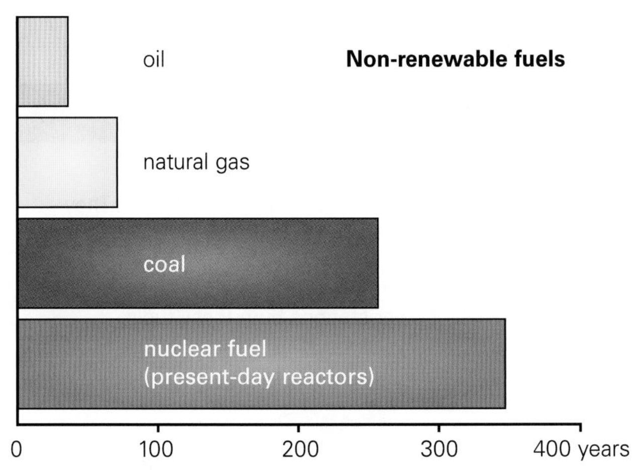

Time that world supplies will last at present rates of consumption from known reserves

Efficiency

When fuels burn, much of their energy is wasted as heat. In a typical power station for example, for every 100 joules of energy in the fuel, only 35 joules ends up as electrical energy. The power station has an ***efficiency*** of 35%.

You can use this equation to work out efficiency:

$$\text{efficiency} = \frac{\text{useful energy output}}{\text{energy input}} \times 100\%$$

Here are some typical efficiency values:

For every **100 J** of **Energy input** ⇨	**Useful energy output**	**Efficiency**
petrol engine	25 J	25%
diesel engine	35 J	35%
fuel-burning power station	35 J	35%
human engine	15 J	15%

Low efficiency is not because of poor design. When an engine is working, some energy has to be wasted. Burning fuel gives particles extra energy. But some of this becomes so spread out that it cannot be used to produce motion. Instead, it is lost as heat.

In some power stations, wasted heat is used to supply the local district with hot water. This is just one way of being less wasteful with fuel so that the world's reserves last longer.

To answer the following, you will need information from Spread 4.21.

1 Some fuels are non-renewable. What does this mean? Give *two* examples.

2 Describe *two* problems which can be caused by gases from burning fuels.

3 Give *two* ways of generating electricity in which no fuel is burnt and the energy is renewable.

4 The energy in petrol originally came from the Sun. Explain how it got into the petrol.

5 A power station has an efficiency of 25%. Explain what this means.

6 Why it is important for engines and power stations to have the highest possible efficiency?

4.21 How the world gets its energy

Solar panels
These absorb energy radiated from the Sun. They use it to heat water.

Solar cells
These use the energy in sunlight to produce small amounts of electricity.

The Sun
The Sun radiates energy because of nuclear reactions deep inside it. The energy is released when nuclei of hydrogen atoms collide and join to form nuclei of helium – a process called nuclear fusion.

Energy in food
We get energy from the food we eat. The food may be from plants ***(biomass)*** or from animals which fed on plants.

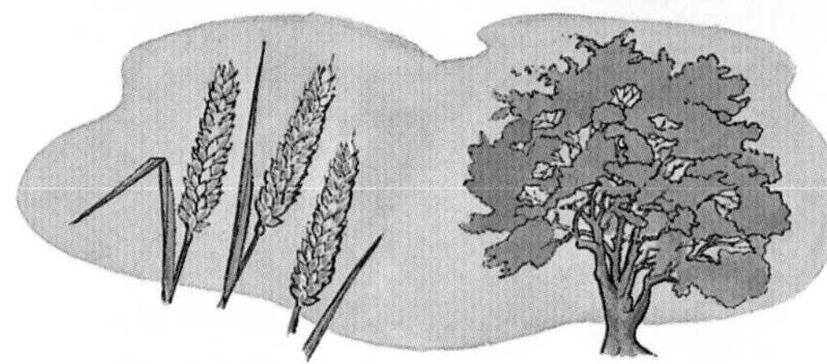

Energy in plants
Plants take in energy from sunlight falling on their leaves. They use it to turn water and carbon dioxide from the air into new growth. Animals eat plants to get the energy stored in them.

Biofuels (biomass) from plants
Wood is still an important fuel in many countries. When wood is burnt, it releases energy which the tree once took in from the Sun. In some countries, sugar cane is grown and fermented to make alcohol. This can be used as a fuel instead of petrol.

Fossil fuels
Oil, natural gas, and coal are called fossil fuels. They were formed from the remains of plants and tiny sea creatures which lived many millions of years ago. Industrial societies rely on fossil fuels for most of their energy. Many power stations burn fossil fuels.

Biofuels from waste
Rotting animal and plant waste can give off methane gas (as in natural gas). This can be used as a fuel. Marshes, rubbish tips, and sewage treatment works are all sources of methane. Some waste can also be used directly as fuel by burning it.

Fuels from oil
Many fuels can be extracted from oil (crude). These include: petrol, diesel fuel, jet fuel, paraffin, central heating oil, bottled gas.

Batteries
Some batteries (e.g. car batteries) have to be given energy by charging them with electricity. Others are manufactured from chemicals which already store energy. But energy is needed to produce the chemicals in the first place.

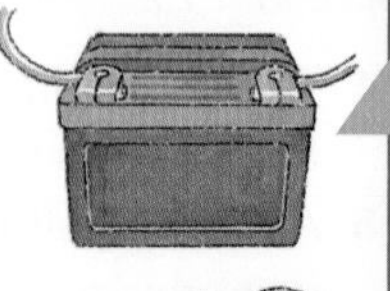

The tides
The gravitational pull of the Moon (and to a lesser extent, the Sun) creates gentle bulges in the Earth's oceans. As the Earth rotates, different places have high and low tides as they pass in and out of the bulges.

Tidal energy
In a tidal energy scheme, an estuary is dammed to form an artificial lake. Incoming tides fill the lake; outgoing tides empty it. The flow of water in and out of the lake turns generators.

The atom
Some atoms have huge amounts of nuclear energy stored in their nuclei (centres). Radioactive materials have unstable atoms which release energy slowly. Nuclear reactors can release energy much more quickly.

Nuclear energy
In a reactor, nuclear reactions release energy from nuclei of uranium atoms. This produces heat which is used to make steam for driving generators.

Weather systems
These are driven by the heating effect of the Sun. Hot air rising above the equator causes belts of wind around the Earth. Heat and winds lift water vapour from the oceans and bring rain and snow.

Geothermal energy
Deep underground, the rocks are hotter than they are on the surface. The heat comes from radioactive materials naturally present in the rocks. It can be used to make steam for heating buildings or driving generators.

Wave energy
Waves are caused by the wind (and partly by tides). Waves cause a rapid up-and-down movement on the surface of the sea. This movement can be used to drive generators.

Hydroelectric energy
An artificial lake forms behind a dam. Water rushing down from this lake is used to turn generators. The lake is kept full by river water which once fell as rain or snow.

Wind energy
For centuries, people have been using the power of the wind to move ships, pump water and grind corn. Today, huge wind turbines are used to turn generators.

4.22 Work, energy, and power

By the end of this spread, you should be able to:
- *calculate work, power, GPE, and KE.*

Work

Work is done whenever a force makes something move. Like energy, work is measured in ***joules*** (J).

One joule of work is done when a force of 1 newton (N) moves an object a distance of 1 metre (m).

To calculate work, you can use this equation:

work done	= force	× distance moved*	* in direction
in J	in N	in m	of force

For example, if a force of 3 N moves an object a distance of 2 m: work done = 3 × 2 = 6 J

Work and energy are linked. If, say, 6 J of work is done, then 6 J of energy is spent.

Power

If one engine has more ***power*** than another, it can do work at a faster rate.

Power is measured in ***watts (W)***. A power of 1 watt means that work is being done (energy is being spent) at the rate of 1 J per second. So 1 W = 1 J/s.

$$\text{power} = \frac{\text{work done}}{\text{time taken}} \quad \text{or} \quad \text{power} = \frac{\text{energy spent}}{\text{time taken}}$$

Typical power outputs

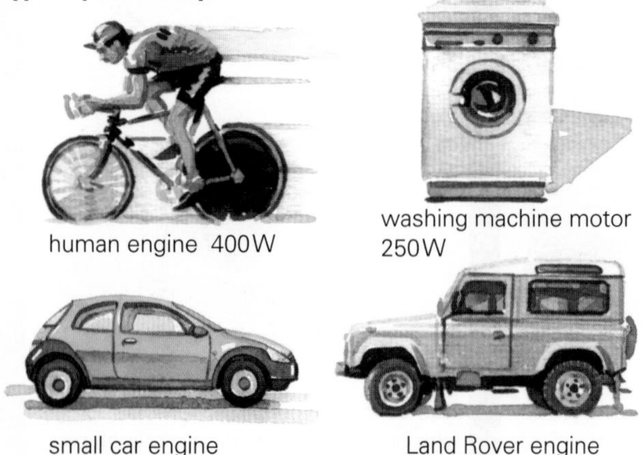

1 kilowatt (kW) = 1000 watts(W)

Gravitational potential energy (GPE)

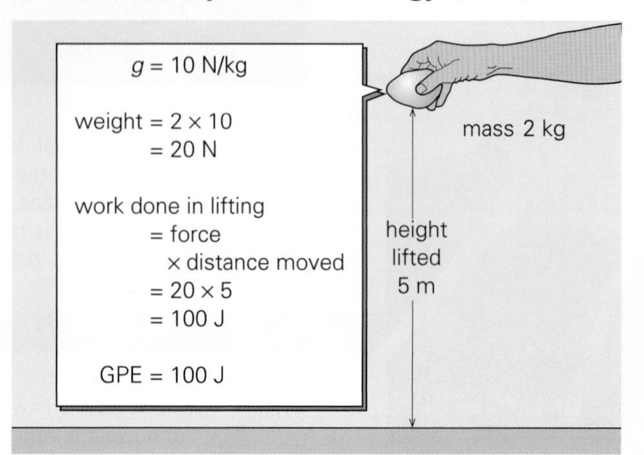

Above, someone has lifted a stone above the ground. The equation work = force × distance moved has been used to calculate the work done.

The stone has gained ***gravitational potential energy (GPE)*** equal to the work done in lifting it. So:

GPE	= weight	× height lifted
in J	in N	in m

In symbols:

$$GPE = mgh$$

m = mass in kg
g = 10 N/kg (see 4.12)
h = height in m

Kinetic energy (KE)

Moving things have kinetic energy (KE). Scientists have worked out an equation for calculating it:

$$KE = \tfrac{1}{2}mv^2$$

KE = kinetic energy in J
m = mass in kg
v = speed in m/s

Below, you can see how to use this equation.

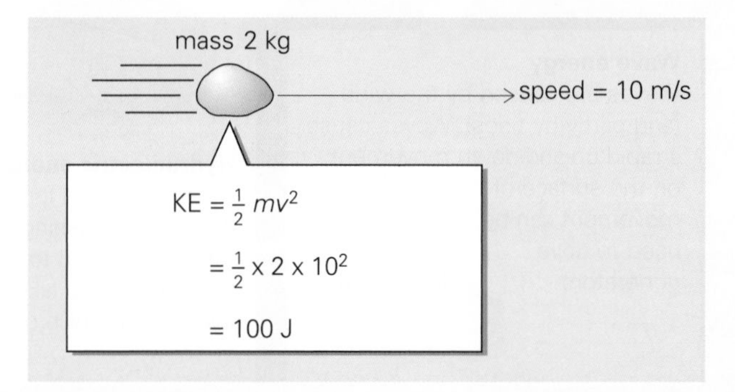

GPE and KE problem

The stone on the right has a mass of 2 kg. It is held 5 metres above the ground and then dropped. As the stone is about to hit the ground, what is (a) its kinetic energy? (b) its speed? (Assume that $g = 10$ N/kg and that no energy is wasted because of air resistance.)

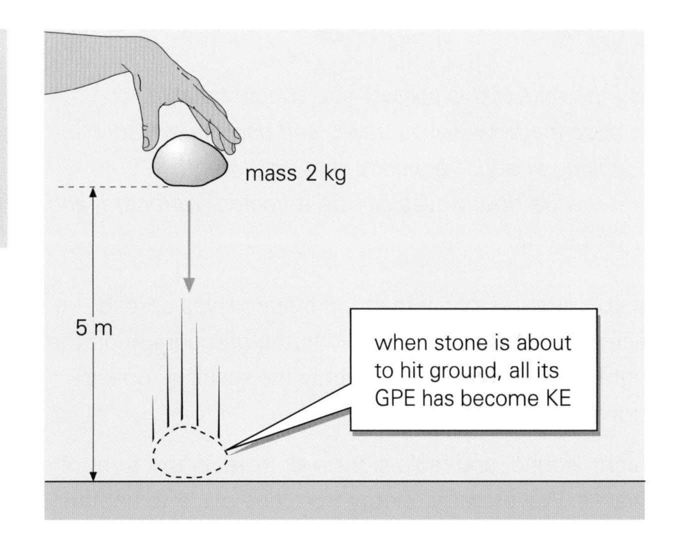

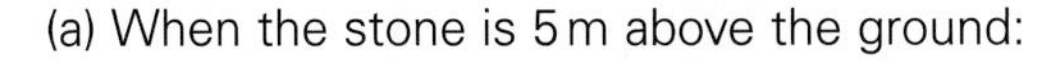

(a) When the stone is 5 m above the ground:

GPE = $mgh = 2 \times 10 \times 5 = 100$ J

When the stone is about to hit the ground, all its GPE has been changed into KE. So its KE is 100 J.

(b) The stone's KE = 100 J, so $\frac{1}{2}mv^2 = 100$

But $m = 2$, so $\frac{1}{2} \times 2 \times v^2 = 100$

so $v = 10$

So, the stone hits the ground at a speed of 10 m/s.

Power problem

The model crane on the right lifts a mass of 4 kg through a height of 3 metres in 10 seconds. What is its power output? (Assume that $g = 10$ N/kg)

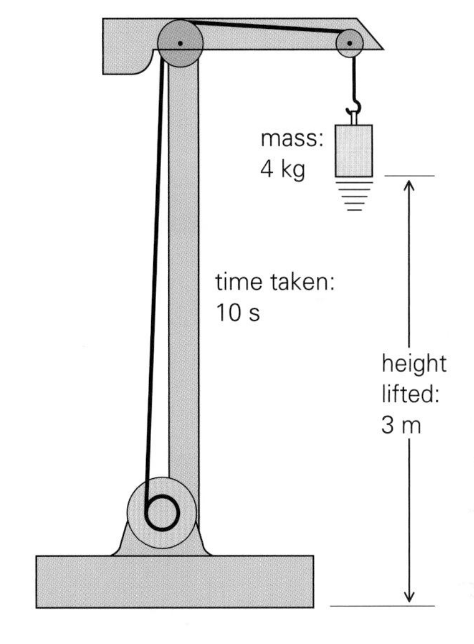

First, calculate the work done. This is the same as the energy spent, and is equal to the GPE gained by the mass.

work done = gain in GPE = mgh
$= 4 \times 10 \times 3$
$= 120$ J

Next, use the power equation:

$$\text{power} = \frac{\text{work done}}{\text{time taken}} = \frac{120}{10} = 12\text{ W}$$

So the crane's power output is 12 W.

Assume that $g = 10$ N/kg, and that no energy is wasted because of air resistance.

1 If you use a force of 20 N to move a wheelbarrow 5 metres, how much work (in J) do you do?

2 If, in question 1, it takes you 10 seconds to move the wheelbarrow, what is your power output?

3 Look at the typical power outputs shown on the opposite page. Convert these into kilowatts.

4 If you lift a mass of 20 kg through a height of 4 metres in 8 seconds **(a)** how much work do you do? **(b)** what is your power output?

5 A rock of mass 3 kg is 20 m above the ground. Find the following:
(a) Its gravitational potential energy (GPE).
(b) Its PE when it has fallen half way to the ground.
(c) Its kinetic energy (KE) when it has fallen half way to the ground.
(d) Its KE just before it hits the ground.
(e) Its speed just before it hits the ground.

4.23 Moving waves

By the end of this spread, you should be able to:
- *describe how waves travel, and use the equation linking speed, frequency, and wavelength*
- *describe how waves can be reflected, refracted, and diffracted.*

If you drop a stone into a pond, tiny waves spread across the surface, as shown in the photograph on the right. The moving wave effect is the result of up-and-down motions in the water.

Light, sound, and radio signals all travel in the form of waves. Waves carry energy from one place to another, but without any material being transferred.

There are two main types of wave. You can demonstrate them with a stretched spring as shown below. When a coil oscillates (moves to and fro), it makes the next one oscillate a fraction of a second later...and so on. This gives the moving wave effect.

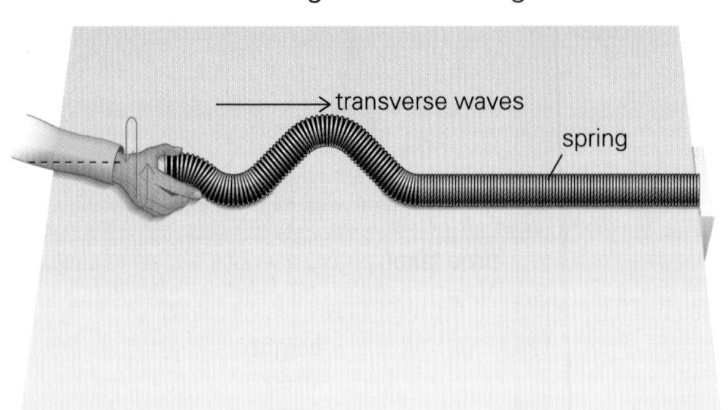

▲ **Transverse waves** have oscillations which are from side to side (or up and down). Light and radio waves travel like this (see Spread 4.28).

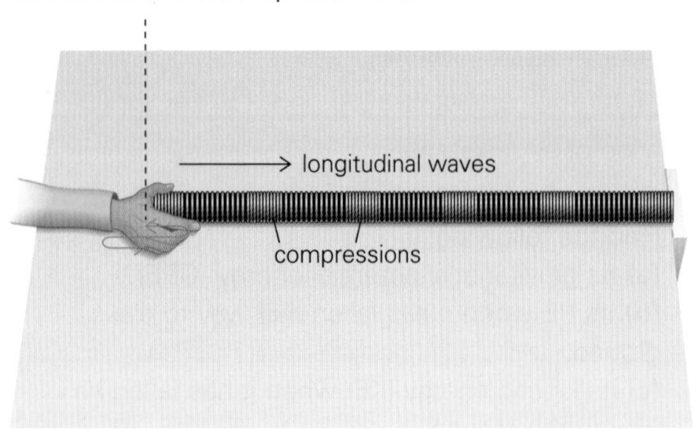

▲ **Longitudinal waves** have oscillations which are backwards and forwards. Sound waves travel like this (see Spread 4.24).

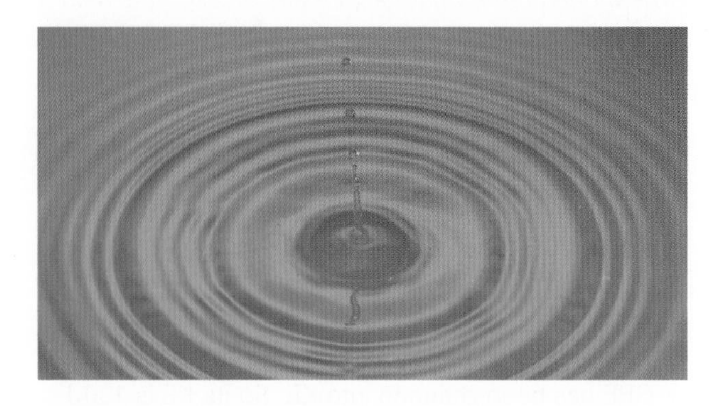

Describing waves

Below, waves are travelling across water. Here are some of the terms used to describe them:

Speed This is measured in metres per second (m/s).

Frequency This is the number of waves per second. It is measured in ***hertz (Hz)***. Below, 3 waves pass the post every second, so the frequency is 3 Hz.

Wavelength This is the distance from one point on a wave to the matching point on the next wave.

Amplitude This is the distance shown below:

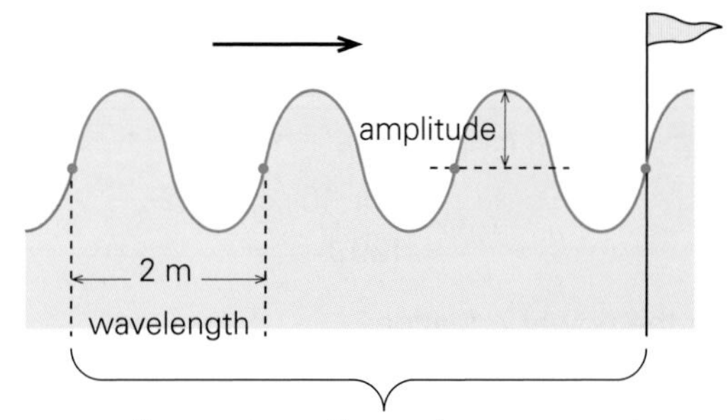

3 waves pass the post every second

Above, 3 waves pass the post every second. Each wave is 2 m long. So the waves travel 6 m every second – their speed is 6 m/s. You could work out this result with an equation, which is true for all waves:

$$\text{speed in m/s} = \text{frequency in Hz} \times \text{wavelength in m}$$

As 3 waves pass the post every second, the time between one wave peak and the next is $\frac{1}{3}$ second. This is called the ***period*** of the wave motion:

$$\text{period} = \frac{1}{\text{frequency}}$$

Wave effects

You can use a ***ripple tank*** to study how waves behave. Ripples are sent across the surface of some water in a shallow tank. Here are some of the effects:

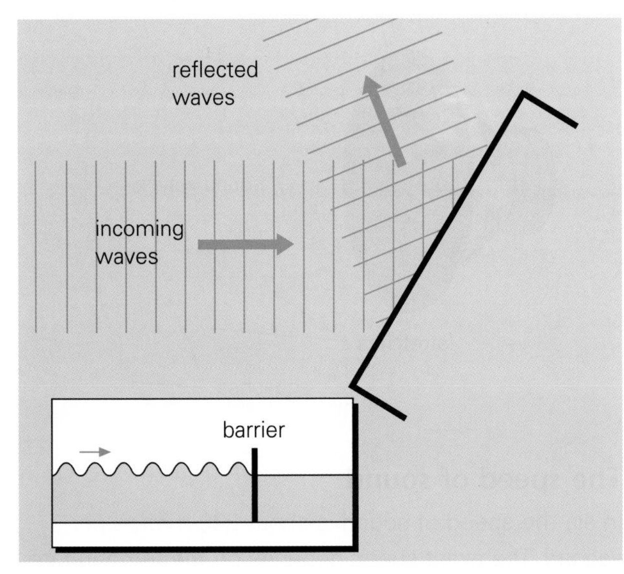

▲ **Reflection** A vertical barrier is put in the path of the waves. The waves are reflected from the barrier at the same angle as they strike it. (For more on the reflection of light, see Spread 4.27.)

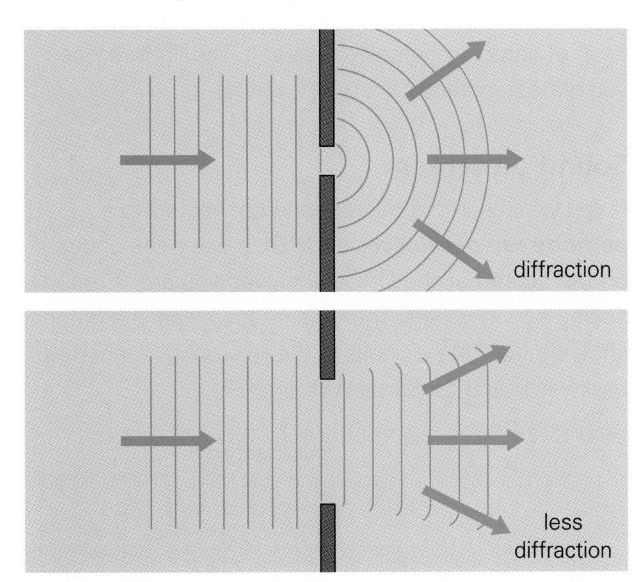

▲ **Diffraction** Waves bend round obstacles, or spread as they pass through a gap. This is called ***diffraction***. It works best if the width of the gap is about the same as the wavelength. Wider gaps cause less diffraction.

Sound waves diffract round large obstacles, so you can hear round corners. Light waves are much shorter, so gaps have to be very tiny to diffract them.

Sound, light, and radio signals can be reflected, refracted, and diffracted. This is evidence that they travel as waves.

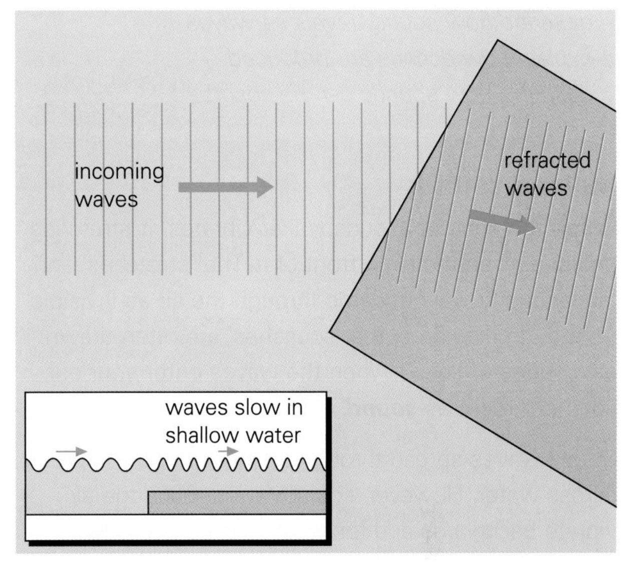

▲ **Refraction** A flat piece of plastic makes the water more shallow. This slows the waves down. When they slow, they change direction. This is called ***refraction***. (For more on the refraction of light, see Spread 4.27.)

1 What is the difference between longitudinal and transverse waves? Give an example of each.

2 If some waves have a speed of 60 m/s, and a frequency of 20 Hz, what is their wavelength?

3 If some waves have a speed of 60 m/s, and a frequency of 10 Hz, what is their wavelength?

4 Below, waves are moving towards a harbour.
(a) What will happen to waves striking the harbour wall at A?
(b) What will happen to waves passing through the harbour entrance at B?
(c) If the harbour entrance were wider, what difference would this make?

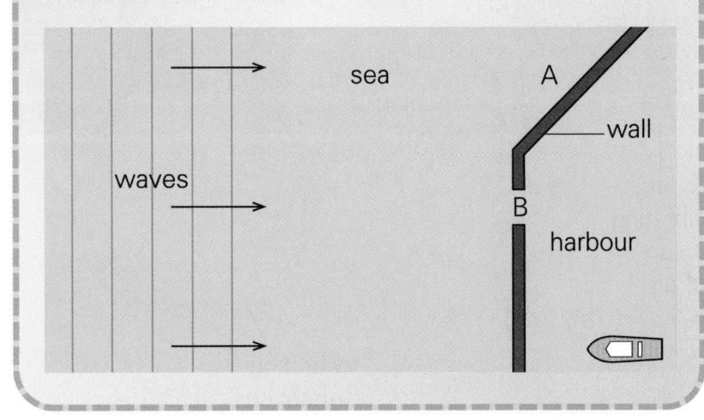

4.24 Sound waves

By the end of this spread, you should be able to:
- *explain what causes sound*
- *describe how sound travels as waves*
- *explain how echoes are produced*
- *describe how sounds can be absorbed.*

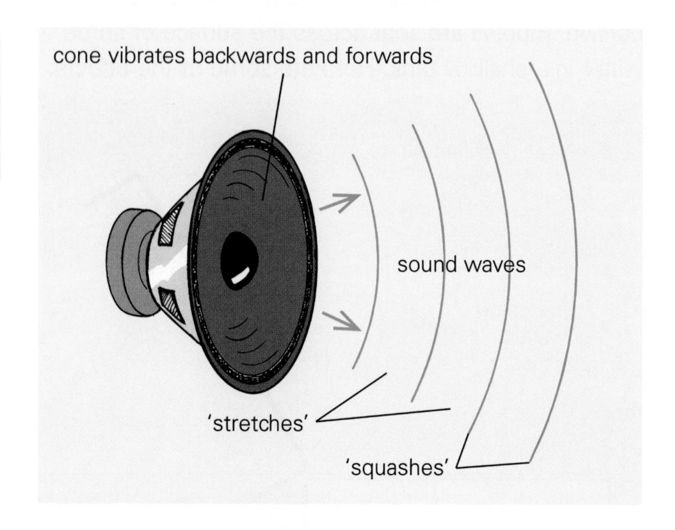

Making sounds

When the cone of a loudspeaker vibrates, it stretches and squashes the air in front of it. The 'stretches' and 'squashes' travel outwards through the air as invisible waves. (In diagrams, the 'squashes' are often drawn as a series of lines.) When the waves enter your ear, you hear them as ***sound***.

Sound waves spread through air just as ripples spread across water. However, sound waves make the air vibrate backwards and forwards, not up and down.

Sound needs a material to travel through

Sound waves can travel through solids and liquids, as well as gases. But they cannot travel through a vacuum (empty space). If there is nothing to stretch and squash, sound waves cannot be made.

Sound is caused by vibrations

The vibrations can be produced in different ways.

You can see some examples below:

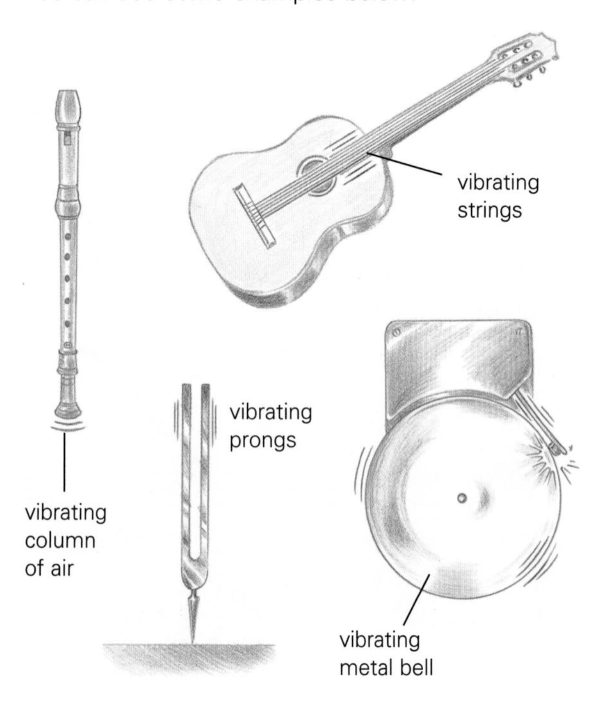

The speed of sound

In air, the speed of sound is about 330 metres per second. The exact speed depends on the temperature.

Sound travels faster through water than it does through air, and even faster through most solids.

Sound is much slower than light, which travels at 300 000 kilometres per second. That is why you see a flash of lightning before you hear it. The light reaches you almost instantly.

Sound on screen

If you whistle into a microphone connected to a ***cathode ray oscilloscope (CRO)***, a wavy line appears on the screen of the CRO. However, you aren't really seeing sound waves. The up-and-down line is a graph showing how the air next to the microphone vibrates backwards and forwards with time.

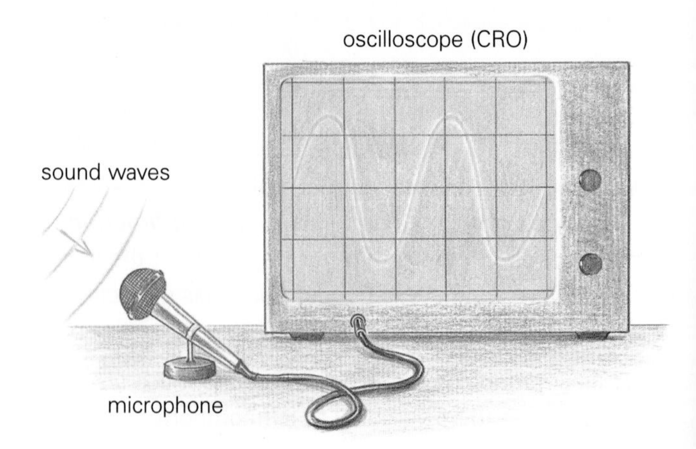

Echoes

Hard surfaces, such as walls, reflect sound waves. When you hear an ***echo***, you are hearing a reflected sound a short time after the original sound.

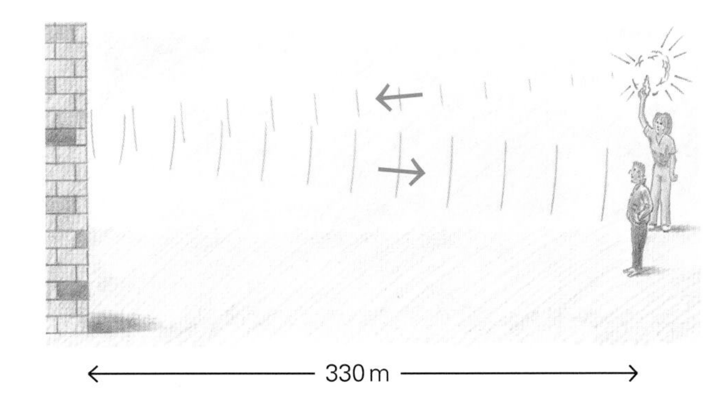

Finding the speed of sound You can use echoes to work out the speed of sound. The girl above is stood 330 metres from a wall. She fires a starting pistol. Her friend hears the echo 2 seconds later.

The sound has travelled a distance of 2 × 330 metres.

The time taken is 2 seconds. So:

$$\text{speed of sound} = \frac{\text{distance travelled}}{\text{time taken}} = \frac{2 \times 330}{2}$$

$$= 330\,\text{m/s}$$

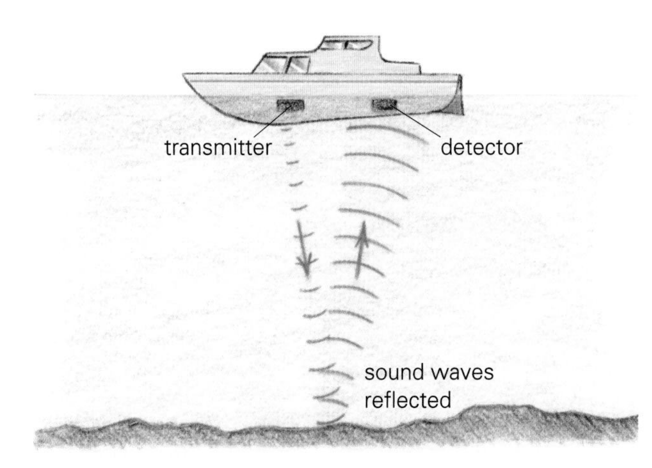

Echo-sounding Boats can use echo-sounding equipment to work out the depth of water beneath them. A sound pulse is sent down through the water. The time for the echo to return is measured. The longer the time, the deeper the water. A microchip can work out the depth and display it on a screen.

Absorbing sounds

Empty rooms sound echoey. The walls reflect the smallest sound, and it may take several seconds for the wave energy to be absorbed so that the sound dies away.

Echoes can be a nuisance in concert halls. Soft materials like carpets and curtains help to absorb sound waves, as do the clothes of the audience. Many large concert halls have specially designed sound absorbers on the ceiling to make them less echoey.

The 'mushrooms' on the ceiling of this concert hall are sound absorbers to reduce unwanted echoes.

Assume that the speed of sound in air is 330 m/s.

1 How could you show that sound travels in the form of waves?

2 Give a reason for each of the following:
(a) You can hear sound coming from the next room, even though all the doors and windows are tightly shut.
(b) Sound cannot travel through a vacuum.

3 **(a)** Why do you hear lightning after you see it?
(b) if you hear lightning 2 seconds after you see it, how far away is the lightning?

4 Chris shouts when he is 110 metres from a wall. When will he hear his echo?

4.25 Sound and ultrasound

By the end of this spread, you should be able to:
- *describe what features of sound waves affect loudness and pitch*
- *describe some of the uses of ultrasound.*

Sounds different

Some sounds are louder than others. Some sounds are higher than others. To see how different sounds compare, you can use a microphone and CRO, as explained in the previous spread.

Amplitude and loudness The height of a peak or trough on the screen is called the ***amplitude***. The higher the amplitude, the more energy the waves carry, and the louder the sound will be.

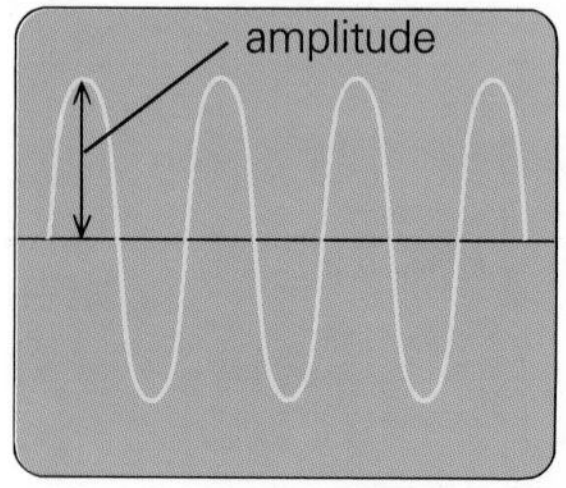

This sound is louder...

amplitude

...than this

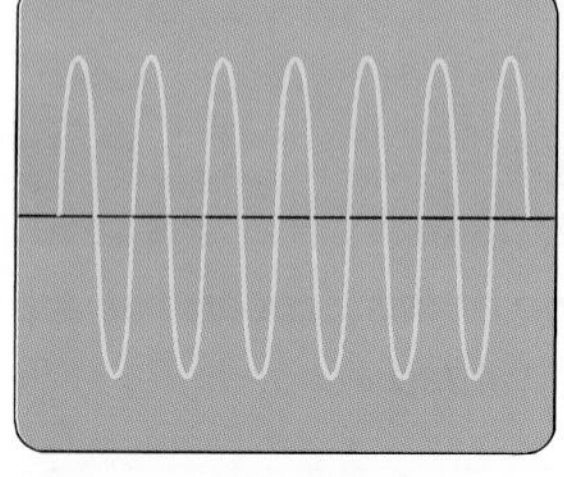

This sound has a higher pitch (and frequency)...

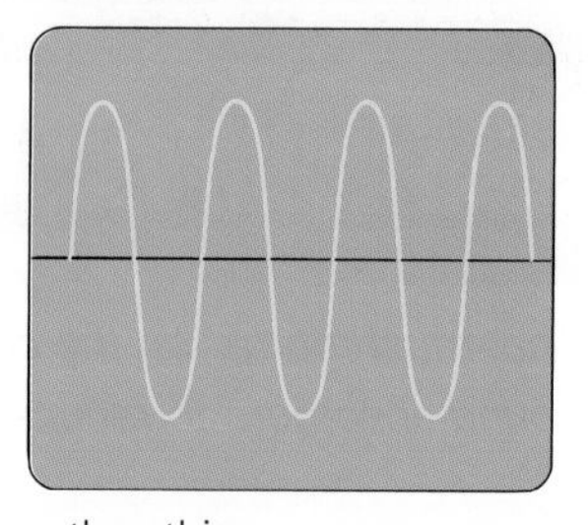

...than this

Frequency and pitch The ***frequency*** of a sound is the number of sound waves being sent out per second.

Frequency is measured in ***hertz (Hz)*** (see Spread 4.23). If a sound has a frequency of, say, 100 Hz, then 100 sound waves are being sent out every second.

The higher the frequency, the higher the note sounds. Musicians say that it has a higher ***pitch***.

If the frequency increases, you see more waves on the CRO screen. The waves are closer together.

		ultrasound	
high frequency	20 000 Hz	highest note heard by human ear	high pitch
	10 000 Hz	whistle	
	1000 Hz	high note from singer	
	100 Hz	low note from singer	
low frequency	20 Hz	drum	low pitch

The human ear can detect sounds up to a frequency of about 20 000 Hz. Sounds above the range of human hearing are called ***ultrasonic sounds***, or ***ultrasound*** (see the next page).

The ***kilohertz (kHz)*** is a useful unit for measuring higher frequencies:

1 kilohertz (kHz) = 1000 Hz

For example: 20 kHz means 20 000 Hz.

The wave equation

The following equation applies to all waves, including sound waves (see also Spread 4.23):

$$\underset{\text{in m/s}}{\text{speed}} = \underset{\text{in Hz}}{\text{frequency}} \times \underset{\text{in m}}{\text{wavelength}}$$

For example, if a sound has a frequency of 110 Hz, and the speed of sound is 330 m/s, you could use the above equation to work out its wavelength:

$$330 = 110 \times \text{wavelength}$$

So the wavelength is 3 m

The higher the frequency, the shorter the wavelength.

Using ultrasound

Here are some examples of how ultrasound is used:

Cleaning Delicate machinery can be cleaned using ultrasound. The machinery is immersed in a tank of liquid, and then the vibrations of high-power ultrasound are used to dislodge the bits of dirt and grease.

Stone breaking In hospitals, concentrated beams of ultrasound can be used to break up kidney stones and gall stones without patients needing surgery.

Echo-sounding The echo-sounders in boats (see Spread 4.24) send and receive pulses of ultrasound.

This bat uses ultrasound to locate insects and other objects. It sends out ultrasounds pulses, then uses its specially shaped ears to pick up the reflections.

1 A sound has a frequency of 200 Hz. What does this tell you about the sound waves?
2 What difference will you hear in a sound if there is an increase in
(a) the amplitude? **(b)** frequency?
3 What is ultrasound?
4 Give *two* examples of the medical use of ultrasound.
5 A boat is fitted with an echo-sounder which uses ultrasound with a frequency of 35 kHz.
(a) What is the frequency in Hz?
(b) If the speed of sound in water is 1400 m/s, what is the wavelength of the ultrasound?
(c) If the frequency were doubled, how would this affect the wavelength?

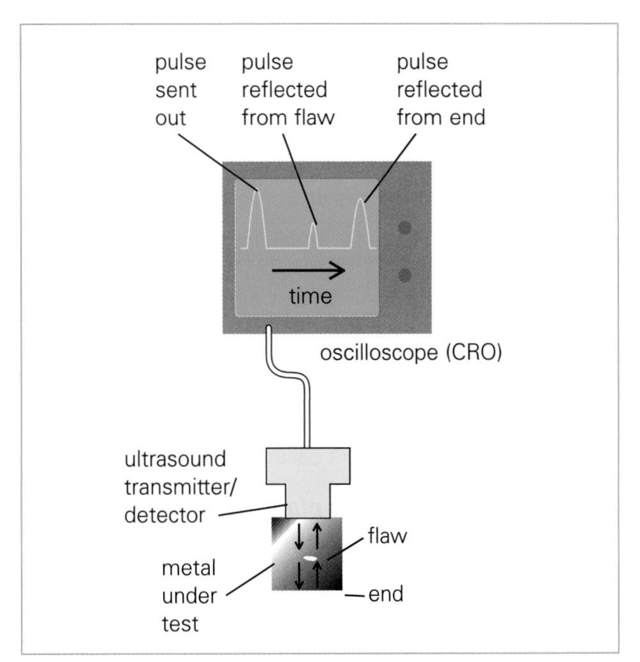

▲**Metal testing** Metals can be tested for flaws using the echo-sounding principle. A pulse of ultrasound is sent through the metal as above. If there is a flaw (tiny gap) in the metal, two reflected pulses are picked up by the detector. The pulse reflected from the flaw returns first, followed by the pulse reflected from the far end of the metal. The pulses can be displayed using an oscilloscope.

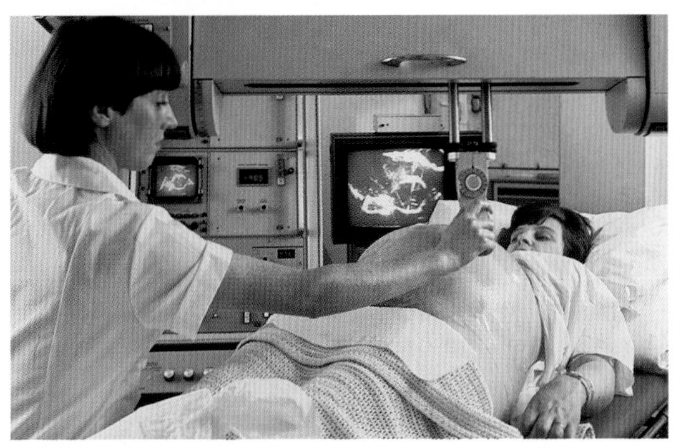

▲**Scanning the womb** The pregnant mother above is having her womb scanned by ultrasound, using the echo-sounding principle. A transmitter sends pulses of ultrasound into the mother's body. The transmitter also acts as a detector and picks up pulses reflected from the baby and different layers inside the body. The signals are processed by a computer, which puts an image on the screen. The method is safer than using X-rays because X-rays can damage cells inside a growing baby.

4.26 Plates and seismic waves

By the end of this spread, you should be able to:
- *describe the theory of plate tectonics*
- *explain what can be learnt from seismic waves.*

To find out what is happening inside the Earth, scientists study the waves from earthquakes.

Beneath the Earth's ***crust*** (outer layer), there is a deep zone of hot, flexible rock called the ***mantle*** (see Spread 3.30). In some parts of the mantle, the rock is molten. The molten rock is called ***magma***.

According to the theory of ***plate tectonics***, the crust (and upper mantle) is made up of huge sections called ***plates*** which 'float' on the denser material beneath. Convection currents in the lower mantle make the plates move very slowly – just a few centimetres per year. The process is driven by heat released by radioactive materials in the mantle.

Plates meet at ***plate boundaries***:

Constructive boundaries These are mainly under oceans. Plates move apart and get bigger as magma wells up between them, and then cools and solidifies to form new crust. The effect is called ***sea-floor spreading*** and it produces ***oceanic ridges***.

Destructive boundaries Plates move together so that one is ***subducted*** (carried down) under the other. This causes intense heat and volcanic activity. If the sliding is not smooth, there are earthquakes.

Conservative boundaries Plates slide past each other, so their shape is 'conserved' – it does not change. Sometimes, the plates catch on each other. When they jerk free, there may be big earthquakes.

Evidence for plate movements
- The radioactivity of materials trapped in rocks weakens with time. From this, scientists can work out when the rock first solidified (see Spread 4.31). Results show that the age of the crust increases as you move further out from an oceanic ridge – evidence of outward movement.
- By comparing the magnetism in rocks, scientists can work out which rocks were once close. This provides evidence that plates move.

Constructive boundary

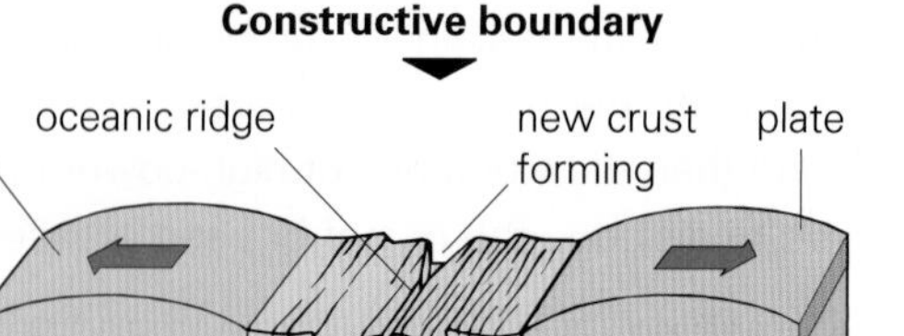

Destructive boundary

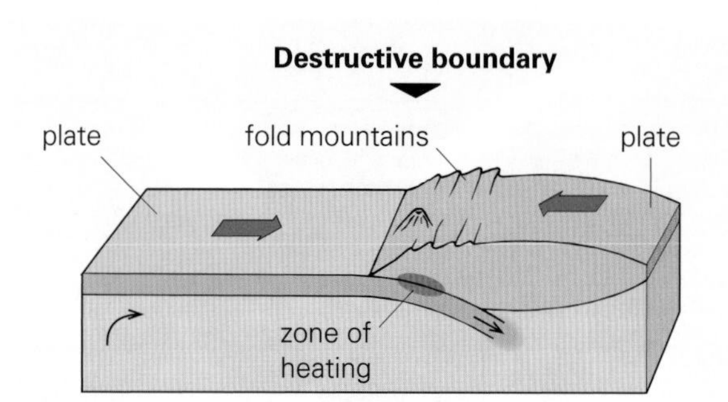

Conservative boundary

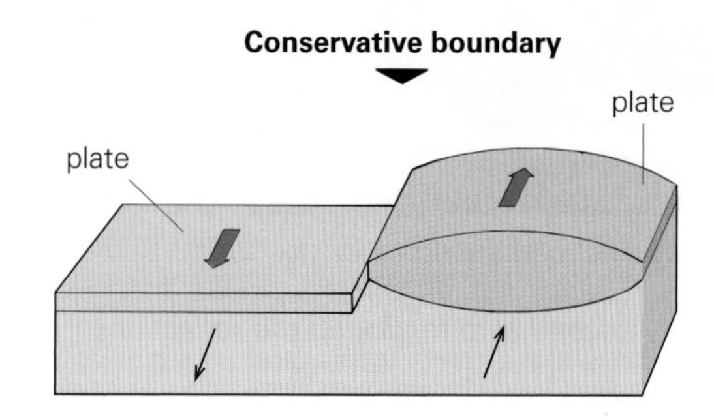

The Earth

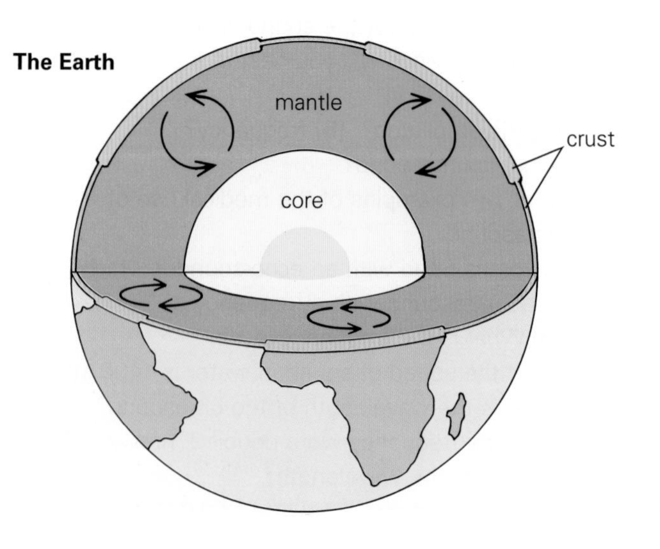

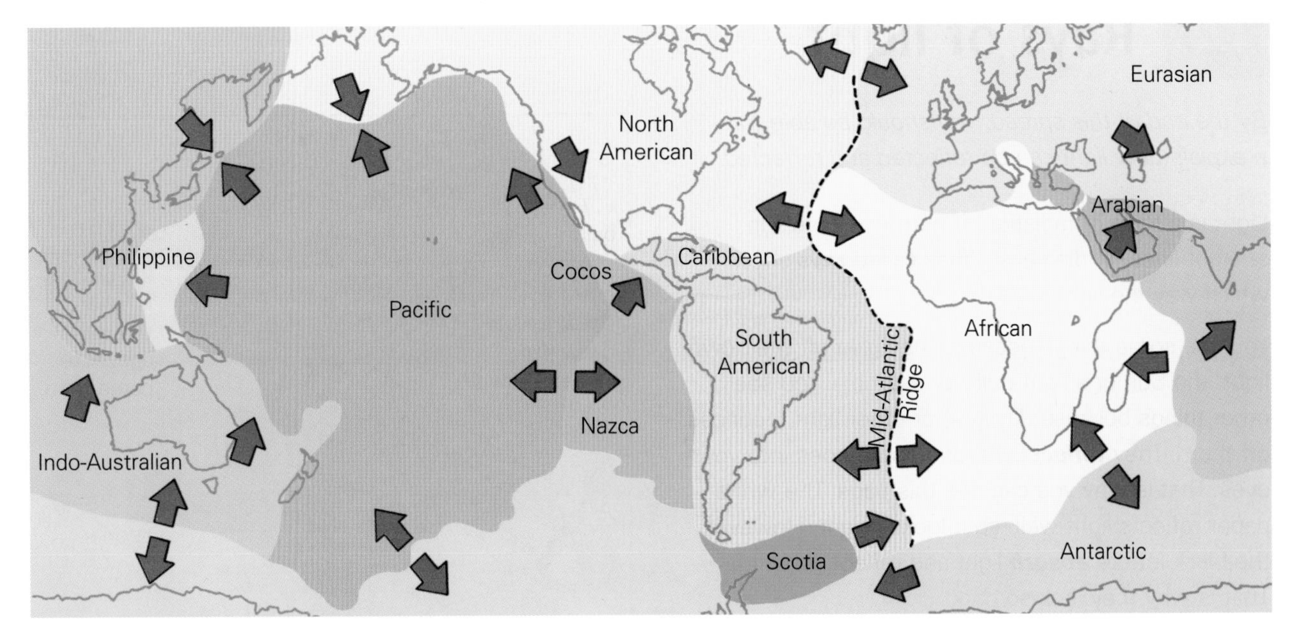

The main plates and directions of movement

Waves from earthquakes

Earthquakes are caused by sudden rock movements in the ground. When they occur, vibrations called ***seismic waves*** travel outwards from the ***focus*** (the site of the earthquake):

P-waves are 'push and pull' waves (longitudinal waves, see Spread 4.23). They can travel through solids and liquids deep in the Earth.

S-waves are 'up and down' waves (transverse waves, see Spread 4.23). They are slower than P-waves, and cannot travel through liquids such as molten rock.

L-waves produce a rolling motion. They are the most destructive, but only travel through surface rocks. They are slower than S-waves.

A ***seismometer*** can detect and measure seismic waves which have travelled many thousands of kilometres from their source. By analysing travel times, scientists can work out what routes the waves must have followed through the Earth. Their results give clues about the Earth's inner structure:

- As rock gets more dense, seismic waves speed up. When their speed changes, they refract (bend) rather like light waves. The curved paths of the waves and the time they take to travel suggest that the Earth has a core which is very dense.
- No S-waves travel through the core. This suggests that the outer part of the core must be liquid.

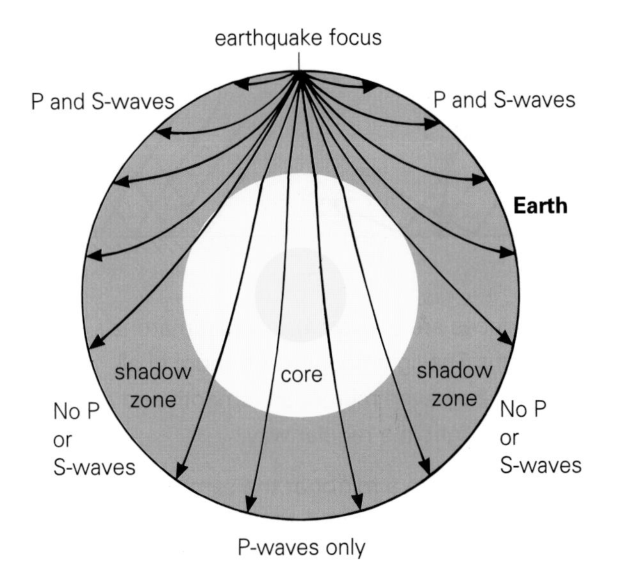

1. Why do earthquakes and volcanoes mainly occur near plate boundaries?
2. What is the difference between a destructive boundary and a constructive boundary?
3. Which will a seismometer detect first: P-waves or S-waves? Why?
4. What evidence is there that part of the Earth's core is liquid?

4.27 Rays of light

By the end of this spread, you should be able to:
- *explain how light can be reflected and refracted.*

Light is a form of radiation. It normally travels in straight lines. In diagrams, lines called ***rays*** show which way the light is going.

You see some things because they give off their own light: the Sun or a light bulb for example. You see other things because daylight, or other light, bounces off them. They ***reflect*** light, and some goes into your eyes. That is why you can see this page. The white paper reflects light well, so it looks bright. However, the black letters ***absorb*** light and reflect very little. That is why they look so dark.

Reflection and mirrors

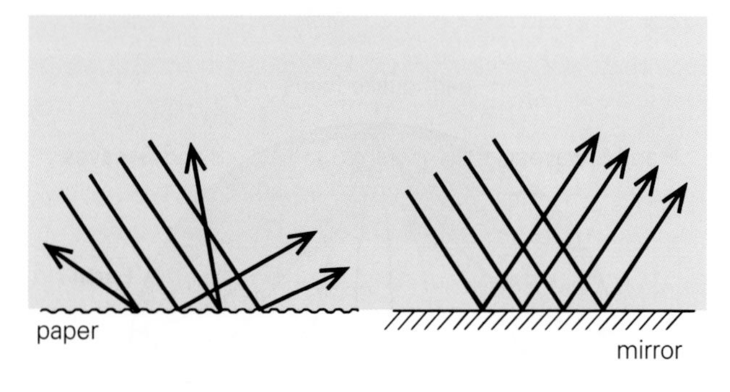

Most surfaces are uneven, or contain materials which scatter light. The light bounces off them in all directions. However, mirrors are smooth and shiny. They reflect light in a regular way.

Light reflects from a mirror at the same angle as it strikes it, as shown in the diagram below:

angle of incidence = angle of reflection

A line at right-angles to the surface of a mirror is called a ***normal***.

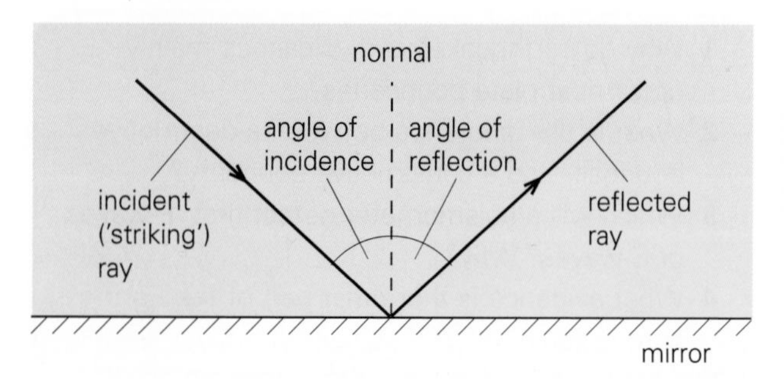

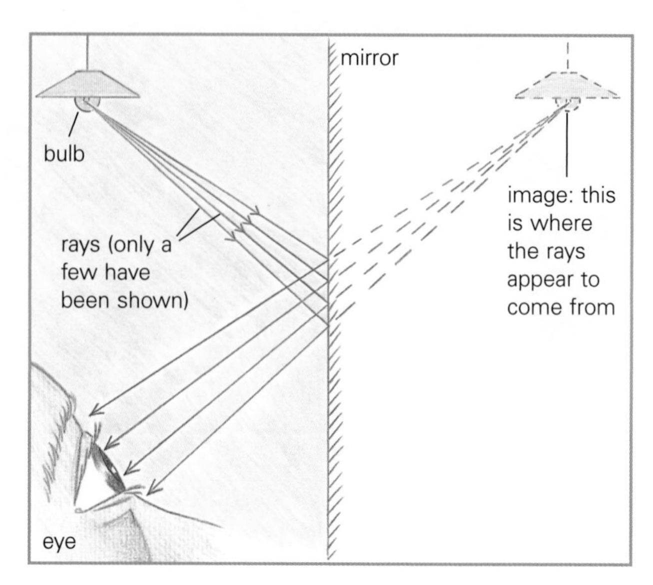

Above, light from a bulb is being reflected by a mirror. Some is reflected into the girl's eye. To the girl, the light seems to come from a point behind the mirror. She sees an ***image*** of the bulb in that position. The image is the same size as the original bulb, and the same distance from the mirror. However it is ***laterally inverted*** (back-to-front).

Refraction

The light passing through the glass block below has been bent. The bending is called ***refraction***: It happens with other transparent materials as well.

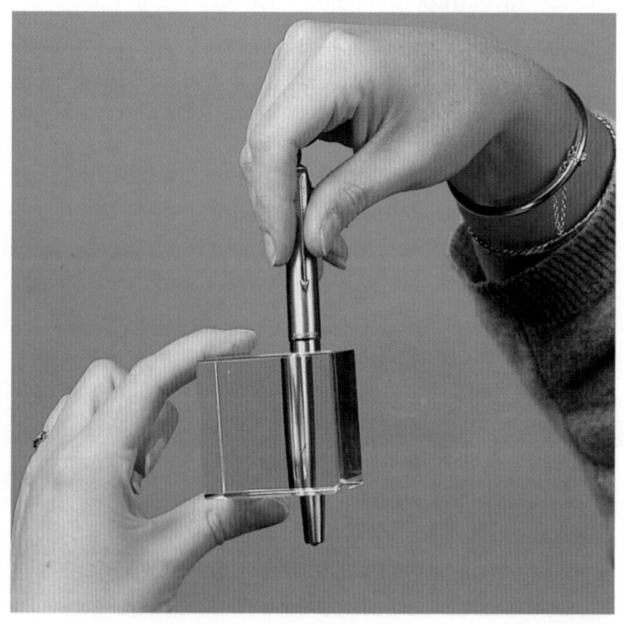

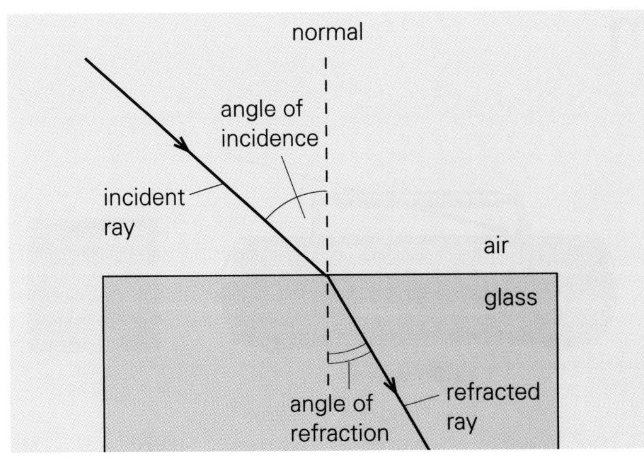

When a ray of light enters a glass block as above, it is refracted towards the normal. Water or clear plastic would have a similar effect. Below, you can see how scientists explain refraction. Light is made up of tiny waves (see 4.23 and 4.28). These travel more slowly in glass than in air. One side of the light beam is slowed before the other. This makes the light waves bend.

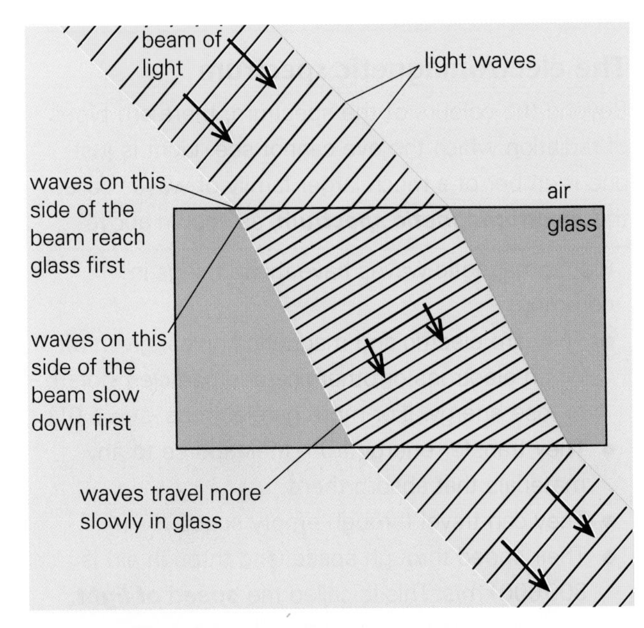

Internal reflections

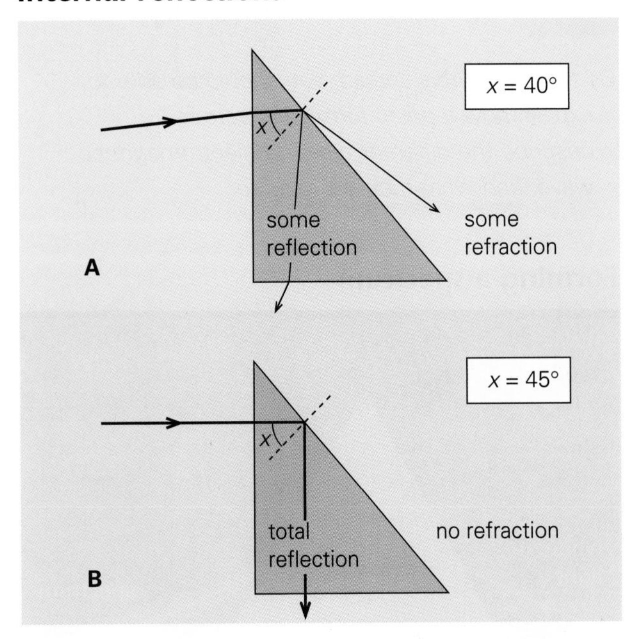

In diagram A above, a ray of light passes through a triangular glass block called a ***prism***. Some light is reflected at point P and some is refracted.

In diagram B, angle x is larger than before. This time, all the light is reflected at P and none is refracted. The inside face of the prism is acting like a perfect mirror. The effect is called ***total internal reflection***.

For glass, total internal reflection only occurs if angle x is more than 41°. This is the ***critical angle*** for glass. The value is different for other materials:

Critical angle

glass (crown)	41°	acrylic plastic	42°
water	49°	diamond	24°

Total internal reflection is used in ***optical fibres*** (see Spread 4.29).

1 A wall reflects light. So does a mirror. What difference is there in the way they reflect light?

2 If you hold a pencil 50 cm in front of a flat mirror, where do you see an image?

3 **(a)** Copy and complete the diagrams on the right to show what happens to each ray.
(b) In which diagram does total internal reflection occur?

4 Why does a beam of light bend when it enters glass at an angle?

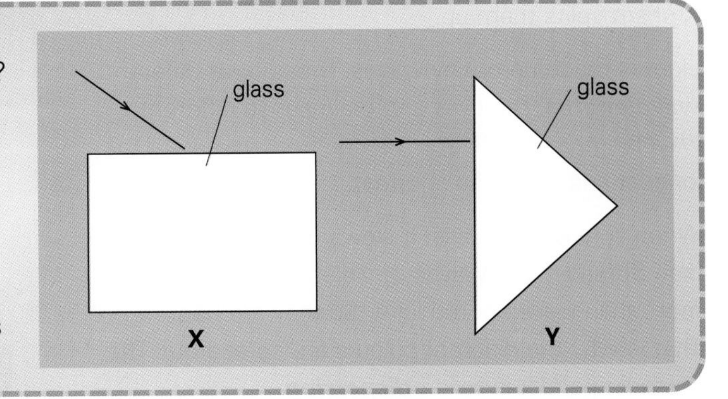

4.28 Across the spectrum

By the end of this spread, you should be able to:
- explain how a prism forms a spectrum
- describe the different types of electromagnetic wave, and what they are used for.

Forming a spectrum

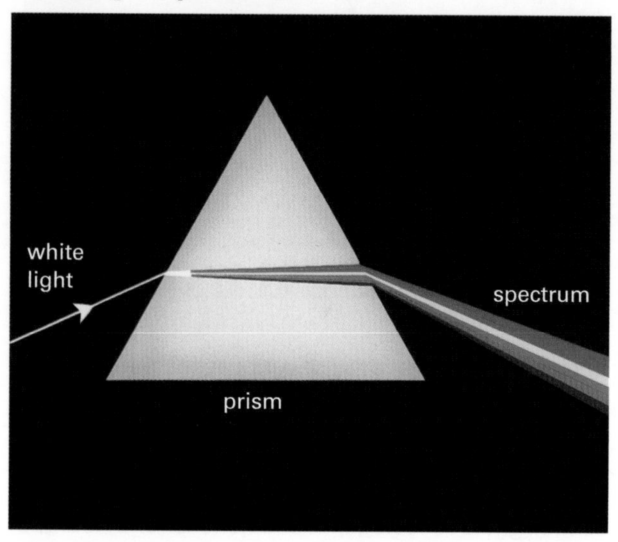

A narrow beam of white light enters a triangular glass block called a ***prism***. The light is refracted (bent) as it goes into the prism, and again as it comes out. The refracted light spreads slightly to form a range of colours called a ***spectrum***.

Most people think they can see six colours in the spectrum: red, orange, yellow, green, blue, and violet.

But really, there is a continuous change of colour from one end to the other.

How a spectrum is formed

White is not a single colour, but a mixture of colours. A prism splits them up.

Light is made up of tiny waves. These have different ***wavelengths***. The eyes and brain sense different wavelengths as different colours. Red waves are the longest and violet the shortest.

When light enters glass, it slows down and bends (see Spread 4.27). Waves of violet light slow down more than waves of red light. So they are bent more. That is why the different colours are spread out. The spreading effect is called ***dispersion***.

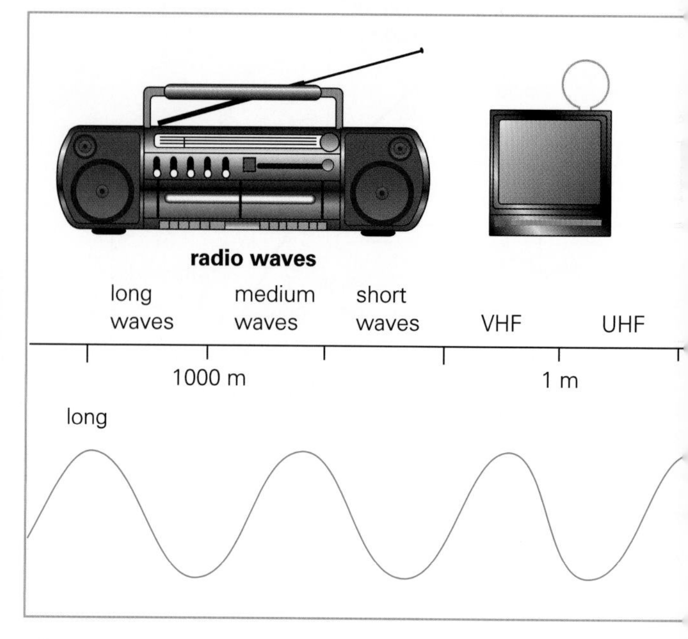

The electromagnetic spectrum

Beyond the colours of the spectrum, there are types of radiation which the eye cannot see. Light is just one member of a much larger family of waves called the ***electromagnetic spectrum***, as shown above.

Electromagnetic waves have these things in common:

- They are electric and magnetic ripples, given off when electrons or other charged particles vibrate or lose energy. (For more on electrons, see 4.01.)
- They transfer energy from their source to any materials that absorb them.
- They can travel through empty space.
- Their speed through space (and through air) is 300 000 km/s. This is called the ***speed of light***.
- As with other types of wave, this equation applies:

  speed (in m/s) = frequency (in Hz) × wavelength (in m) (see 4.23)

  So, the higher the frequency of the waves, the shorter their wavelength.

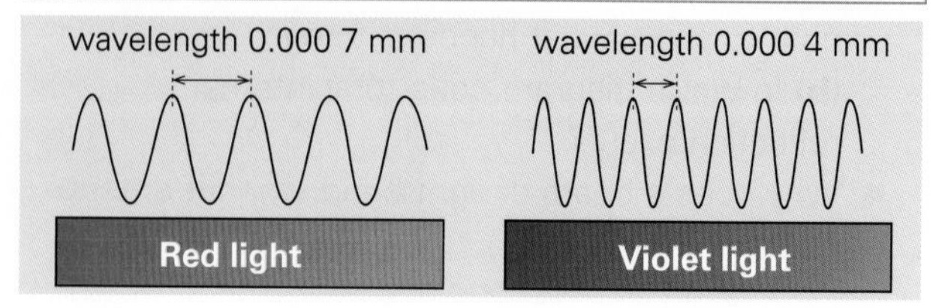

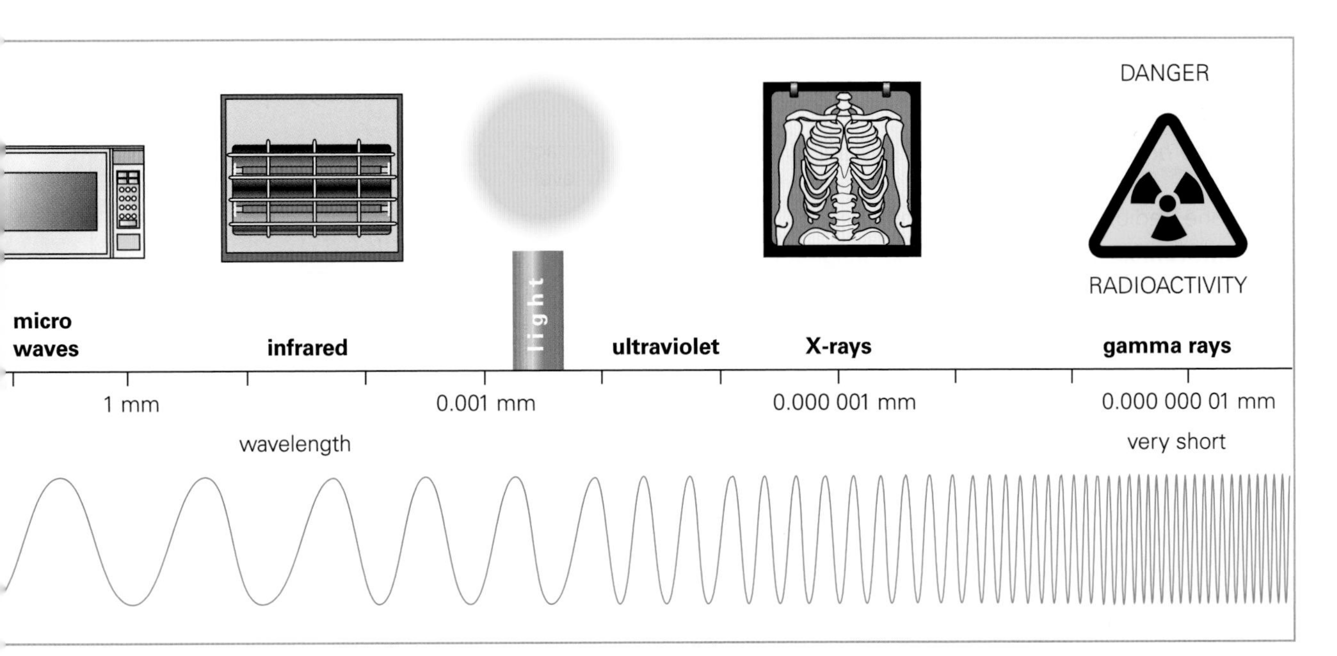

Radio waves can be produced by making electrons vibrate in an aerial. They can carry signals which tell a radio or TV what sounds or pictures to make.

Microwaves are radio waves of very short wavelength. They are used for radar and in telecommunications (see the next spread, 4.29). Some microwaves are absorbed by food. This makes the food hot. The idea is used in microwave ovens.

Infrared Hot things like fires and radiators all give off infrared radiation. In fact, everything gives off some infrared. If you absorb it, it heats you up.
TV remote controllers transmit instructions using pulses of infrared. In telecommunications, optical fibres carry pulses of infrared (see Spread 4.29).

Light This is the visible part of the spectrum – the only part which the eye can detect.

1 Which colour of light **(a)** is refracted most by a prism? **(b)** has the longest wavelength?

2 *gamma rays radio waves microwaves X-rays light ultraviolet infrared*
Which of the above **(a)** can be detected by the eye? **(b)** are used for communications **(c)** are used in cooking? **(d)** can pass through flesh? **(e)** can damage cells deep in the body? **(f)** have the highest frequency?

Ultraviolet Some lamps produce ultraviolet. In fluorescent lamps, the tube is coated with a white powder which glows and gives off light when it absorbs ultraviolet from 'electrified' gas in the tube.

Sunlight contains ultraviolet. In people like this it produces a tan, but too much can cause skin cancer.

X-rays Shorter wavelengths can penetrate dense metals. Longer wavelengths can pass through flesh, but not bone. So they can be used to take 'shadow' photographs of bones. Only brief bursts of X-rays must be used for this because X-rays can kill living cells deep in the body or make them cancerous.

Gamma rays come from radioactive materials (see Spread 4.30) and have the same effects as X-rays. In hospitals, they can be used for sterilizing medical instruments because they kill germs. Concentrated beams are used to kill cancer cells.

4.29 Sending signals

By the end of this spread, you should be able to:
- *explain how information can be sent using wires, optical fibres, or radio waves*
- *describe how analogue and digital signals differ.*

Telephone, radio, and TV are all forms of ***telecommunication*** – ways of transmitting (sending) information long distances. The information may be sounds, pictures, or computer data.

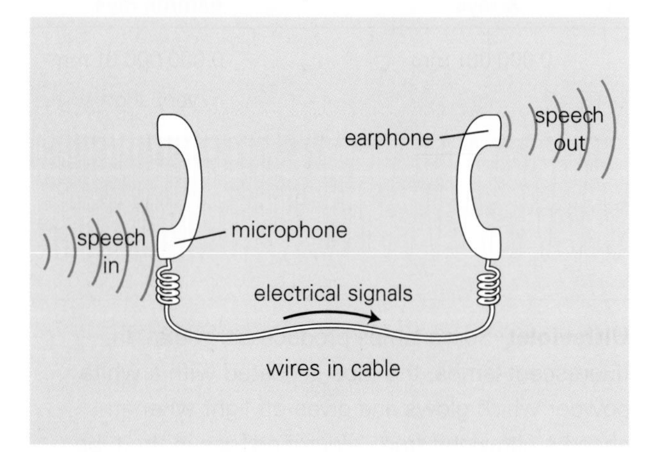

The diagram shows a simple telephone system. The microphone turns the incoming information (speech) into a changing electric current. The changes, called ***signals***, pass along wires to the earphone. This turns them back into useful information (speech). In a real system, ***amplifiers*** are used to boost the strength of the electrical signals.

The signals used by telecommunications systems can be changes in current, changes in the intensity of a beam of light, or changes in the strength or frequency of radio waves. They may be transmitted using wires, optical fibres, or radio waves (see the next page).

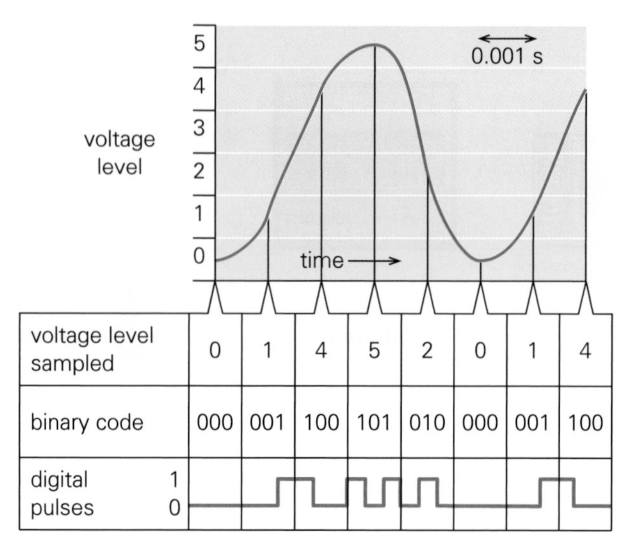

voltage level sampled	0	1	4	5	2	0	1	4
binary code	000	001	100	101	010	000	001	100
digital pulses 1 0								

Analogue and digital signals

The sound waves entering a microphone make the current through it vary – as shown in the graph above. A continuous variation like this is called an ***analogue signal***. The table shows how it can be converted into ***digital signals*** – signals represented by numbers. The original signal is ***sampled*** electronically thousands of times per second. In effect, the height of the graph is measured repeatedly, and the measurements changed into ***binary codes*** (numbers using only 0s and 1s). These are transmitted as a series of pulses and turned back into analogue signals at the receiving end.

Advantages of digital transmission Compared with analogue signals, digital signals can carry more sets of information per second along a cable or radio link. Also digital signals deliver better quality. Both types of signal are spoilt by interference as they travel along. However, digital pulses can be 'cleaned up' and their quality restored.

1 Telephone systems sometimes make use of optical fibres:
(a) In what form do the signals travel along the fibre?
(b) Give *two* advantages of sending digital signals rather than analogue ones.
(c) Give *two* advantages of using an optical fibre link rather than a cable with wires in it.

2 Explain why, if you use a radio down in a valley, AM reception may be good, but FM poor.

3 A radio station broadcasts on a frequency of 100 MHz. The speed of radio waves is 300 000 km/s.
(a) What is the frequency in Hz?
(b) What is the speed of the waves in m/s?
(c) Use the wave equation in Spread 4.28 to calculate the wavelength of the radio waves.

Using optical fibres

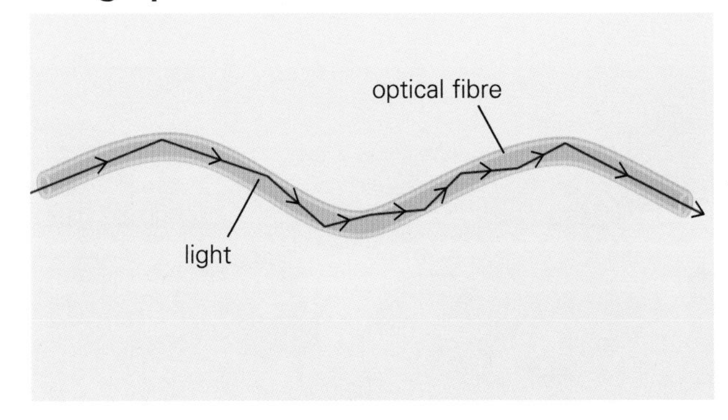

Optical fibres are thin, flexible strands of glass or plastic. They make use of total internal reflection (see Spread 4.27). When light enters one end of a fibre, it is reflected from side to side, until it comes out of the other end. The diagram above shows the principle. High-quality fibres have an outer layer of glass or plastic to protect the core which carries the light.

Telephone networks often use ***optical fibres*** for long-distance transmission. At the sending end, electrical signals are turned into light signals (usually pulses of infrared). These travel along the fibre. At the receiving end, they are turned back into electrical signals.

Optical fibre cables are thinner and lighter than electric cables. They can carry more signals, with less loss of power. Most of the telephone calls between Britain and the United States are carried by an optical fibre cable under the Atlantic Ocean.

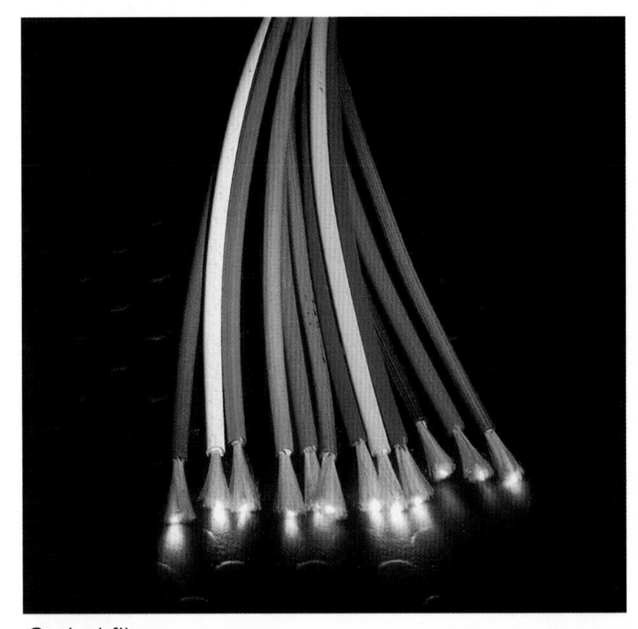

Optical fibres

Further information

Frequency

1 kilohertz (kHz) = 1000 waves per second

1 megahertz (MHZ) = 1000 000 waves per second

For more about...	*See Spread...*
frequency, wavelength, diffraction	4.23
total internal reflection	4.27
radio waves, light, and infrared	4.28
communications satellites	4.33

Using radio waves

If a high-frequency alternating current is fed to an aerial, radio waves are given off. Signals from an ***aerial*** connected to the ***transmitter*** are picked up by another aerial connected to the ***receiver***. Radio signals can be used for transmitting pictures as well as sounds, so the receiver could be a radio, TV, or mobile phone.

Here are examples of radio waves and their uses:

Typical frequency	Description and typical uses
200 kHz 1 MHz	long wave: AM radio medium wave: AM radio *Long and medium waves diffract (bend) round hills. Some medium waves are also reflected by the ionosphere – a layer of charged particles in the Earth's upper atmosphere. This increases their range.*
100 MHz 500 MHz	VHF (very high frequeny): FM radio UHF (ultra high frequency): terrestrial TV *VHF and UHF are blocked by hills, so there must be an unbroken straight line from the transmitter to the receiver.*
10 000 MHz	microwaves: satellite communication, telephone links *Microwaves are beamed between dish aerials.*

Satellite communication

Communications satellites are used to relay (pass on) radio signals. They pick up signals from dish aerials on the ground, amplify them, and beam them back to a different area of the Earth.

4.30 Radioactivity

By the end of this spread, you should be able to:
- *explain why some substances are radioactive*
- *describe the three main types of nuclear radiation and their effects.*

Isotopes

Atoms of the same element are not all alike. Elements can exist in different versions, with different numbers of neutrons in the nucleus. These different versions are called ***isotopes*** (see Spread 3.05). For example:

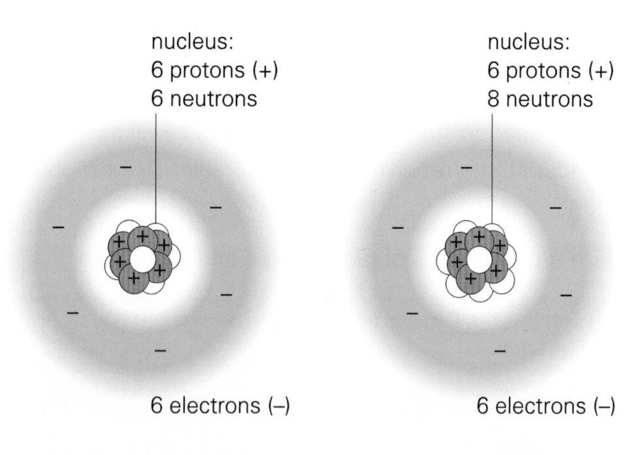

Most carbon atoms have 6 protons and 6 neutrons in the nucleus. This common isotope is called carbon-12 (12 is the total of protons plus neutrons). But some carbon atoms have 6 protons and 8 neutrons. This rare isotope is carbon-14.

Nuclear radiation

Some isotopes have atoms with unstable nuclei. In time, the nucleus breaks up and shoots out a tiny particle and often a burst of wave energy as well. These 'radiate' from the nucleus. They are ***nuclear radiation***. If an isotope gives out nuclear radiation, scientists say that it is ***radioactive***. It is known as a ***radioisotope*** (or ***radionuclide)***

Some of the materials in nuclear power stations are highly radioactive. But nuclear radiation come from many natural sources as well, as shown on the right. This means that there is a small amount of ***background radiation*** around us all the time.

Containers for radioactive waste must be strong enough to withstand crashes like this.

Ionizing effect

Nuclear radiation can remove electrons from atoms in its path. In other words it can make ions (see Spread 3.07): it has an ***ionizing*** effect. Ionizing radiation can be very dangerous. It may stop cells in vital organs working properly. It can also damage the chemical instructions in normal cells so that the cells grow abnormally and cause cancer. However, in ***radiotherapy*** treatment, carefully directed radiation (gamma rays) is used to kill cancer cells.

Isotopes		
Stable	*Unstable, radioactive*	*Found in...*
carbon-12 carbon-13	carbon-14	air, plants, animals
potassium-39 potassium-41	potassium-40	rocks, plants, sea-water
	uranium-234 uranium-235 uranium-238	rocks

Alpha, beta, and gamma

There are three main types of nuclear radiation: ***alpha*** particles, ***beta*** particles, and ***gamma*** rays. They can be detected by a ***Geiger–Müller tube (GM tube)***, connected to an electronic counter or meter.

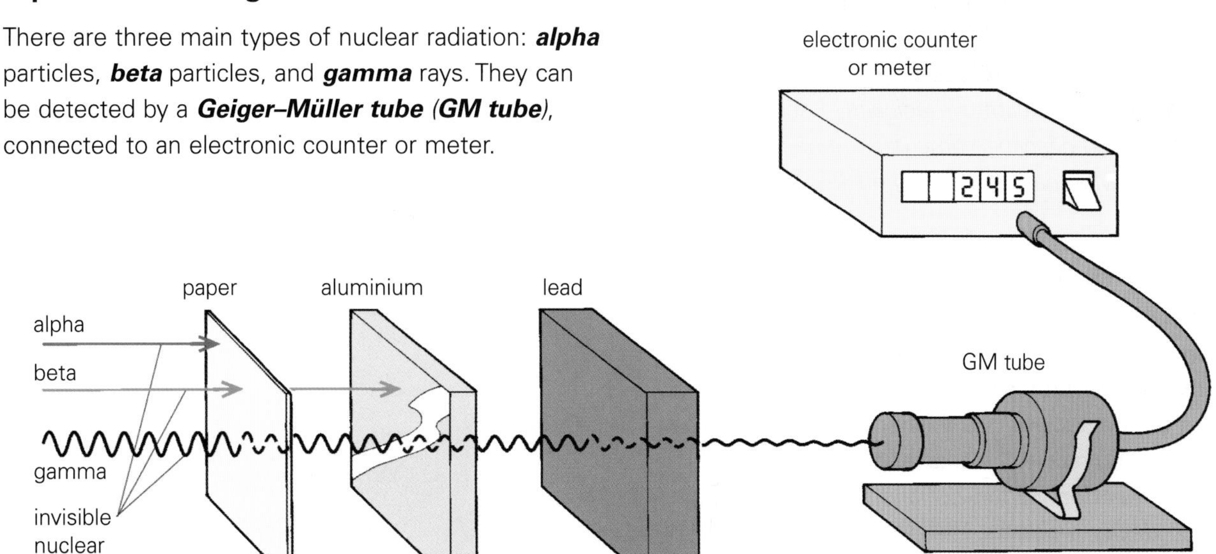

Nuclear radiation	**Alpha particles**	**Beta particles**	**Gamma rays**
	Each particle is 2 protons + 2 neutrons	Each particle is an electron (formed when the nucleus breaks up)	Electromagnetic waves similar to X-rays (see Spread 4.28)
Electric charge	+	–	No charge
Ionizing effect	Strong	Weak	Very weak
Penetrating effect	Not very penetrating: stopped by thick piece of paper, or skin	Penetrating: stopped by thick sheet of aluminium	Highly penetrating: never completely stopped, though lead and very thick concrete reduce strength

1 Comparing atoms of *carbon-12* and *carbon-14:*
(a) What do the numbers '12' and '14' tell you?
(b) In what ways are the atoms the same?
(c) How are the atoms different?

2 If a substance is *radioactive*, what does this mean?

3 Nuclear radiation has an *ionizing effect*. What does this mean?

4 Why can ionizing radiation be dangerous?

5 What are the three main types of nuclear radiation?

6 Which type of radiation is stopped by skin or thick paper?

7 Which type of radiation can penetrate lead?

8 Which type of radiation is most ionizing?

9 Explain what is meant by *background radiation.*

4.31 Decay and fission

By the end of this spread, you should be able to:
- *explain what is meant by 'radioactive decay' and 'half-life' and describe some uses of radioactivity*
- *describe what happens during nuclear fission.*

Radioactive decay

The break up of unstable nuclei is called ***radioactive decay***. When radium-226 decays (see right) it loses 2 neutrons and 2 protons as an alpha particle. So it becomes the nucleus of a different element, the new isotope being radon-222. this is an example of a ***nuclear reaction***. Radon-222 and the alpha particle are the ***decay products***. The reaction can also be written as a nuclear equation. Can you work out what each number and symbol in the equation stands for?

In radioactive decay, energy is released, and the products move faster than the original atoms. For example, it is the decay of radioactive materials in the rocks which keeps temperatures high underground.

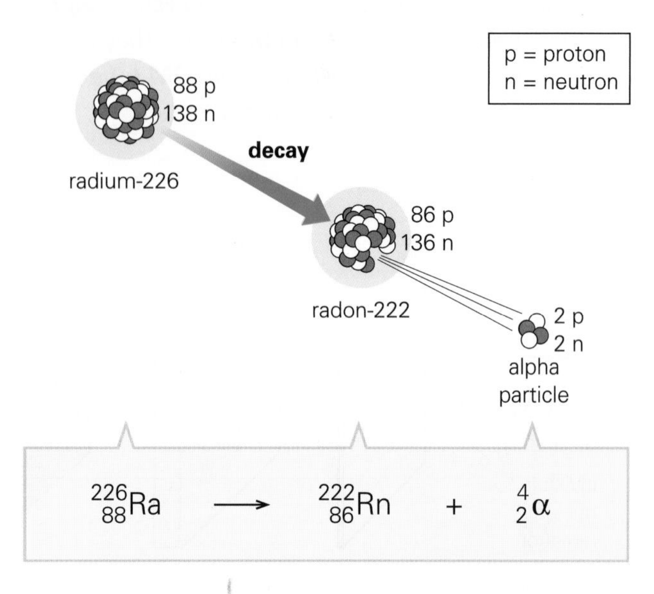

$$^{226}_{88}\mathrm{Ra} \longrightarrow \, ^{222}_{86}\mathrm{Rn} + \, ^{4}_{2}\alpha$$

Half-life

Radioactive decay is a random process. You cannot tell which nucleus is going to break up next, or when. But some types of nucleus are more unstable than others. They decay at a faster rate.

The graph on the right shows the decay of a sample of iodine-128. The number of nuclei decaying per second is called the ***activity***. It is measured in ***becquerel (Bq)***. As time goes on, there are fewer and fewer unstable nuclei left to decay, so the activity gets less and less. After 25 minutes, half the unstable nuclei have decayed, so the activity has halved. After another 25 minutes, the activity has halved again...and so on. Iodine-128 has a ***half-life*** of 25 minutes.

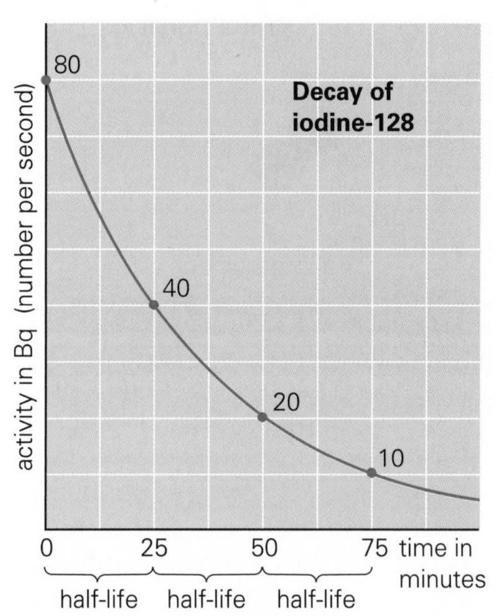

Radioisotope	Half-life
radon-222	3.8 days
strontium-90	28 years
radium-226	1602 years
carbon-14	5730 years
plutonium-239	24 400 years
uranium-235	710 000 000 years

Fission

Natural uranium is a dense, radioactive metal. It is mainly a mixture of two isotopes: uranium-238 (over 99%) and uranium-235 (less than 1%). Both isotopes decay slowly. However, nuclei of uranium-235 can be split by neutrons. This process is called ***fission***. It can release energy very quickly, like this:

A neutron strikes a uranium-235 nucleus, making it split it into two roughly equal parts. Two or three neutrons are shot out as well. If these hit other uranium-235 nuclei, they split and give out more neutrons...and so on in a ***chain reaction***.

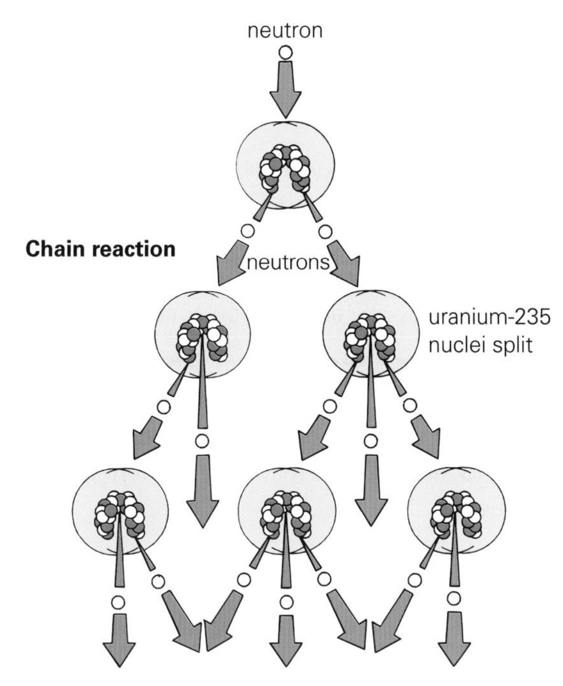

In an uncontrolled chain reaction, huge numbers of nuclei are split in a fraction of a second. The heat builds up so rapidly that the material bursts apart in an explosion. This happens in a nuclear bomb.

In a controlled chain reaction, the flow of neutrons is regulated so that there is a steady output of heat. This happens in a nuclear reactor. The power station below uses a ***nuclear reactor*** as the heat source for its boilers (see Spread 4.20). There is no burning fuel to pollute the atmosphere. But a small amount of radioactive waste is produced. Some of this waste will need to be stored for hundreds of years before its activity falls to a safe level.

Using radioactivity

Radioactive dating can be used to find the age of rocks. When rocks are formed, natural radioisotopes become trapped in them – along with their decay products. For example, in mica rock, over millions of years, the amount of radioactive potassium-40 decreases as the amount of its decay product, argon-40, increases. By measuring the ratio of potassium to argon, the age of the rock can be estimated.

Radioactive tracers are small (and safe) amounts of radioisotopes whose movement can be tracked. For example, to check how well the thyroid gland takes up iodine, a patient can be given a drink with iodine-131 in it. Over the next day, the activity of the tracer is measured to see if it is building up in the thyroid.

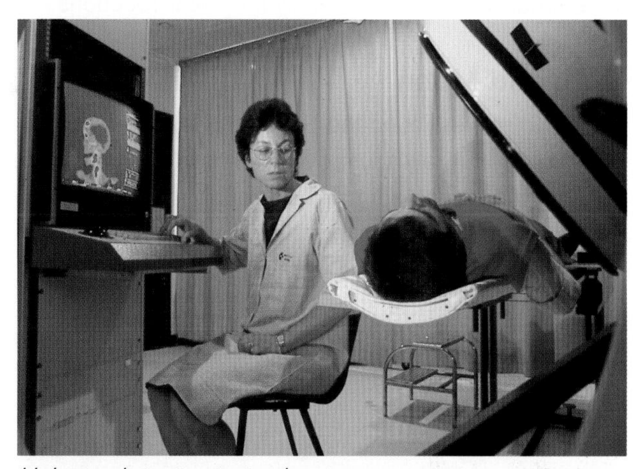

Using a detector to track a tracer

1 Radium-226 decays by emitting an alpha particle. **(a)** What happens during the decay? **(b)** Why is a different element formed?

2 Strontium-90 has a *half-life* of 28 years. What does this mean?

3 Here are some measurements of the activity of a small sample of iodine-131:

Time in days	0	4	8	12
Activity in Bq	240	170	120	85

(a) What is the half-life of iodine-131?
(b) After how many days would you expect the activity of the sample to be 30 per second?

4 Why do you think that radioisotopes used as tracers should have short half-lives?

5 Give an example of a *chain reaction*. Where is a controlled chain reaction used?

4.32 Sun, Earth, Moon, and sky

By the end of this spread, you should be able to:
- *explain why the Earth has day and night*
- *describe how stars and planets appear from Earth*
- *describe the orbits of the Moon and Earth*
- *describe the effects of gravity.*

The Sun

The Sun is a huge, hot, brightly glowing ball of gas, called a ***star***. It is 150 000 000 kilometres away from us. Its diameter, 1.4 million kilometres, is more than a hundred times the Earth's. The Sun is extremely hot: 6000 °C on the surface, rising to 15 000 000 °C in its core, where the heat comes from nuclear reactions.

Earth and Moon

The Earth moves round the Sun in a path called an ***orbit***. One orbit takes just over 365 days, which is the length of our year. As it moves through space, the Earth spins slowly on its axis once a day. This gives us day and night as we move from the sunny side facing the Sun to the dark side away from it.

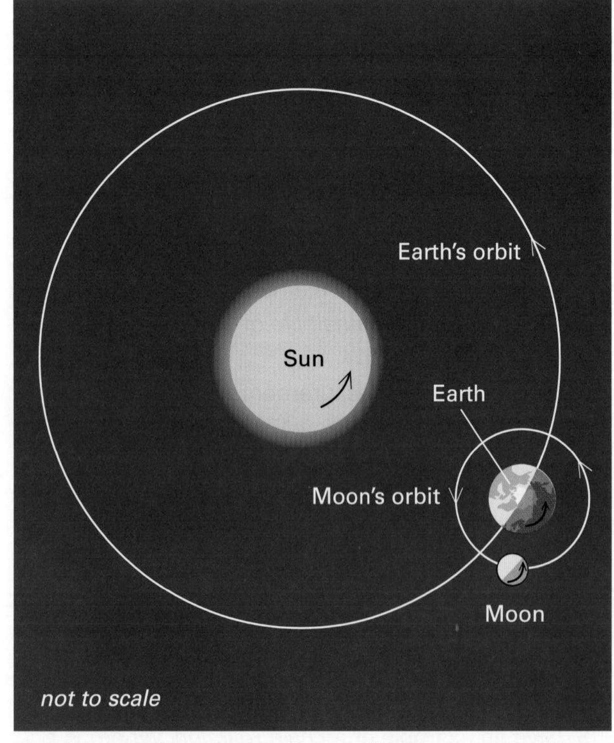

The Moon orbits the Earth, and the Earth orbits the Sun. The orbits are near-circular.

Looking at the sky

On a clear night, the sky is full of tiny points of light. Most of these are stars. Like the Sun, they appear to move across the sky as the Earth rotates. However, relative to each other, their positions hardly change by any amount you could notice.

The brightest stars seem to form patterns – the different groups are called ***constellations***. However, the stars in a constellation aren't really grouped together. Some may be much further away than others.

A few of the dots in the night sky do appear to change position (see Spread 1.03). These are ***planets***, which, like the Earth, are in orbit around the Sun.

During the daytime, you can't see stars or planets because their light is completed swamped by sunlight.

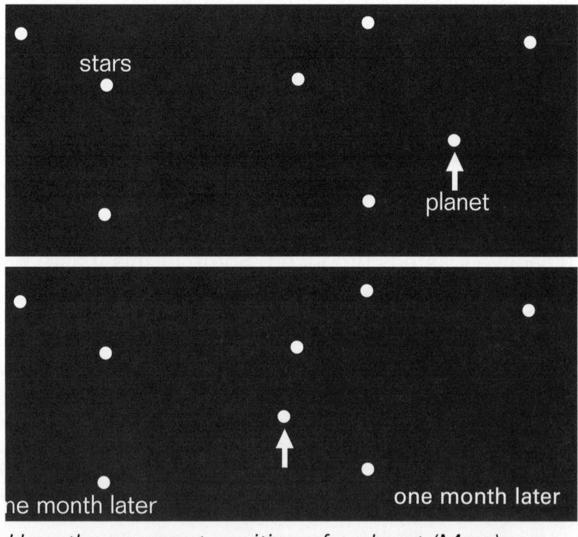

How the apparent position of a planet (Mars) can change relative to the stars in the background.

The ***Moon*** is in orbit around the Earth and 380 000 kilometres away from us. It is smaller than the Earth and has a rocky, cratered surface. We only see the Moon because its surface reflects sunlight. And we don't see the part that is in shadow.

The Moon takes about 28 days to orbit the Earth. It also takes the same time to turn once on its axis, so it always keeps the same face toward us.

Looking at the Moon, we only see the sunlit part.

Gravity in action

We are pulled to the Earth by the force of gravity. No one knows what causes gravity. But scientists know that there is a gravitational force between all masses:

- The force is always a force of attraction.
- The greater the masses, the greater the force.
- The force weakens with distance. Isaac Newton found that it obeys an ***inverse square law***: doubling the distance between two masses reduces the force between them to a quarter...and so on.

The gravitational force between everyday things is far too weak to detect. It only becomes strong if one of the things has a huge mass, like the Earth.

The gravitational force between the Earth and the Sun holds the Earth in orbit around the Sun. The gravitational force between the Moon and the Earth holds the Moon in orbit around the Earth.

Bombarded from space

Millions of pieces of rock, ice, and dust orbit the Sun. Some have elliptical orbits which can bring them close to the Earth. When high-speed dust particles hit the Earth's atmosphere, they burn up, causing streaks of light called ***meteors*** ('shooting stars'). Bigger lumps of material may reach the surface as ***meteorites***.

A large meteorite caused the huge ***impact crater*** above, in Arizona, USA. It is 800 metres across and was probably formed about 50 000 years ago.

Today, impacts by large meteorites are very rare. They were much more common 4000 million years ago, when the Earth was young.

Many of the craters on the Moon's surface were caused by meteorites, though some are due to past volcanic activity. The Moon has no atmosphere or oceans, so its craters have not been eroded away like those on Earth.

1 About how long does it take for
(a) the Earth to turn once on its axis?
(b) the Earth to orbit the Sun?
(c) the Moon to orbit the Earth?

2 If the Moon and Earth were further apart, how would this affect the gravitational pull between them?

3 **(a)** Copy the diagram on the right. Shade in the part of the Earth that is in shadow.
(b) Is it daytime or night in Britain?

4 What are constellations?

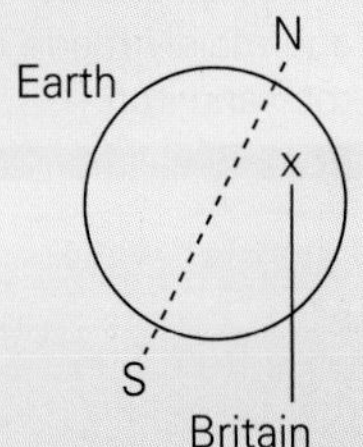

Not to scale

5 Without a telescope, how could you tell whether a dot in the night sky was a star or a planet?

4.33 Planets, stars, and galaxies

By the end of this spread, you should be able to:
- *explain what the Solar System is, and its place in the Universe.*

The Solar System

The Earth is one of many ***planets*** in orbit around the Sun. The Sun, planets, and other objects in orbit are together known as the ***Solar System***.

Planets are not hot enough to give off their own light. We can only see them because they reflect the Sun's light. From Earth, they look like tiny dots in the night sky. Without a telescope, it is difficult to tell whether you are looking at a star or a planet.

Most planets move in near-circular orbits around the Sun. Many have smaller ***moons*** in orbit around them.

Comets are collections of ice, gas, and dust which orbit the Sun and reflect its light. They have highly elliptical orbits which bring them close to the Sun and then far out in the Solar System.

Saturn is a gassy giant. Its 'rings' are billions of orbiting bits of ice and rock which reflect light.

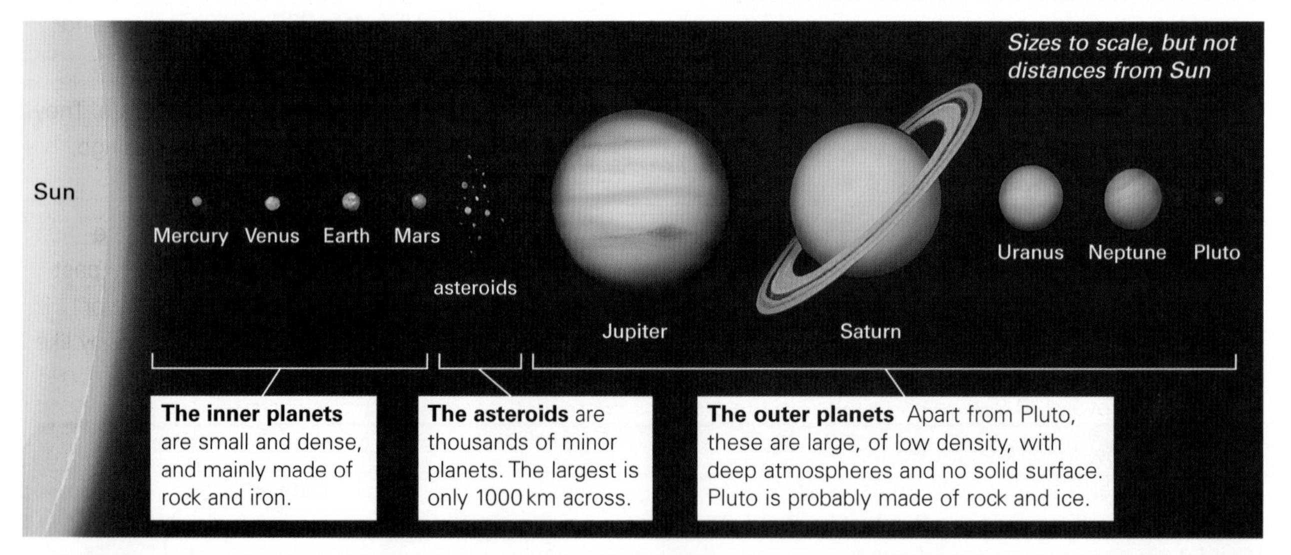

	Mercury	Venus	Earth	Mars	Jupiter	Saturn	Uranus	Neptune	Pluto
Average distance from Sun in million km	58	108	150	228	778	1427	2870	4490	5900
Time for one orbit in years	0.24	0.62	1	1.88	11.86	29.46	84.01	164.8	247
Diameter in km	4900	12 100	12 800	6800	143 000	120 000	51 000	49 000	2300
Average surface temperature	350 °C	480 °C	22 °C	−23 °C	−150 °C	−180 °C	−210 °C	−220 °C	−230 °C
Number of moons	0	0	1	2	52	30	21	8	1

Stars and galaxies

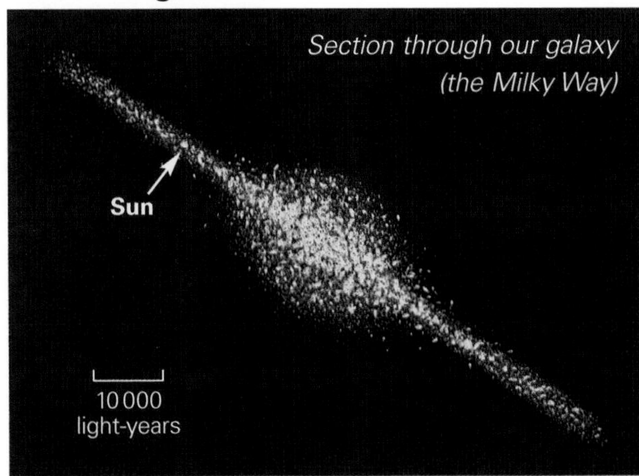

Our Sun is just one star in a huge star system called a ***galaxy***. This contains over 100 billion stars. It is so big that a beam of light, travelling at 300 000 kilometres per second, would take 100 000 years to cross it! Scientists say that the galaxy is a distance of 100 000 ***light-years*** across.

Ours is not the only galaxy. In the whole ***Universe***, there are over 100 billion galaxies.

Our galaxy is called the ***Milky Way***. You can see the edge of its disc as a bright band of stars across the night sky. The Milky Way is a member of a local cluster of about 30 galaxies. The other major member is the ***Andromeda Galaxy***.

Exploring space

People have stood on the Moon, unmanned spacecraft have landed on Mars and Venus, and space probes have passed close to most of the outer planets. But travelling further into space is a problem. The *Voyager 2* probe took 12 years to reach Neptune. At that speed, it would take over 100 000 years to reach the nearest star! To find out more about stars and galaxies, we have to rely on the light and other forms of radiation picked up by telescopes.

The Andromeda Galaxy is 2 million light-years away.

Life in the Universe

Earth seems to be the only planet in the Solar System where life has evolved. It has a rare combination of factors to make this possible, including liquid water, the correct temperature range, an ozone layer which absorbs the Sun's ultraviolet, a magnetic field which shields the Earth from other types of radiation, and a 'bodyguard', Jupiter, to pull away some of the comets and meteorites that might have hit the Earth.

So is the Earth unique? No one knows. The materials needed to produce the molecules of life are present elsewhere – for example, scientists haven't ruled out the possibility of finding simple bacteria in the water on Europa, one of Jupiter's Moons. There are planets around other stars, and as there are billions of stars in each galaxy, and billions of galaxies, it seems possible that some of these planets might be able to support life. But there is no evidence one way or the other.

1 Why do planets give out light?
2 Which planets are smaller than the Earth?
3 In the table on the left, what link can you see between the time for a planet's orbit and its distance from the Sun?
4 **(a)** Which planets are colder than the Earth? **(b)** Why do you think they are colder?
5 Carbon dioxide in Venus' atmosphere produces a greenhouse effect (global warming). What clues are there for this in the table on the left?
6 What is **(a)** galaxy? **(b)** a light-year?
7 Look at the Andromeda Galaxy in the photograph. **(a)** How long does its light take to reach us? **(b)** Why are humans unlikely to visit it?

4.34 Action in orbit

By the end of this spread, you should be able to:
- *describe what satellites can be used for*
- *describe the features of a comet's orbit around the Sun.*

Satellites at work

There are hundreds of satellites in orbit around the Earth. Here are some of the jobs they do:

Communications satellites These are used to pass on signals for telephones and TV from one part of the Earth to another (see Spread 4.29).

Research satellites Some of these carry telescopes for looking at stars and planets. Above the atmosphere, they get a much clearer view.

Monitoring satellites These study conditions down on the Earth. For example, weather satellites send pictures back to Earth (using radio signals) so that forecasters can see what the weather is doing.

Navigation satellites These send out radio signals which ships, aircraft, or people on the ground can use to find their position.

This GPS (Global Positioning System) receiver picks up signals from a network of satellites, uses the data to calculate its position, then shows the result.

If your TV is connected to a dish aerial, it gets its signals from a satellite like this.

The Hubble Space Telescope *uses radio signals to send pictures of distant stars and galaxies back to Earth.*

Putting satellites into orbit

Most satellites are in circular orbits around the Earth. They are launched by rockets whose engines are so powerful that they can only work for ten minutes or so before running out of fuel.

For a low orbit, just above the Earth's atmosphere, a speed of about 29 000 km per hour (18 000 mph) is required. A satellite travels fastest in a low orbit. In a high orbit, it travels at a lower speed. However, the launch rocket must leave the Earth at a higher speed, otherwise it will not be able to 'coast' far enough out into space when its engines shut down.

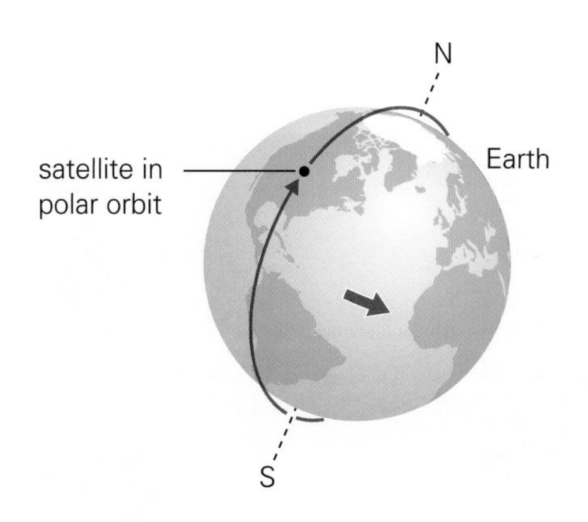

Satellites which survey the Earth are often put into a low ***polar orbit*** – one that passes over the North and South Poles. As the Earth turns beneath them, they can scan the whole of its surface.

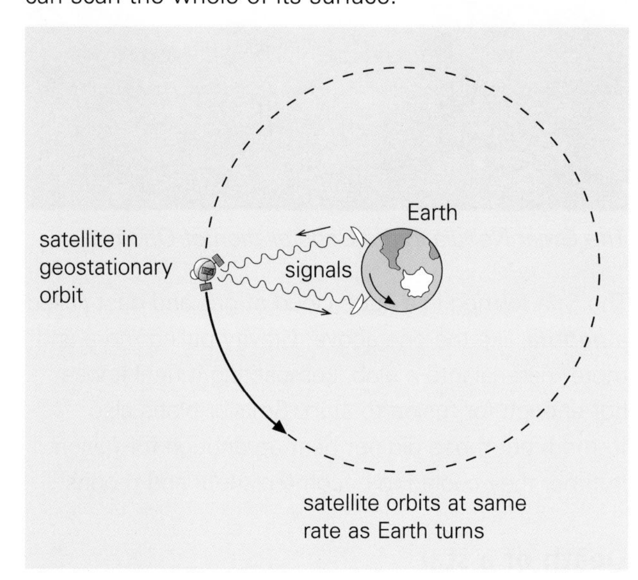

Communications satellites are normally put into a ***geostationary orbit***, as shown above. The satellite orbits at exactly the same rate as the Earth turns, so it appears to stay in the same position in the sky. Down on the ground, the dish aerials sending and receiving the signals can point in a fixed direction. For a geostationary orbit, a satellite must be 35 900 km above the equator and travelling at a speed of about 11 000 km per hour (6900 mph).

Out of orbit

If a rocket leaves the Earth at more than 11 200 m/s (25 000 mph), it will 'coast' so far out into space that it is never pulled back to Earth again. This speed is called the ***escape velocity*** for the Earth. Space probes sent to other planets need to reach this velocity.

Orbit of a comet

A comet (see Spread 4.33) has a highly elliptical orbit around the Sun. It has least speed when furthest from the Sun. That is also when the gravitational pull is weakest. As it 'falls' closer to the Sun, the force of gravity increases. Also, the comet speeds up. It reaches its maximum speed when closest to the Sun.

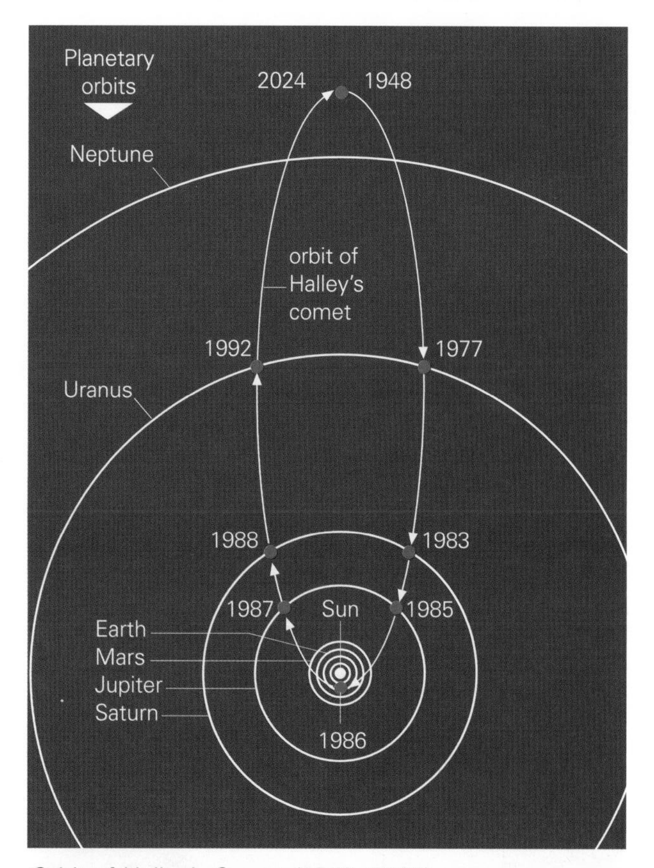

Orbit of Halley's Comet (1948–2024)

1 A survey satellite is in a low polar orbit.
(a) What is the advantage of this type of orbit?
(b) Give three other uses of satellites.

2 Satellites that transmit TV pictures are in orbit and moving. Yet, down on the ground, the dish aerials that receive the signals point in a fixed direction. How is this possible?

3 Satellites A and B are identical, but A has been put into a higher orbit around the Earth than B.
(a) Which satellite has the greatest gravitational force on it?
(b) Which satellite has the greatest speed?
(c) Which takes the longest to orbit the Earth?

4.35 Birth and death

By the end of this spread, you should be able to:
- *describe how the Solar System was formed*
- *describe the life cycle of a star*
- *explain why scientists think that the Universe is expanding, and started with the big bang.*

Birth of a star

The Sun started to shine about 4500 million years ago. Hydrogen is its 'fuel'. Deep in its core, the intense heat and pressure make hydrogen nuclei fuse (join) together to form helium nuclei, releasing energy as they do so. The process is called ***nuclear fusion***.

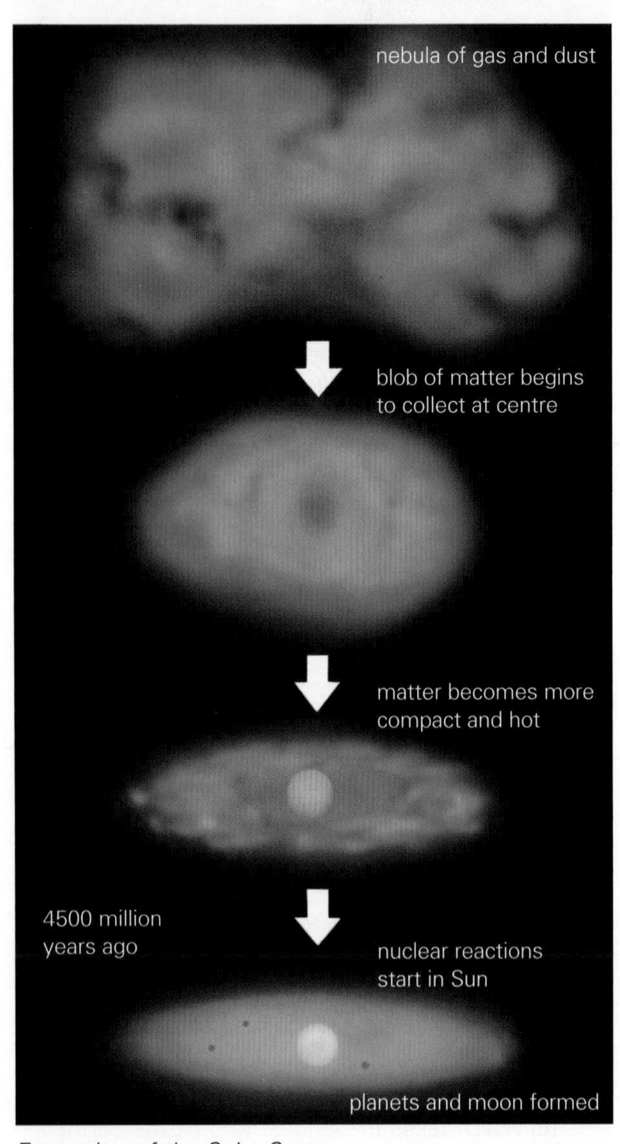

Formation of the Solar System

The Great Nebula in the Constellation of Orion

The Sun formed in a huge cloud of gas and dust called a ***nebula***, like the one above. Gravity pulled more and more material into a blob, compacting it until it was hot enough for fusion to start. Smaller blobs also formed, but these did not heat up enough for fusion. In time, they cooled to become planets and moons.

Death of a star

At present, the Sun is stable. But in about 6000 million years time, it will have converted all of its hydrogen to helium. Then it will swell and its outer layer will cool to a red glow. It will have become a ***red giant***. Eventually, its outer layer will drift into space, exposing a hot, dense core called a ***white dwarf***. This tiny star will use helium as its nuclear fuel. When this runs out, the star will cool and fade for ever.

In each galaxy, new stars are forming and old ones are dying. The most massive stars have a different fate from that of the Sun. Eventually, they blow up in a gigantic nuclear explosion called a ***supernova***. This leaves a dense core called a neutron star. If this has enough mass, it continues to collapse under its own gravity. Nothing can resist the pull. Even light cannot escape. The star becomes a ***black hole***.

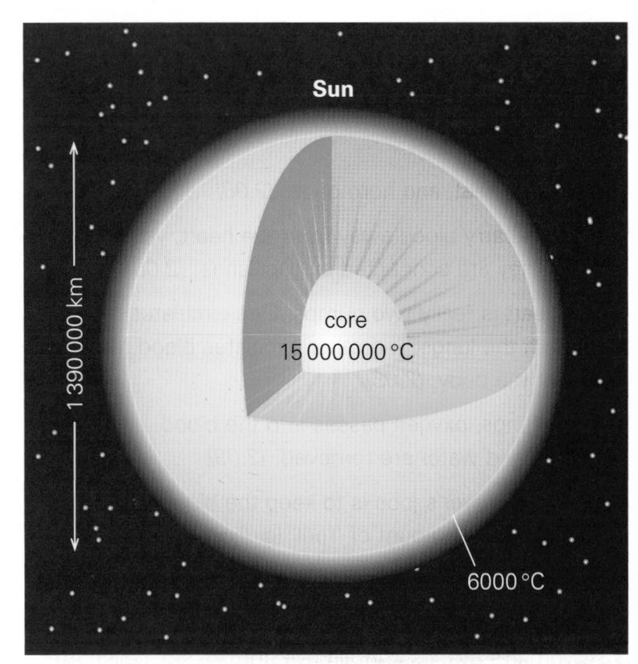

Nuclear fusion takes place in the Sun's core. The high temperatures create outward forces which stop the Sun collapsing under the pull of its own gravity.

Made from stardust

In stars, fusion reactions change lighter elements into heavier ones – hydrogen changing into helium is an example of this. However, to make very heavy elements (gold and uranium for example), the extreme conditions which create a supernova are needed. The Sun and inner planets contain very heavy elements. This suggests that the nebula in which they formed included 'stardust' from an earlier supernova. In other words, the Sun is a ***second-generation star***.

1. By what process does the Sun gets its energy?
2. The Sun formed in a nebula.
   (a) What is a nebula?
   (b) What made matter in the nebula collect together in a blob?
   (c) About how long ago was the Sun formed?
   (d) What evidence is there that the nebula in which the Sun formed contained material from an earlier supernova?
3. What is **(a)** a red giant? **(b)** a white dwarf?
4. How is a black hole formed?
5. What evidence is there that the Universe may have started with a big bang?

The expanding Universe

The light waves we receive from distant galaxies are 'stretched out' – their wavelengths are longer. This is called the ***Doppler effect***. Scientists think that it is occurs because the galaxies seem to be rushing apart at high speed. We are living in an expanding Universe.

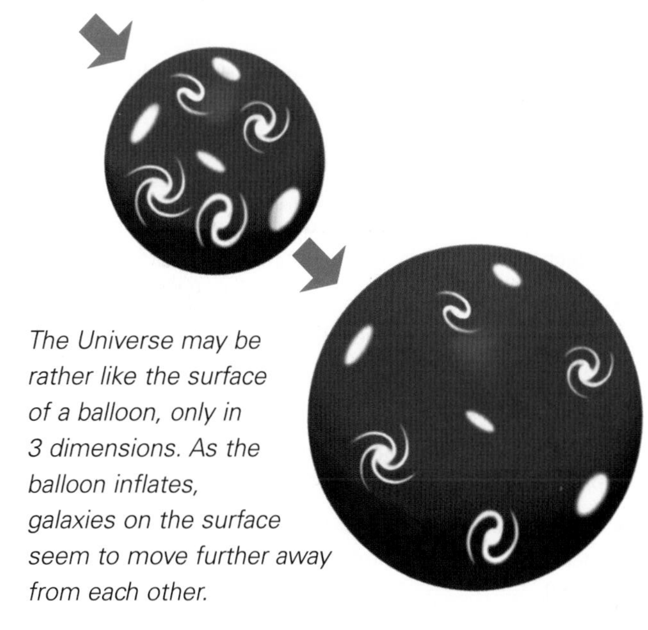

The Universe may be rather like the surface of a balloon, only in 3 dimensions. As the balloon inflates, galaxies on the surface seem to move further away from each other.

The Universe may have been created in a gigantic burst of energy called the ***big bang***. Here are two pieces of evidence to support this idea:

- Radio telescopes have picked up background microwave radiation coming from every direction in space. This may be an 'echo' of the big bang.
- As the galaxies seem to be moving apart, they may all have come from the same tiny volume of space.

Scientists have estimated the rate at which the galaxies seem to be moving apart. It is thought to be about 1 km per year for each 14 000 million km of separation. They call this the ***Hubble constant***. Using this value, it is possible to work out that the galaxies must have started to separate about 14 000 million years ago. If so, that is when the big bang occurred.

The future of the Universe is unknown. Until recently, it was thought that the gravitational pull between galaxies might be slowing the expansion down. But new evidence suggests that the expansion may actually be increasing.

Summary

The spread numbers in brackets tell you where to find more information

- Animals and plants feed, respire, excrete, grow, move, reproduce, and show sensitivity. *(2.01)*
- Unlike animal cells, plant cells have a cell wall and chloroplasts. *(2.01)*
- Plants absorb the Sun's energy. They use it to make their own food, mainly from carbon dioxide and water. The process is called photosynthesis. *(2.02)*
- Respiration is the process in which cells get energy from their food. *(2.02)*
- Photosynthesis produces glucose, which is turned into starch for storage in leaves, stems, and roots. *(2.03)*
- The rate of photosynthesis depends on light intensity, water supply, carbon dioxide concentration, and temperature. *(2.03)*
- Plants need water and minerals from the soil. These enter roots by diffusion and are pushed up through the plant by active transport. *(2.03)*
- Flowers contain sex cells. For a seed to develop, a male sex cell must combine with a female sex cell. *(2.04)*
- Plants use chemicals called hormones to control how different parts work. For example, auxin stimulates growth. Artificial hormones can be used to help roots develop, to control ripening, and as weedkillers. *(2.04)*
- The human body is a collection of organs. These are made of cells. *(2.05)*
- Blood carries food, oxygen, and other substances to cells and carries away their waste products. *(2.05)*
- As food passes along the gut, it is digested and then absorbed into the blood. *(2.06)*
- Food is a mixture of carbohydrates, fats, proteins, minerals, vitamins, fibre, and water. *(2.06)*
- Your blood absorbs digested food through villi on the inner surface of the small intestine. *(2.07)*
- During aerobic respiration, oxygen is used up. However, muscles can deliver energy for short periods without oxygen. In this case, the respiration is anaerobic. *(2.07)*
- Enzymes speed up vital chemical reactions in the body. *(2.07)*
- Blood carries oxygen, digested food, water, waste products, heat, and hormones. *(2.08)*
- Arteries carry blood away from the heart. Veins carry it back. They are connected by capillaries. *(2.08)*
- The heart is really two pumps. One circulates blood through the lungs. The other circulates blood round the rest of the body. *(2.08)*
- In the lungs, oxygen passes into the blood, and carbon dioxide and water are removed. *(2.09)*
- One of the liver's jobs is to keep the blood topped up with the right amount of 'fuel' (glucose). *(2.09)*
- The kidneys filter unwanted substances from the blood. These pass out of the body as urine. *(2.09)*
- A woman's ovaries contain ova (female sex cells). An ovum is released about every 28 days as part of the menstrual cycle. *(2.10)*
- A baby grows from a cell formed when a sperm from the father fertilizes an ovum inside the mother. *(2.10)*
- Your body is controlled by the central nervous system. Sensory nerve cells carry information to it; motor nerve cells carry instructions from it. *(2.11)*
- In the eye, a lens system forms an image on the retina. This contains millions of light-sensitive cells that send nerve impulses to the brain. *(2.11)*
- The body uses hormones to control some processes. Hormones are made in the endocrine glands. *(2.12)*
- The menstrual cycle is controlled by hormones. *(2.12)*
- Artificial hormones include steroids, contraceptive pills, and fertility drugs. *(2.12)*
- In your body, temperature, amount of water, and blood sugar are three factors which must be kept steady. Maintaining the balance is called homeostasis. *(2.13)*
- In the kidneys, nephrons remove unwanted substances from the blood and maintain water balance. *(2.13)*
- Insulin is a hormone that controls how much glucose the liver changes into glycogen for storage. *(2.13)*
- Microbes that cause disease are known as germs. They include some bacteria and viruses. They can be transferred by air, contact, animals, and food. *(2.14)*
- Defending against disease is the job of your immune system. Antibiotics and vaccines can also be given to help the immune system. *(2.14)*

- Skin is an oily, waterproof barrier that makes it difficult for germs to get into the body. *(2.15)*
- Your immune system uses white blood cells to deal with invading germs. Some digest the germs. Others make toxins that destroy germs. *(2.15)*
- Smoking, excess alcohol, and solvent or drug abuse will damage your health. *(2.15)*
- You inherit some of your characteristics. Others depend on your environment. *(2.16)*
- The chemical instructions for inherited characteristics are called genes. Genes are small sections of the chromosomes in your cells. *(2.16)*
- By selecting the parents carefully, animals and plants can be bred for particular characteristics. *(2.16)*
- Alleles are different versions of a gene. Alleles operate in pairs. The dominant allele controls the characteristic. *(2.17)*
- A characteristic you can see is called a phenotype; the combination of alleles producing it is the genotype. *(2.17)*
- Sex cells (sperms and ova) are formed by a type of cell division called meiosis. Each has half the number of chromosomes in the original cell. *(2.17)*
- Human cells each have two sex chromosomes: XX in females and XY in males. *(2.17)*
- For growth, cells must multiply. They do this by a type of cell division called mitosis. *(2.18)*
- In cells, there are molecules of DNA. These carry genetic information in the form of a chemical code that controls what proteins are made. *(2.18)*
- In asexual reproduction, cells divide to form a new organism genetically identical to the parent. *(2.18)*
- If a plant or animal is homozygous (pure bred) for a characteristic, the two alleles are the same. If it is heterozygous (hybrid), they are different. *(2.19)*
- In simple cases, you can predict the results of crossing organisms. *(2.19)*
- Making genetically identical copies of a plant or animal is called cloning. *(2.19)*
- Inherited (genetic) diseases include cystic fibrosis, sickle-cell anaemia, and haemophilia. *(2.19)*
- In genetic engineering, or genetic modification, genes are transferred from one organism to another. *(2.19)*
- According to Darwin's theory of natural selection, some variations in a species help in the struggle for survival. Survivors reproduce and pass on their characteristics to later generations. So the species becomes adapted to its environment. *(2.20)*
- Mutations are chance genetic changes. They produce variation and so contribute to natural selection. *(2.20)*
- Fossils provide evidence of how species have evolved – or have died out – in the past. *(2.21)*
- The science of genetics grew from Mendel's work. Without it, features of variation and evolution were difficult to explain. *(2.21)*
- Conditions for animals and plants change daily and seasonally, and habitats are shared with other organisms. These factors affect populations. *(2.22)*
- As animals eat other animals, and plants, a change in the size of one population will affect others. *(2.22)*
- The size of a population is limited by predators, competitors, and the food and shelter available. *(2.23)*
- Humans change land use and cause pollution. Our use of resources needs to be sustainable. *(2.23)*
- The burning of fossil fuels is putting extra carbon dioxide into the atmosphere. This is trapping more of the Sun's heat and may be causing global warming. *(2.24)*
- Other environmental problems caused by human activity include acid rain, destruction of the ozone layer, forest destruction, and river pollution. *(2.24)*
- Plants are food for animals, which are food for other animals, and so on in a food chain. Where food chains are interconnected, they form a food web. *(2.25)*
- Pyramid charts can be used to show the numbers or the masses of the organisms in a food chain. *(2.25)*
- Decomposers are microbes that feed on the remains of dead plants and animals. They put useful chemicals back into the soil. *(2.25)*
- Carbon and nitrogen are recycled by living things. The recycling depends on microbe activity. *(2.26)*
- Energy is lost at every stage of a food chain. This limits the number of levels in the chain. *(2.27)*
- The rate at which plant or animal matter decomposes depends on temperature and the availability of water and air for the microbes. *(2.27)*
- If polluting chemicals pass along a food chain, they become more and more concentrated. *(2.27)*
- Food production from the land and sea needs to be carefully controlled to avoid damage to the environment, animal welfare, and human health. *(2.27)*

Summary

The spread numbers in brackets tell you where to find more information

- Everything is made from about 100 simple substances called elements. There are two main types of element: metals and non-metals. *(3.01)*
- Atoms of different elements can join together in chemical reactions to form new substances called compounds. *(3.01)*
- Unlike compounds, mixtures have components of varying amounts which can be separated easily. *(3.01)*
- Solids, liquids, and gases are made up of tiny particles that are constantly on the move. *(3.02)*
- Particles of one material may spread through another. This is called diffusion. *(3.02)*
- $\text{density (kg/m}^3) = \dfrac{\text{mass (kg)}}{\text{volume (m}^3)}$ *(3.02)*
- If one substance (the solute) dissolves in another (the solvent), the result is a solution. *(3.03)*
- The solubility of most solids increases with temperature. For gases, it decreases. *(3.03)*
- Crystallization, chromatography, and distillation are three methods of separating the components of solutions. *(3.03)*
- Chemical changes form new substances and are usually difficult to reverse. Physical changes do not form new substances and are easy to reverse. *(3.04)*
- Exothermic reactions give out energy (heat). Endothermic reactions take in energy. *(3.04)*
- Atoms have a nucleus of protons (+ charge) and neutrons (no charge), with electrons (– charge) moving around it. *(3.05)*
- Elements exist in different versions called isotopes. These have the same atomic number but different mass numbers. *(3.05)*
- The relative atomic mass scale compares the masses of atoms with each other. The standard is an atom of carbon-12, with a mass of 12.000 units. *(3.05)*
- In the periodic table, elements in the same column have similar properties and similar arrangements of outer shell electrons. *(3.06)*
- The noble gases, in the last column of the table, have full outer shells and are unreactive. *(3.06)*
- Ions are atoms (or groups of atoms) that have gained or lost electrons. *(3.07)*
- In an ionic crystal, such as sodium chloride, strong forces of attraction bind positive and negative ions tightly together. *(3.07)*
- Some atoms bond together by sharing electrons. This is called covalent bonding. A molecule is a group of atoms joined by covalent bonds. *(3.08)*
- In water, neighbouring molecules are held together by hydrogen bonds. *(3.08)*
- In giant molecules, such as diamond, the atoms are rigidly fixed by strong covalent bonds. *(3.09)*
- Metals are made up of positive ions surrounded by a 'sea' of electrons. *(3.09)*
- Balanced chemical equations show how atoms change partners during reactions. *(3.10)*
- Balanced ionic equations for a reaction show only the ions that change. *(3.10)*
- One mole of any gas contains 6×10^{23} particles (e.g. molecules) and occupies 24 dm^3 (litres) at normal room temperature and pressure. *(3.10)*
- In any chemical reaction, energy is absorbed in order to break down the bonds between the atoms in the reacting substances. Energy is released by the formation of new bonds in the products. *(3.11)*
- You can calculate the net change of energy during a chemical reaction if you know the bond energies of all the links being broken or made. *(3.11)*
- There are different types of chemical reaction, including synthesis, thermal decomposition, precipitation, and redox. *(3.12)*
- In redox reactions, electrons pass from a reducing agent to an oxidizing agent. *(3.12)*
- Some metals can form more than one type of ion. These ions can be changed from one to another by oxidation or reduction. *(3.12)*
- All acids contain hydrogen. If an acid reacts with a metal, hydrogen gas is given off. *(3.13)*
- A base will neutralize an acid. Alkalis are soluble bases. *(3.13)*
- Acids turn litmus red; alkalis turn it blue. *(3.13)*
- The strength of an acid or alkali is measured on the pH scale. pH1 is a very strong acid; pH14 is a very strong alkali; pH7 is neutral. *(3.13)*

- When a base neutralizes an acid, the result is a solution containing a salt. *(3.14)*
- When a carbonate reacts with an acid, a salt is produced and carbon dioxide is give off. *(3.14)*
- Factors affecting the rate of a reaction include the size of the bits, concentration, temperature, and presence of a catalyst. *(3.15)*
- Catalysts speed up reactions without being used up themselves. *(3.15)*
- Enzymes are biological catalysts. They are very sensitive to temperature and to acid/alkali conditions. *(3.16)*
- Enzymes are used in the production of cheese, yoghurt, alcohol, and bread. *(3.16)*
- In a reversible reaction, the products can recombine to reform the original substances again. *(3.17)*
- Ammonia and sulphuric acid are made by reversible reactions. Manufacturers use conditions that give a reasonable compromise between yield and cost. *(3.17)*
- Ammonia is manufactured by the Haber process. *(3.18)*
- Ammonia is used in the manufacture of nitric acid and Nitram fertilizer (ammonium nitrate). *(3.18)*
- The atoms of the elements in any group of the periodic table get bigger and become more metallic down the group. Across a period the atoms get smaller and more non-metallic. *(3.19)*
- The oxides of the elements in Groups 1 and 2 are basic. The oxides become increasingly acidic from left to right across a period. *(3.19)*
- The elements in Group 7, the halogens, include chlorine, bromine, and iodine. Their bleaching actions, and their reactions with hydrogen, show that there is a decrease in reactivity down the group. *(3.20)*
- Chlorine is used to kill germs in water and in the manufacture of plastics and chemicals. *(3.20)*
- Many of the transition metals have similar properties. Most are hard, dense, and strong, form more than one type of ion, and have coloured compounds. *(3.21)*
- Transition metals and their compounds are often used as catalysts. *(3.21)*
- Metals can be arranged in a reactivity series by observing how well they react with air, water, and dilute acid. The alkali metals are the most reactive. Gold is unreactive. *(3.22)*
- Some metals corrode when their surface is in contact with air and/or water. *(3.22)*
- A metal will displace a less reactive metal from a solution of one of its salts. *(3.23)*
- Reactive metals dissolve in dilute acids by displacing hydrogen. *(3.23)*
- In some reactions, metals compete for oxygen. *(3.23)*
- The more reactive a metal, the more difficult it is to separate from its ore. *(3.24)*
- Iron is separated from its ore by smelting. This is done in a blast furnace. The iron produced is mostly used to make steel. *(3.24)*
- Aluminium is separated from its ore by electrolysis. Electrolysis is also used to anodize aluminium and to purify copper. *(3.25)*
- Sodium chloride is the raw material for many important chemicals and processes. *(3.26)*
- The electrolysis of sodium chloride solution produces chlorine, hydrogen, and sodium hydroxide. *(3.26)*
- Limestone is used to make cement, quicklime, and slaked lime and in steel production. *(3.26)*
- The simplest compounds of carbon are the hydrocarbons. They include the alkanes, such as methane, ethane, propane, and butane. *(3.27)*
- Crude oil is mainly a mixture of alkanes. The different fractions can be separated by distillation. Some are used as fuels, others in making plastics. *(3.27)*
- Long-chain alkanes from oil can be cracked to form shorter-chain molecules. These include alkenes, such as ethene and propene. *(3.28)*
- In addition polymerization, short alkene molecules are joined up to form long-chain molecules. Polythene is produced in this way. *(3.28)*
- Today, the Earth's atmosphere is mainly nitrogen (78%) and oxygen (21%). Four billion years ago, it was mainly carbon dioxide. It changed to its present composition because of the activities of microbes, plants, and animals. *(3.29)*
- The Earth has a crust, mantle, and core. *(3.30)*
- Igneous rocks are formed when molten magma cools. The cooling rate affects the crystal size. *(3.30)*
- Sedimentary rocks are formed from layers of deposited sediment. *(3.30)*
- Metamorphic rocks are rocks that have been changed by heat or pressure. *(3.30)*
- As movements take place in the Earth's crust and mantle, rocks are recycled. *(3.30)*

SECTION 4

Summary

The spread numbers in brackets tell you where to find more information

- Metals and carbon are good conductors of electricity. Most non-metals are insulators. *(4.01)*
- Like charges repel. Unlike charges attract. *(4.01)*
- The force between charges is used in photocopiers and in some inkjet printers. *(4.01)*
- Current is measured in amperes (A). Voltage is measured in volts (V). *(4.02)*
- Bulbs (or other components) can be connected in series or in parallel. *(4.02)*
- When a current flows through a resistance, energy is given off as heat. *(4.03)*
- 1 kW h is the energy supplied to a 1 kilowatt appliance in 1 hour. Electricity companies set a cost for each Unit (kW h) supplied. *(4.03)*
- At constant temperature, the current through a metal conductor is proportional to the voltage across it. *(4.04)*
- Resistance is measured in ohms (Ω):

  $\text{resistance } (\Omega) = \dfrac{\text{voltage (V)}}{\text{current (A)}}$ *(4.04)*
- Thermistors, light-dependent resistors, and diodes have a resistance that can vary. *(4.04)*
- Power is measured in watts (W):

  power (W) = voltage (V) × current (A) *(4.05)*
- Charge is measured in coulombs (C):

  charge (C) = current (A) × time (s) *(4.05)*
- Energy is measured in joules (J):

  energy (J) = charge (C) × voltage (V) *(4.05)*
- With series components, the current is the same but the voltage is shared. With parallel components, the voltage is the same but the current is shared. *(4.06)*
- Electromagnets are used in magnetic relays and some circuit breakers. *(4.06)*
- There is a force on a current-carrying wire if it lies across a magnetic field. The effect is used in electric motors. *(4.07)*
- Generators use electromagnetic induction to produce a current. *(4.07)*
- In UK mains plugs, the wires are coloured brown (live), blue (neutral), and yellow/green (earth). *(4.08)*
- Switches and fuses are fitted in the live wire. *(4.08)*
- Mains current is alternating current (AC). Batteries supply direct current (DC). *(4.08)*
- Transformers step AC voltages up or down:

  $\dfrac{\text{output voltage}}{\text{input voltage}} = \dfrac{\text{output turns}}{\text{input turns}}$ *(4.09)*
- Mains power is transmitted across country at high voltage to reduce the current in the cable. *(4.09)*
- $\text{speed (m/s)} = \dfrac{\text{distance travelled (m)}}{\text{time taken (s)}}$ *(4.10)*
- The force of friction provides grip. But in machines, it opposes motion and wastes energy. *(4.10)*
- Factors affecting a car's stopping include the speed and the driver's reaction time. *(4.10)*
- Velocity is speed in a particular direction. *(4.11)*
- $\text{acceleration (m/s}^2\text{)} = \dfrac{\text{change in velocity (m/s)}}{\text{time taken (s)}}$ *(4.11)*
- Motion can be represented using distance–time and speed–time graphs. *(4.11)*
- Force is measured in newtons (N). *(4.12)*
- Weight is the force of gravity. *(4.12)*
- force (N) = mass (kg) × acceleration (m/s^2) *(4.12)*
- The Earth's gravitational field strength, *g*, is 10 N/kg. *g* is also the acceleration of free fall 10 m/s^2. *(4.12)*
- For every a force (action) there is an equal but opposite force (reaction) on another object. *(4.12)*
- Unbalanced forces produce acceleration. *(4.13)*
- If an object is at rest, or has a steady velocity, the forces on it are balanced. *(4.13)*
- An object falling through air reaches a terminal velocity. *(4.13)*
- Pressure is measured in N/m^2, or pascals (Pa):

  $\text{pressure (Pa)} = \dfrac{\text{force (N)}}{\text{area (m}^2\text{)}}$ *(4.14)*
- Hydraulic machines use liquid pressure. *(4.14)*
- If a material obeys Hooke's law, its extension is proportional to the stretching force. *(4.15)*
- For a fixed mass of gas at constant temperature, pressure × volume stays the same. *(4.15)*
- moment = force × distance from turning point. *(4.16)*

- If something is balanced, the total left-turning moment is equal to the total right-turning moment. This is the law of moments. *(4.16)*
- Energy can change into different forms, but it cannot be made or destroyed. *(4.17)*
- Conduction and convection are two ways in which the energy may travel because of a temperature difference. *(4.18)*
- Some surfaces are better at emitting and absorbing thermal radiation than others. *(4.19)*
- Most power stations use steam from a heat source to turn their generators. *(4.20)*
- Fuels like coal, oil, and natural gas are non-renewable. Energy sources like the wind, and water behind dams, are renewable. *(4.20)*
- If a power station has an efficiency of 35%, then 35% of its fuel's energy is changed into electrical energy. The rest is wasted as heat. *(4.20)*
- Nearly all of our energy comes from the Sun. *(4.21)*
- Work, like energy, is measured in joules (J):
- work done (J) = force (N) × distance moved (m) *(4.22)*
- Power is measured in watts (W):

  $$\text{power (W)} = \frac{\text{work done (J)}}{\text{time taken (s)}} \quad \text{or} \quad \frac{\text{energy spent (J)}}{\text{time taken (s)}}$$
- gravitational potential energy (J)

  = weight (N) × height lifted (m) *(4.22)*
- kinetic energy = $\frac{1}{2}$ × mass × velocity^2 *(4.22)*
- Waves can be transverse or longitudinal. *(4.23)*
- If 3 waves are sent out per second, the frequency is 3 hertz (Hz). *(4.23)*
- speed (m/s) = frequency (Hz) × wavelength (m) *(4.23)*
- Waves can be reflected, refracted, and diffracted. *(4.23)*
- Sound waves can travel through solids, liquids, and gases, but not through a vacuum. *(4.24)*
- Echoes are reflected sounds. *(4.24)*
- The higher the amplitude, the louder the sound. The higher the frequency, the higher the pitch. *(4.25)*
- Sound above the limit of human hearing is called ultrasound. Its uses include cleaning, metal testing, and scanning the womb. *(4.25)*
- The Earth's crust and upper mantle is made up of huge tectonic plates. Their movements cause earthquakes, volcanoes, and fold mountains. *(4.26)*
- Seismic waves provide evidence that the Earth has a layered structure, with a partly liquid core. *(4.26)*
- Light is reflected from a mirror at the same angle as it strikes it. *(4.27)*
- Light bends when it enters a transparent material such as glass. This is called refraction. *(4.27)*
- Above a certain angle, light is totally reflected from the inside face of a transparent material. *(4.27)*
- A prism will split white light into a spectrum of all the colours of the rainbow. *(4.28)*
- The electromagnetic spectrum is made up of radio waves, microwaves, infrared, light, ultraviolet, X-rays, and gamma rays. *(4.28)*
- Analogue signals vary continuously. Digital signals are pulses representing numbers. Digital signals can be transmitted with less loss of quality. *(4.29)*
- Digital signals can be sent along optical fibres as pulses of light or infrared. *(4.29)*
- Radioactive materials give out nuclear radiation. The main types of radiation are alpha particles, beta particles, and gamma rays (waves). *(4.30)*
- Ionizing radiation can damage or kill cells. *(4.30)*
- The break-up of unstable nuclei is called radioactive decay. The half-life is the time it takes for half the unstable nuclei to decay. *(4.31)*
- The splitting of nuclei (such as uranium-235) by neutrons is called fission. It releases energy. *(4.31)*
- Uses of radioactivity include dating rocks and in tracers. *(4.31)*
- The Earth orbits the Sun, held by the force of gravity. The Moon orbits the Earth. *(4.32)*
- The Solar System is mainly made up of the Sun and its planets. *(4.33)*
- The Sun is one star in a vast galaxy of 100 billion stars. There are more than 100 billion galaxies in the whole Universe. *(4.33)*
- Satellites are used for communications, research, monitoring, and navigation. *(4.34)*
- Comets have elliptical orbits around the Sun. *(4.34)*
- The Sun and planets formed in a huge cloud of gas and dust called a nebula. *(4.35)*
- The Sun is powered by nuclear fusion. When its fuel runs out, it will swell, then fade. *(4.35)*
- There is evidence that the Universe started with a big bang and has been expanding ever since. *(4.35)*

Exam-style questions

1 a If a person has heart disease, it may be necessary for them to have a heart operation. During this operation, the heart is stopped and their blood is passed through a heart–lung machine.
(i) Suggest a reason why the heart is stopped during the operation.
(ii) Describe *two* jobs that the heart–lung machine must do for the person during the operation.

b Heart rates can be recorded on graph paper or on a screen. The line drawn is called an ECG. Here is an example.

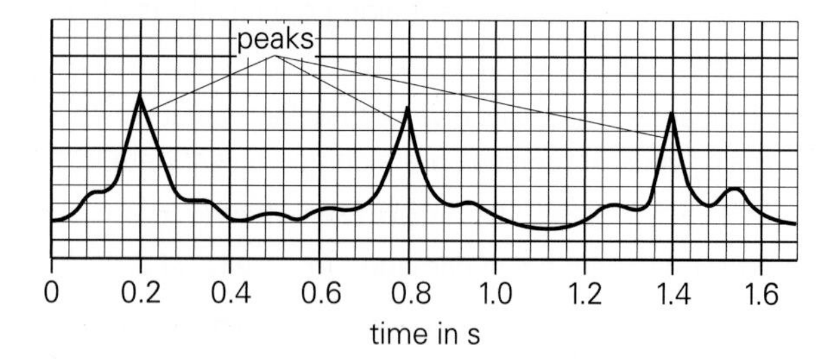

The time taken for one heart beat is represented by the distance between two peaks. Work out the heart rate in the ECG above in beats per minute. Show all your working.

2 The diagram below shows a plant.

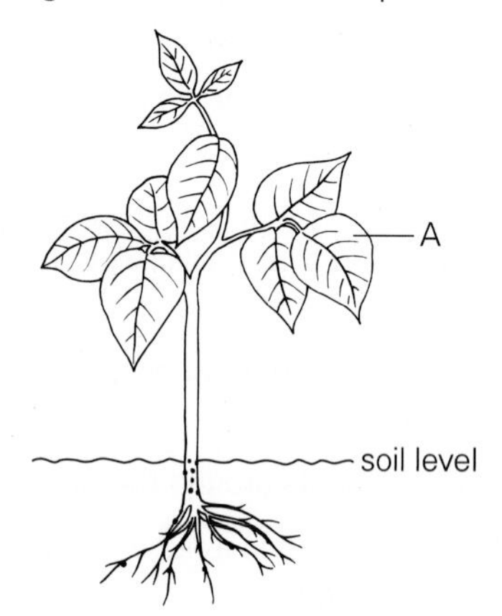

a What job is done by structure A.

b The next diagram shows a cell from the leaf of this plant. How does the structure of this cell enable it to make food for the plant?

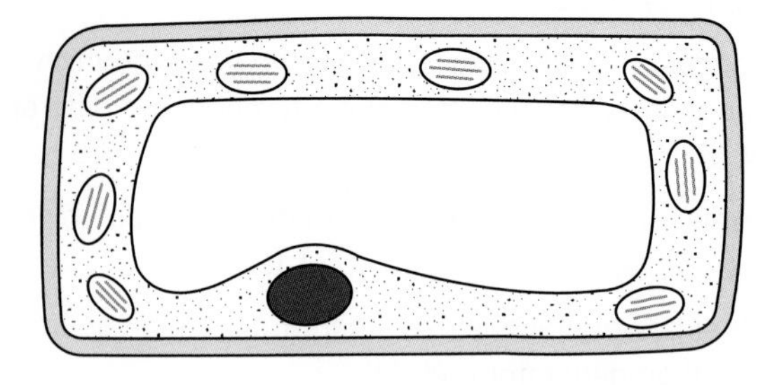

c (i) Copy and complete the following word equation to show what happens during photosynthesis:

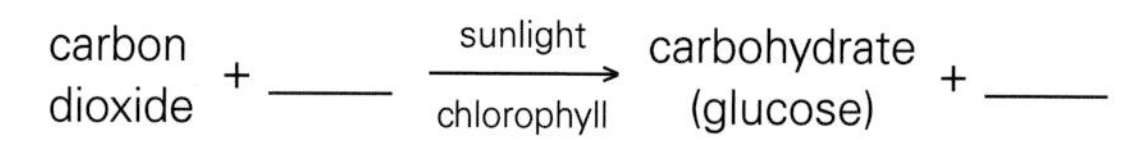

(ii) How does carbon dioxide enter a leaf for photosynthesis?

d In Autumn many plants die. Describe the part played by microbes in re-converting the carbohydrate from the plant into carbon dioxide.

e Sugar is a word that can be used to describe several carbohydrates. Here are two examples:

glucose – chemical formula $C_6H_{12}O_6$

maltose – chemical formula $C_{12}H_{22}O_{11}$

(i) Name the elements contained in sugars.
(ii) When sugars are heated, they melt at low temperature. When sodium chloride is heated, it does not melt until a much higher temperature. Explain this in terms of the structures of these compounds.

3 a The cell in the diagram below has four chromosomes. With the help of diagrams, explain how this cell can divide by mitosis.

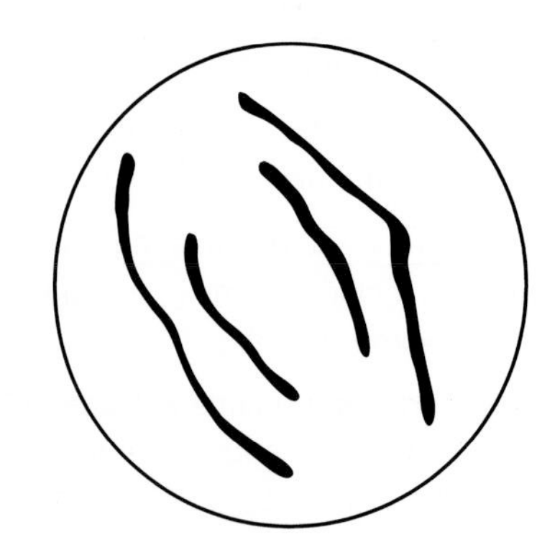

b The black panther is black variety of the spotted leopard. It is commonly found in dense tropical rain forests. However, it is much rarer in areas of open grassland where the normal (spotted) form is more common.

The diagram below shows the results of a cross between a spotted leopard and a black panther. It also shows the results of a cross between two of their offspring (the F_1 generation).

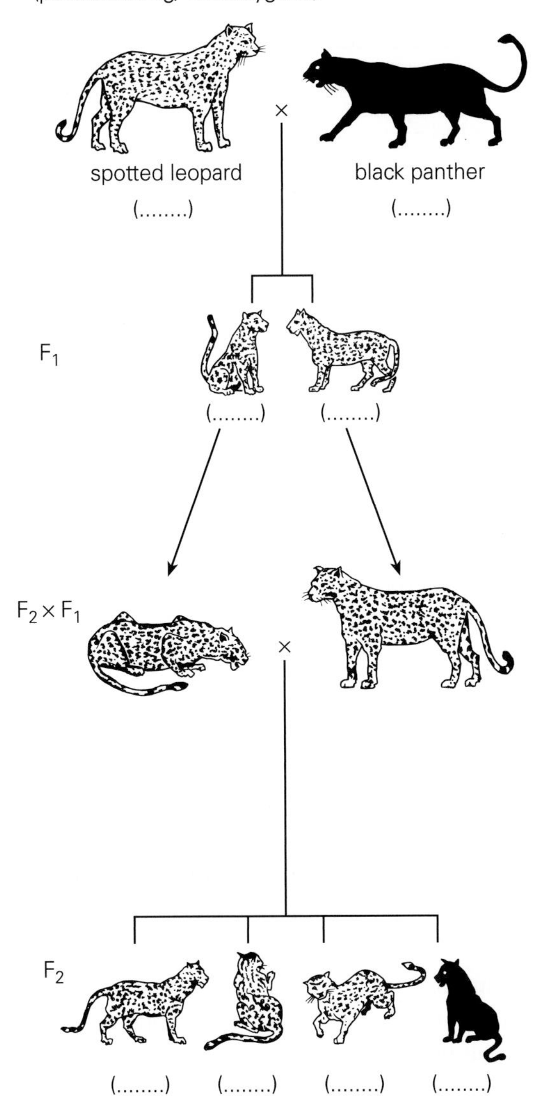

Assume that coat colour is controlled by a single gene with two alleles (H and h). One of these is responsible for the spotted coat, the other for the black coat.

(i) Which one of the alleles H (dominant) or h (recessive) controls black coat colour? Explain your answer.

(ii) Make a simplified copy of the diagram. Write hh, Hh, or HH in the spaces between each of the brackets in your own copy of the diagram to show the possible alleles possessed by each of the animals shown.

(iii) What is the expected ratio of homozygous to heterozygous individuals in the F_2 generation?

(iv) Black panthers are far more common in the rain forests than in open grassland. Explain how natural selection could cause this.

4 Mendeleev published his periodic table in 1869. The following shows how groups 1 and 7 looked. These were the *outer* two groups in his table:

	Group 1	Group 7
Period 1	H	
Period 2	Li	F
Period 3	Na	Cl
Period 4	K Cu	Mn Br
Period 5	Rb Ag	I

Use the information from this and the modern periodic table (p.224) to answer these questions:

a Explain what is meant by period 1, 2, etc.

b (i) How does Group 1 differ in the modern periodic table?
(ii) Why is the modern table more appropriate for this group?
(iii) What name is given to the Group 1 elements?

c (i) What name is given to the group 7 elements?
(ii) Why is Mn out of place in Group 7?

Below Mn is one of several places where Mendeleev left a gap in his table.

d Suggest why Mendeleev left these gaps.

e Why is there no group to the right of Group 7 in Mendeleev's table?

5 Ammonium sulphate is used as a fertilizer. The flow diagram below shows the main stages in making it:

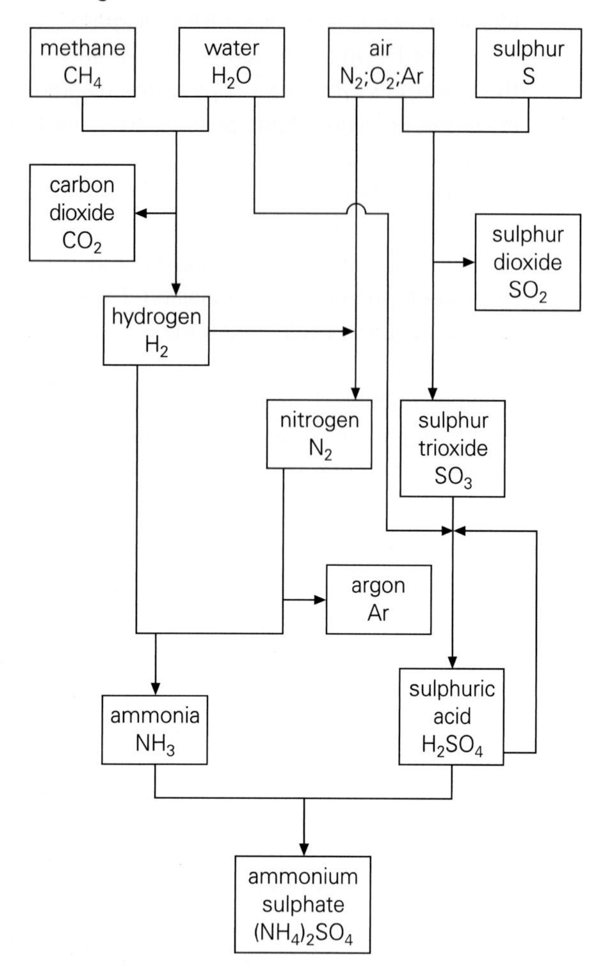

a Several of the above stages use a catalyst. What is the reason for using a catalyst?

b The by-products of the various processes include carbon dioxide, sulphur dioxide, and argon. Which of these by-products is harmful and which useful? Explain your answer.

c Ammonia is produced by the Haber process.
(i) What two gases are used in making ammonia?
(ii) The reaction is reversible. What problems does this create when manufacturing ammonia?

d (i) Give a balanced symbolic equation for the reaction between ammonia and sulphuric acid.
(ii) What mass of sulphuric acid (in tonnes) is needed to exactly neutralize 340 tonnes of ammonia? (The relative atomic masses are H 1, N 14, O 16, and S 32.)
Show all your working clearly.

6 The diagram below shows a section through a volcano.

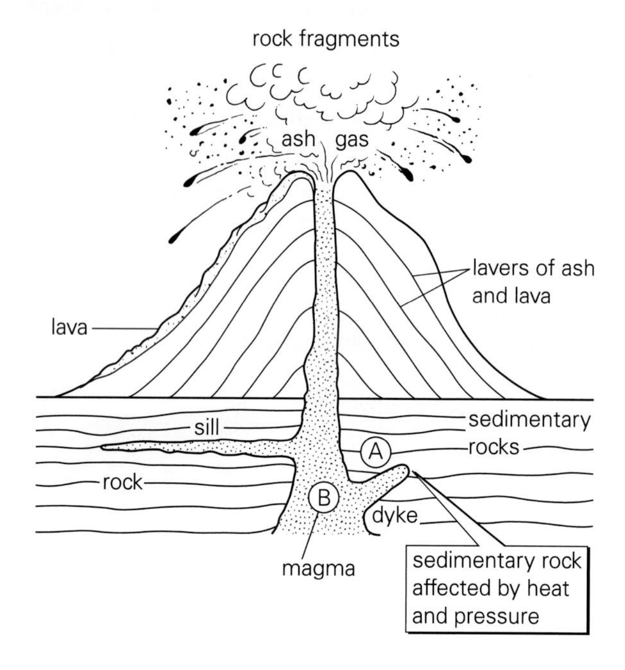

a What type of rock you would expect to find at A?

b (i) What type of rock will eventually be formed at B?
(ii) Explain how this type of rock (B) is formed.

7 Ethene and bromine react in an addition reaction to form dibromoethane:

C_2H_4 *(g)*	+	Br_2 *(l)*	→	$C_2H_4Br_2$ *(l)*
ethene		bromine		dibromoethane
colourless		red/brown		colourless

a Draw a displayed formula for ethene. (It contains a carbon–carbon double bond.)

b Draw a displayed formula for dibromoethane. (It contains only single bonds.)

c Explain why it is called an addition reaction.

d What colour change would tell you that a reaction had occurred?

An approximate value for the energy change in this reaction can be found using the bond energies.

e Use the bond energies in the table at the top of the next page to calculate the energy
(i) to break the bonds in ethene;
(ii) to break the bonds in bromine;
(iii) to make the bonds in dibromoethane.

Bond	Bond energy (kJ)
C—C	347
C=C	612
C—H	413
Br—Br	193
C—Br	290

f Calculate the energy difference in bond breaking and bond making.

g Is this reaction exothermic or endothermic? Explain your answer.

8 The diagram below represents the Earth and the Sun.

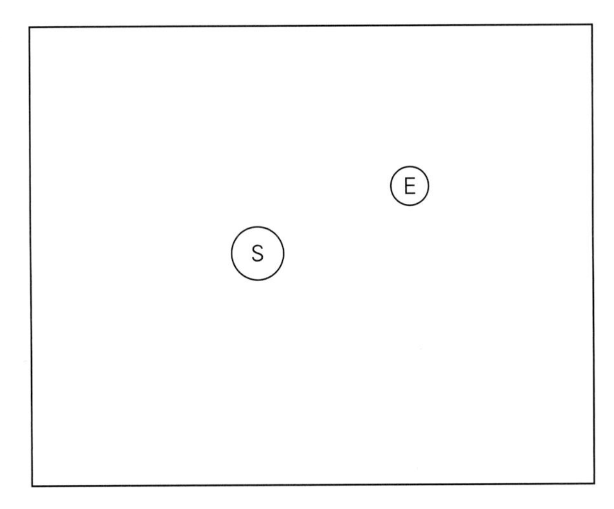

a Copy the diagram. Draw a line to show how the Earth will move in the next six months.

b The orbit time of Venus is 0.6 years and that of Mars is 1.9 years. Add a 'V' and an 'M' to your diagram to show the approximate positions of Venus and Mars.

9 a The diagram below shows a circuit with a lamp and buzzer.

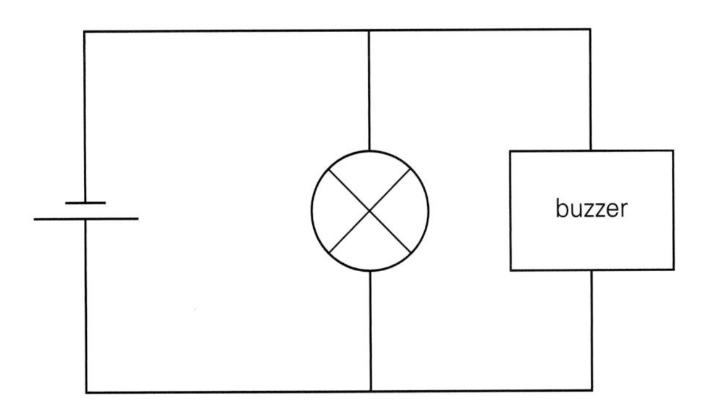

(i) Redraw the circuit, adding a single switch to control both the lamp and the buzzer.

(ii) Redraw the circuit with two switches. One switch should control only the lamp and the second switch should control only the buzzer.

b (i) A student wants to be able to control the brightness of a lamp. Draw and label a circuit diagram to show how this can be done.

The student adds an ammeter and a voltmeter to the circuit. He predicts that to double the current he will need to double the voltage. The table below shows the readings he takes.

lamp	voltage (V)	current (A)
dim	4.0	1.0
normal	7.0	1.5
bright	11.0	2.0

(ii) Using the data in the table, explain whether his prediction is correct.

(iii) Calculate the resistance of the lamp filament when it is bright and when it is dim. What can you deduce about the resistance?

10 The diagram below shows a hydroelectric power station.

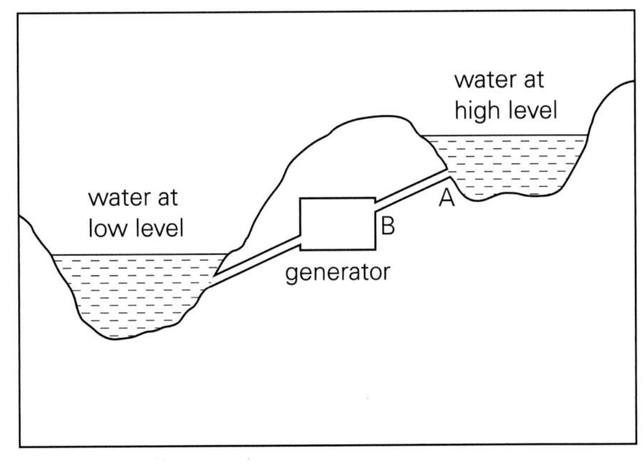

a What form of energy does the water have at the position marked A on the diagram?

b What forms of energy does the water have at B?

c (i) Why will the energy delivered by the generator be less than the change in energy of the water?

(ii) What happens to the missing energy?

d Give *three* advantages of a hydroelectric power station over a coal-fired power station.

11 The diagram below shows a hot-water storage tank. The water is heated electrically.

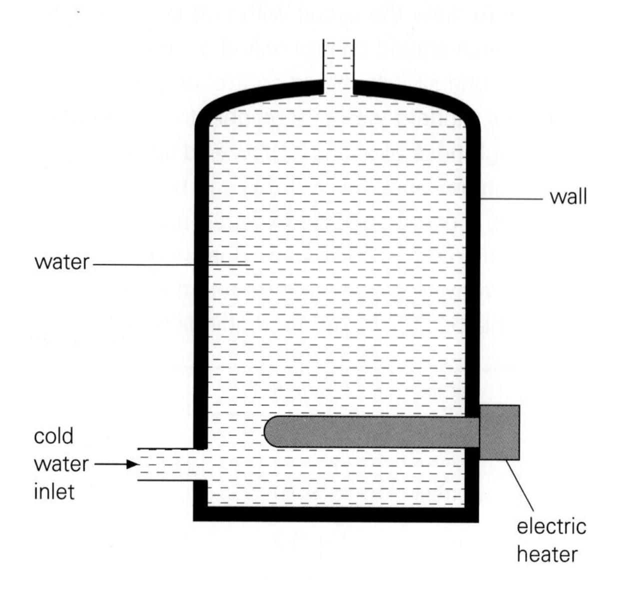

a Explain why the heater is placed at the bottom of the tank rather than at the top.

b Explain how you would reduce the energy loss through the walls of the tank, naming any materials you might use.

12 The diagram below shows a microwave oven. Microwaves are absorbed by water in food.

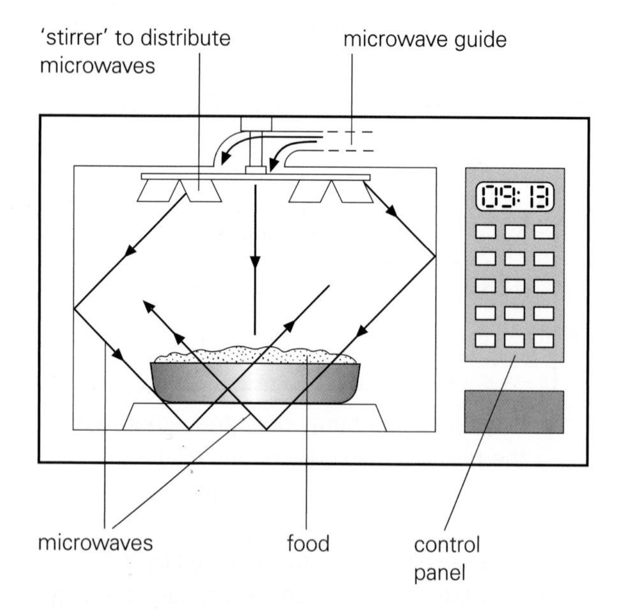

a Why does the food get hot?

b The output power of the oven is rated at 650 W. How much energy does the cooker transfer from the electricity supply in two minutes?

c The microwaves produced by the oven have a frequency of 2450 MHz and a speed of 300 000 km per second. Calculate their wavelength.

Microwaves are part of the electromagnetic spectrum.

d (i) Name another part of the electromagnetic spectrum which can be used to heat up food.
(ii) Do the waves in your last answer have a *longer* or *shorter* wavelength than microwaves?

e Name two parts of the electromagnetic spectrum which can cause cancer.

f Name two parts of the electromagnetic spectrum which can be used for communication.

13 A bicycle and rider roll down a ramp onto level ground and are then stopped by the brakes. The graph below represents the motion.

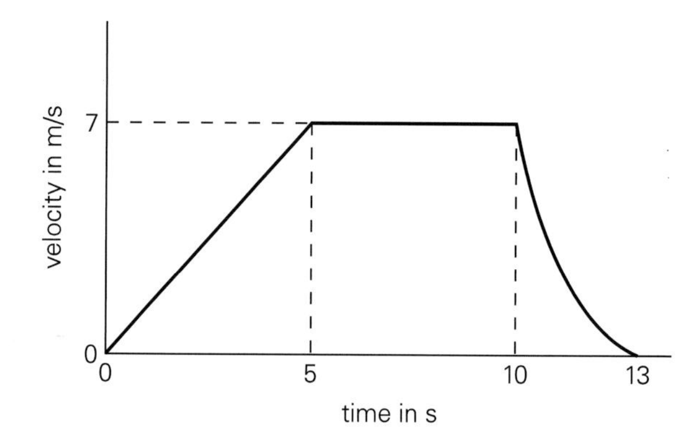

a Calculate
(i) the acceleration during the first 5 seconds;
(ii) the distance travelled in the first 10 seconds.

b The total mass of the bicycle and rider is 90 kg. The gravitational field strength of the Earth is 10 N/kg (10 N weight per kg).
Calculate
(i) the accelerating force on the bicycle and rider during the first 5 seconds;
(ii) the kinetic energy of the bicycle and rider after 5 seconds;
(iii) the thermal energy (heat) produced in the brakes between 10 seconds and 13 seconds;
(iv) the average power developed in the brakes.

Answers to questions on spreads

2.01
1 a chloroplast **b** cellulose **c** cytoplasm **d** membrane **e** nucleus **2** Living things feed, respire, excrete, grow, move, reproduce, show sensitivity, are made of cells **3** For energy, growth **4 a** Have nucleus, cytoplasm, membrane **b** Have cell wall, sap, chloroplasts **5 a** Groups of similar cells **b** Organ

2.02
1 a Carbon dioxide **b** Oxygen **2** Glucose (sugar) **3** No light **4** Healthy growth (making body proteins) **5** Stoma **6** Stream of water (and dissolved minerals) moving up through plant **7 a** Oxygen **b** Carbon dioxide **8** Plants make oxygen

2.03
1 a No change **b** Rate of photosynthesis limited by amount of light available **2** Osmosis **3** Control stomata size, hence water loss and gas flow in/out of leaf; swell up by osmosis **4** Used in respiration, changed into starch, used in making proteins

2.04
1 a Ovules (in carpels) **b** Pollen (in stamens) **2** Combined with male sex cell **3** In cross-pollination, pollen transferred to flower on different plant **4** Auxin collects away from light, stimulates growth that side, so shoot bends **5 a** Make plants grow too quickly, run out of food, and die **b** Control of ripening **6** Shortage of nitrate or potassium ions

2.05
1 a brain **b** heart **c** stomach **d** lung **e** lung **f** liver **g** kidney **2** Oxygen, water, food **3** Through lungs and kidneys

2.06
1 Change food into simpler liquid form **b** Absorbed into blood **c** Leaves body through anus **2 a** growth **b** bones and teeth; **a** fish, bread **b** cheese, milk **3** Helps food pass along the gut **4 a** Carbohydrates, fats

2.07
1 a 2200 kJ **b** 950 kJ **c** 1650 kJ **2** If energy not needed, food turned into fat **3** Bumps on surface of small intestine; absorb digested food into blood **4** Not enough oxygen available; oxygen not used in anaerobic respiration **5** Speed up reactions **6** Kill bacteria, create acid conditions for digestive enzymes **7** Gall bladder; aids digestion of fat

2.08
1 a Arteries **b** Veins **2** Fine tubes; carry blood near cells **3** To get food and oxygen, and get rid of waste products **4** Plasma **5** White **6** Red; haemoglobin **7** Lungs **8** So blood only flows one way **9** Passes through lungs, heart, other parts of body, heart

2.09
1 a Oxygen **b** Carbon dioxide (+ water) **2** For blood to absorb/release gases **3** Ribs move out, diaphragm down **4** Glycogen, iron, vitamins **5** Removes old red blood cells, produces heat **6** Filtering/cleaning blood **7** Go to bladder **8** They remove unwanted substances from body **9** Lungs

2.10
1 Egg cell; release of egg cell; cycle of egg cell release, uterus lining growth, period **2** 28 days **3** Passes out of uterus **4** Must meet sperm **5** Skin forms round ovum B In testicles **7** Condom, diaphragm

2.11
1 Respond to stimuli **2** Carry signals **a** from receptors to central nervous system **b** from central nervous system to muscles/other organs **3** blinking, coughing, sneezing **4** Ear sends signals to brain, brain sends signals to muscles **5 a** retina **b** lens **c** iris B optic nerve

2.12
1 In the blood **2** See table on p. 42 **3 a** Adrenal glands **b** Prepares body for rapid energy output **4 a** Pituitary gland **b** Causes egg to mature, increases oestrogen production **c** Reduces FSH production, increases LH production **d** Causes egg release **5** Steroids (heart disease), fertility drugs (multiple births)

2.13
1 a Drink, food, respiration **b** urine, sweat **2** Filter liquid from blood, then cause useful substances and water to be reabsorbed **3** Reduces nephrons' water output by increasing reabsorption **4** Kidneys not working properly; same job as nephrons **5** Makes liver change glucose (sugar) into glycogen **6** Glands in pancreas produce more insulin which affects liver as in Q5. **7 a** Increased **b** decreased **8** Sweating

2.14
1 Contact, animals, droplets in air **2** To avoid contaminating food **3** Some white blood cells digest germs, others make antibodies which kill germs **4** Contain dead or harmless germs which make immune system produce antibodies **5** Stops it working **6** Sexual contact, blood-to-blood contact, infected mother to unborn child **7** Bacterial cell has no nucleus

2.15
1 Barrier, with oils that kill germs **2 a** Traps dust and germs **b** Swallowed **3** Put microbes out of action or make them easier for phagocytes to digest **4** Chemicals which destroy toxins **5** Destroy their toxins, digest them **6 a** Body becomes dependent on them **b** Nicotine, heroin **7** See p. 49

2.16
1 Height, weight **2** Tongue-rolling, blood group **3** Chemical instructions for a characteristic; in nucleus of cells **4** One from each parent **5 a** Same set of genes **b** Characteristics depend partly on environment **6** Weight depends on eating habits **7** Racehorses, wheat **8** Preserve genes

2.17
1 Phenotype is characteristic which is seen (e.g. black hair), genotype is combination of alleles producing it (e.g. Hh) **2 a** Meiosis **b** Alleles **c** Gametes **3 a** 23 **b** 46 **4 a** Female **b** Male **5** 1 in 2

2.18
1 a mitosis **b** meiosis **2 a** In chromosomes **b** Carrying genetic information **3** Protein **4** 'Unzips', then unattached bases join on to produce two identical molecules **5** Genetically **a** different **b** the same

2.19
1 The same **2** Making genetically identical copies; growing plants from cuttings **3** Females have two X chromosomes, so chances of both having harmful allele are low; mother can pass on harmful allele to child **4** Pieces of DNA from human cells inserted into DNA of bacteria **5** All F_1 are Ss, no wrinkled seeds; F_2 can be SS, Ss, or ss, 1/4 have wrinkled seeds

2.20
1 Reptiles, dinosaurs, birds, mammals **2** Height, weight **3** Help in struggle for survival, so animals live to reproduce **4** Chance change in genes; dark wing colouring (peppered moth) **5** Radiation, chemicals **6** Dark one; better camouflage **7** Fewer dark moths eaten by birds, so more reproduce **8** Members start to live and breed in separate groups

2.21
1 a Giraffes stretch necks when reaching food; longer necks inherited by offspring **b** Giraffes with longer necks reach food, so more likely to survive, breed, and pass on characteristics **2** Genetics shows that acquired characteristics not passed on **3** Implied that humans and animals had common ancestors **4** Non-resistant variations killed, so resistant variations thrive and reproduce

2.22
1 Place where organism lives; group of same organisms; all organisms in one habitat; animal which feeds on other animals **2** Tides cover/uncover beach **3** Temperature change from summer to winter **4** For light and water **5** Has features which help its survival; bear's fur for winter warmth **6 a** Eaten by toads **b** Toads have died off, more plants **7** Fewer toads because fewer slugs for food

2.23
1 Food limited, predators eat rabbits **2** Providing building materials (changes landscape), increasing crop yields (destroys habitats) **3 a** Risk of injury **b** Release of polluting chemicals **4 a** Renewed at same rate as used **b** Softwood; can be regrown **5** Oil spillages cause pollution **6** Dumping of chemical waste, overuse of fertilizers

2.24
1 Released by burning fossil fuels **2** Carbon dioxide traps Sun's heat **3** Stops ultraviolet **4** Sulphur dioxide and nitrogen oxides from burning fossil fuels; attacks stone and steelwork **5** Chemicals cause poisoning; nutrients in sewage encourage algae growth which uses up oxygen; warm water discharge lessens oxygen, encourages weed growth **6** Releases more carbon dioxide; fewer plants to absorb carbon dioxide **7** Leads to erosion; deprives soil of nutrients

2.25
1 Producers (e.g. plants) make food, consumers (e.g. snails) eat it **2** Materials which rot; leaves, paper **3** Octopus, crab, seal, seagull; populations would fall **4** and **5** See pyramid examples on p. 68 (Q5 biomasses: frogs 200 g, worms 10 000 g, leaves 500 000 g)

2.26
1 a Photosynthesis **b** Respiration, burning **2** Burning coal makes carbon dioxide, which is absorbed by plants, which are eaten by animals **3** For making proteins **4** From nitrates in soil **5** By eating plants **6** By microbes taking nitrogen from air and dead organisms, by effect of lightning

2.27
1 Herbivore feeds on plants, carnivore feeds on animals; herbivore; plants are at 1st trophic level **2** Energy lost at each level, so not enough left for more **3** Fewer and fewer organisms contain them **4** Plenty of moisture, air, warmth **5** Return vital elements to soil **6** Land needed to produce beef for one person can produce wheat for 20 **7** Too many young fish caught, so not enough left to breed

3.01
1 a Substances that cannot be broken down **b** Metals, non-metals **2** Al Fe Ca Na K Mg **3** From Latin name (Ferrum) **4** sulphur **5** two or more atoms **6** H_3PO_4 **7** Compounds are new substances made when elements combine; mixtures are made up of variable amounts

3.02
1 Particles in solid vibrate about fixed positions, particles in liquid vibrate but can change positions, particles in gas move about freely **2** Particles colliding with container walls **3** Particles wandering through those of another material **4** Slow down **5** Pure water boils at 100 °C **6** Escaping particles take energy away **7** 2700 kg/m^3 **8** 0.002 m^3

3.03
2 a Substance which dissolves **b** Substance it dissolves in **c** Solution with maximum amount of solute **d** Solution with water as solvent **3 a** 37 g/100 g water; 30 g/100 g water **b** 39 g/100 g water; 180 g/100 g water **4** See table on p. 79 **5** Dissolve in water, then evaporate water **6 a** Solutes in same solution **b** Liquids with different boiling points **7** Distillation

3.04
1 a Physical changes easy to reverse, no new substances **b** Physical – melting; chemical – burning **c** Total mass unchanged **2** 88 g **3** Bubbles of carbon dioxide **4** Changes of state, colour, temperature **5** Exothermic – heat out, endothermic – heat in

3.05
1 Proton (mass 1, charge +1), neutron (1, 0), electron (0, –1) **3** number of protons in atom; number of protons + neutrons; atoms of same element with different mass numbers **4** 3 protons, 4 neutrons, 3 electrons **5** Relative atomic mass

3.06
1 a Gases, unreactive, full outer shells **b** Metals, reactive, one outer shell electron **2** See table on p. 84 **3** See row 2 on p. 224 **4 a** 1 **b** 4 **5** Energy levels holding electrons up to a set number **6 a** 2e, 8e, 1e **b** 2e, 8e, 6e **c** 2e, 8e, 7e **d** 2e, 8e, 8e

3.07
1 Nothing **2** 2e, 8e; 2e, 8e, 8e **3** Charged atom (or group of atoms); positive ion; bonding due to attraction between opposite ions **4** See p. 87 (top right) **5** Billions of ions cluster; very strong bonds **6 a** NaOH **b** $(NH_4)_2SO_4$ **c** Fe_2O_3

3.08
1 Covalent bond (shared electrons) **2** Two atoms sharing electron give full outer shells **4** Electrons not distributed evenly around nuclei **5** Temperature not high enough for bonds to break **6** Form ions which are attracted to end of water molecules; ions are mobile

3.09
1 a Attractions between oppositely charged ions **b** Weak bonds between molecules **c** 'Sea' of free electrons **2** Weak bonds between molecules easily broken by heat **3** Each carbon atom strongly bonded to four others **4** Flat layers of atoms slide over each other **5** Mixture not compound **6** Different sizes of atoms resist sliding of layers

3.10
1 a 2; so that equation balances **b** Solid and gas **2** Relative atomic mass, relative molecular mass; 18 **3** 6×10^{23} molecules **a** 16 g **b** 24 dm^3 **4** 3 moles of solute in 1 dm^3 solution
5 $CH_4(g) + 2O_2(g) \rightarrow CO_2(g) + 2H_2O(l)$; **a** 36 g **b** 48 dm^3

3.11
1 Oxygen; 4 shared electrons **2** Energy needed to start reaction **3** More energy released in making new bonds than absorbed in breaking existing bonds **4** Reverse of 3 **5** Energy needed to break a bond (or energy released when bond is made) **6** 2054 kJ

3.12
1 breaking down of substance due to heat **2** Change from green to black; $CuCO_3(s) \rightarrow CuO(s) + CO_2(g)$ **3** barium sulphate; $Ba(NO_3)_2(aq) + CuSO_4(aq) \rightarrow BaSO_4(s) + Cu(NO_3)_2(aq)$
4 Removal of electrons; addition of electrons **5** Fe^{2+} and Fe^{3+} ions **6 a** Zinc **b** Sodium **c** Iron

3.13
1 Hydrogen **2** Split up into ions **3** Fewer hydrogen ions in weak acid **4** Base will neutralize acid; alkali is soluble base
5 Hydrogen gas; pops when lit **6** Colour depends on acidity/alkalinity **7** Acidic **8** pH7

3.14
1 Water, sodium chloride; $HCl(aq) + NaOH(aq) \rightarrow H_2O(l) + NaCl(aq)$ **2** $H^+(aq) + OH^-(aq) \rightarrow H_2O(l)$ **3** acid + base → water + salt **4** Neutralize copper(II) oxide with sulphuric acid, then evaporate solution **5** nitric acid + calcium carbonate → water + carbon dioxide + calcium nitrate **6** Indigestion tables, treating acid soil

3.15
1 Collisions which break bonds **2 a** Break up chips, increase temperature, increase concentration of acid **b** Fizzing stops
3 Sulphur precipitated **4** Speeds up reaction; forms looser bonds **5** Manganese dioxide; $2H_2O_2(l) \rightarrow O_2(g) + 2H_2O(l)$ **6** See table on p. 115

3.16
1 Biological catalyst; breaks down starch into glucose **2** Curds separated from milk, pressed, enzymes added **3** Glucose changed into ethanol + carbon dioxide **4** Fermentation of starch creates carbon dioxide bubbles **5** Certain molecules fit enzymes which weaken bonds **6** Enzyme destroyed if too high, shut down if too low **7** Treating starch with enzyme in mould **8** See table on p. 105

3.17
1 a Products recombine to give original substances **b** Heating in sealed container; $CaCO_3(s) \rightarrow CaO(s) + CO_2(g)$ **2 a** 150 atm, 450 °C **b** Compromise between speed and good yield **c** 17%
d 500 atm, 380 °C **3** Unchanged but reached more quickly
4 a Pollution, lower yield **b** Using excess oxygen, cooling and removing sulphur trioxide

3.18
1 Nitrogen from air and hydrogen from methane + steam **2** Iron
3 Ammonium nitrate; neutralizing nitric acid with ammonia
4 Nitrogen, phosphorus, potassium **5** Six; make ammonia, nitric acid, Nitram, sulphuric acid, phosphoric acid, NPK blends
6 Anywhere near a port with road/rail links and close to supplies of natural gas and potassium chloride

3.19
1 a Bigger **b** Bigger **2** Form positive ions, ionic chlorides, ionic (basic) oxides **3** High melting point, semiconductor, oxide both acidic and basic; shows some properties of metals as well as non-metals **4** Ionic solid; turns it blue **5** Violently;
$2K(s) + 2H_2O(l) \rightarrow 2KOH(aq) + H_2(g)$ **6** Potassium chloride, doesn't raise it

3.20
1 Salt producer **2** Chlorine – iron glows brightly, bromine – iron glows less brightly, iodine – slow reaction **3 a** Increase
b Increase **c** Decrease **4** Bleaches/removes colour; kills germs
5 Hydrogen burns in chlorine, bromine needs a catalyst to react
6 Precipitation (potassium nitrate) **7** Chlorine displaces bromine, potassium chloride formed

3.21
1 See table on p. 114 **2** See p. 114 bottom left **3 a** 2e, 8e, 8e, 2e **b** 2e, 8e, 9e, 2e **c** 2e, 8e, 18e, 2e **d** 2e, 8e, 18e, 3e
4 Chromium **5** Potassium dichromate(VI); orange to green
6 Platinum-rhodium alloys; change polluting gases into harmless ones

3.22
1 By comparing reactions with air, water, dilute acid **2** Hydrogen
3 a Slowly **b** Slowly **c** Slowly **4** Calcium, gold **5** Oxide layer protects rest of metal **6** Experiment as on p. 117 **7** iron coated with zinc

3.23
1 a Iron displaces copper **b** Iron displaces silver **2** Zinc corrodes instead of iron **3** Used as a reference, shows how metal will react with acid **4** Iron(II) oxide and copper; aluminium oxide and iron **5 a** iron(II) sulphate, copper(II) sulphate **b** Between zinc and iron **c** (i) Slow (in steam) (ii) Steady (iii) Displace copper

3.24
1 Minerals containing metals **2 a** Galena **b** Haematite
c Bauxite **3** Gold unreactive, aluminium reactive **4** Potassium; electrolysis **5 a** Carbon dioxide **b** Iron (and carbon dioxide)
c Slag **6** Blowing oxygen and lime onto surface; steel has less carbon and impurities **7 a** Chromium, nickel **b** Manganese or tungsten

3.25
1 Aluminium ore; aluminium oxide **2 a** Splitting ionic liquid using current **b** Ionic liquid **3 a** Positive **b** Negative **4 a** Graphite
b Graphite **c** Aluminium ore dissolved in cryolite **5** Saves ore and energy **6** aluminium with aluminium oxide coating
7 a Copper passes into solution **b** Copper deposited **c** Ions move through, but no change

3.26
1 Mined **2** Flavouring, treating roads, softening water **3** Making glass, detergents, treating sewage, softening water **4** Saves raw materials and energy **5** Chlorine (purifying water) hydrogen (making ammonia), sodium hydroxide (making soaps) **5** See p. 125 right column

3.27
1 Carbon compound **2 a** Compound of hydrogen and carbon **b** Hydrocarbon with single covalent bond **c** Chemicals with similar properties **3** Methane (CH_4), ethane (C_2H_6), propane (C_3H_8), butane (C_4H_{10}) **4** Mixture of alkanes; underground **5** By fractional distillation; group of alkanes with similar chain lengths **6 a** $C_3H_8(g) + 5O_2(g) \rightarrow 3CO_2(g) + 4H_2O(l)$
b $C_5H_{12}(g) + 8O_2(g) \rightarrow 5CO_2(g) + 6H_2O(l)$

3.28
1 Changing into lighter fractions **2** unsaturated – double carbon–carbon bond **3** Add bromine water, watch for colour loss **4 a** Alkene molecules joining up **b** Small molecules that will link with others in a chain **c** Chain formed when monomers link **5** By pressurizing and heating ethene gas **6 a** propene (carpets, ropes) **b** phenylethene (insulation, disposable cups)

3.29
1 a Rocks dissolved by rain **b** Shells of sea creatures forming **2** Band of O_3 high in atmosphere **3 a** Condensed to form oceans **b** Carbon became locked up in sedimentary rocks **c** Given off by plants when these appeared **4** Act as reservoir for dissolved carbon dioxide

3.30
1 Core (very hot, part molten), mantle (hot, flexible rock), crust (thin outer layer) **2** Molten rock; release of pressure melts hot rock **3 a** Rocks formed when magma cooled at surface **b** Rocks formed when magma cooled underground; extrusive – smaller crystals **4** Deposited fragments compressed and cemented **5** Sea shore **6** Remains or outlines of ancient plants or animals; whether on land or sea **7** Rocks changed by heat and/or pressure

4.01
1 Conductor has free electrons **2** Ball **a** attracted **b** repelled **3 a** positive **b** negative **4** Sparks when refuelling aircraft **5** In electrostatic precipitators, photocopiers

4.02
1 Ammeter; 2.0 **2** Voltmeter **3 a** Brighter **b** Higher **c** Higher (approx. double) **4 a** Reduced to previous brightness **b** Reduced to previous value (2.0) **5** Brightness not reduced; one bulb keeps working if other bulb removed

4.03
1 a Lower resistance **b** Resistance too high, heat given off, current reduced **c** Heating elements; because high resistance gives heating effect **2** Brighter; less resistance, so higher current **3** 2 kW h; 20p **4 a** 80p **b** 10p **5** 160p

4.04
1 a X is 4 Ω, Y is 2 Ω, Z is 0.5 Ω **b** 4 A, 8 A, 12 A **2** Thermistor **b** Diode **3 a** 2 Ω **b** 4 Ω **4** B because current is less for any given voltage

4.05
1 a 2 A **b** 3 A **c** 0.5 A **2 a** 2 Ω **b** 18 W **3 a** 3 C **b** 18 J **4 a** 4 A **b** 2 A **c** 6 A

4.06
1 With a compass **2** More turns on coil, higher current **3** Wrap coil round nail, connect coil to battery **4** Circular **5** Small current switches on electromagnet in relay, relay switches on motor circuit **6 a** Switches off current when this is too high **b** To pull catch when current is high enough

4.07
1 a Connecting outside circuit to coil **b** Reversing current through coil every half turn **2** Stronger magnet, more turns on coil, higher current **3** AC keeps changing direction; **a** DC **b** AC **c** AC **4 a** By turning coil in magnetic field **b** To connect coil to outside circuit **5** Meter needle flicks one way, then other way

4.08
1 a Live; so that wire in cable is not live when switch is off **b** Safety, takes current from metal body to earth so that fuse blows **c** Brown **2 a** 2 A **b** 3 A **c** Lamp current small, so any increase might cause overheating without blowing fuse **3** Smaller wave (lower amplitude) **4** Line nearer centre line

4.09
1 Increasing voltage from power station; reducing voltage in TVs **2** DC doesn't cause changing magnetic field **3 a** Step-down **b** 23 V **c** 46 W **d** 46 W **e** 2 A **4** To reduce current, and energy loss **5 a** 100 A **b** 1 A

4.10
1 10 m/s **2** Head-down position, shaped helmet, smooth and tight-fitting oufit, streamlined bike (e.g. no spokes) **3 a** Brakes, tyres, steering wheel **b** Wheel bearings, moving parts of engine and gearbox, car body moving through air **4 a** 15 m **b** 18 m; at higher speed, car travels further in same time **5 a** 50 m **b** 99 m

4.11
1 a 30 m/s **b** 3 m/s^2 **c** 6 m/s^2 **d** 600 m **2 a** 20 m/s **b** Steady speed (0–20 s), speed drops (20–30 s), speed zero (30–40 s)

4.12
1 a 50 N **b** 10 m/s^2 **2** Double force, but double mass **3** 15 N **4** 0.5 m/s^2 **5** Earth's mass too great for force to have detectable effect **6 a** 8 N **b** 1.6 m/s^2

4.13
1 a Terminal velocity **b** Air resistance **c** Equal **d** Reduced; lower speed needed for air resistance to equal weight **2 a** 1000 N **b** 1.25 m/s^2 **c** Steady velocity

4.14
1 a Larger area so lower pressure **b** Smaller area so higher pressure **2 a** 100 Pa **b** 50 Pa **3 a** 200 Pa **b** 100 N **c** Greater

4.15
1 b Graph is straight line through origin **c** Up to 27 mm extension **d** 2.3 N **2 b** Pressure increases if volume decreases **c** 33 cm^3 **3** 12 m^3

4.16
1 To give turning effect which matches turning effect of load, so that centre of mass of crane and load is over base **2** When load is changed, turning effect of counterbalance must be changed **3** 200 N m **4** 200 N m **5** 0.5 m **6** 200 N (counterbalance is then at maximum distance) **7** Wider base, heavier weight low down on base

4.17
1 a Moving car **b** Battery **c** Stretched spring **2** 10 000 J **3** Chemical; changed to heat + kinetic energy **4** Kinetic → heat **5** Chemical → kinetic → potential → kinetic → heat **6** Energy can change forms, but total amount stays the same

4.18
1 a conduction **b** convection **c** conduction **2 a** Metal conducts, plastic insulates **b** All contain pockets of air, which is an insulator **3 a** B **b** B

4.19
1 a To reflect Sun's radiation **b** Poor radiators of heat **2** Black; better absorber of radiation **3 a** Part-vacuum, stopper **b** Shiny surfaces **4** Reduces heat flow in either direction **5 a** radiation **b** conduction, convection, evaporation

4.20
1 Cannot be replaced; coal, oil **2** Global warming (carbon dioxide), acid rain (sulphur dioxide) **3** Hydroelectric scheme, aerogenerators **4** Ancient sea plants absorbed Sun's energy, ancient sea creatures fed on plants, remains of plants and creatures trapped and crushed by sediment to form oil, petrol extracted from oil **5** 35% of fuel's energy changed into electrical energy, rest wasted as heat **6** Less fuel burnt, so supplies last longer and less pollution

4.22
1 100 J **2** 10 W **3** 0.4 kW, 0.25 kW, 45 kW, 95 kW **4** 100 W **5 a** 600 J **b** 300 J **c** 300 J **d** 600 J **e** 20 m/s

4.23
1 Longitudinal – backwards–forwards vibrations (e.g. sound), transverse – side-to-side vibrations (e.g. light) **2** 3 m **3** 6 m **4 a** Reflected **b** Diffracted **c** Diffracted less

4.24
1 Detect waves with microphone, show waveform on CRO **2 a** Sound waves travel through solids **b** No material to carry vibrations **3 a** Sound much slower than light **b** 660 m **4** 0.67 s after he shouts

4.25
1 200 waves every second **2 a** louder **b** higher pitch **3** Sounds above range of human hearing (above 20 kHz) **4** Scanning womb, breaking up kidney stones **5 a** 35 000 Hz **b** 0.04 m (4 cm) **c** Halved to 0.02 m

4.26
1 Plates move against each other, crust cracked, heat produced **2** Plates move together at destructive boundaries, apart at constructive boundaries **3** P-waves, faster **4** S-waves cannot travel through liquids, no S-waves travel through core

4.27
1 Mirror reflects light in regular way, wall scatters light **2** 50 cm behind mirror **3 b** Y **4** Light waves on one side of beam reach glass first and slow before those on other side, so beam bends

4.28
1 a Violet **b** Red **2 a** Light **b** radio, micro, infrared, light **c** infrared, micro **d** X-rays, gamma **e** X-rays, gamma **f** gamma

4.29
1 a Pulses of light or infrared **b** More information carried, less loss of quality **c** Can carry more signals, less power loss **2** AM uses longer waves which will diffract into valley **3 a** 100 000 000 Hz **b** 300 000 000 m/s **c** 3 m

4.30
1 a Total of protons + neutrons **b** Same number of protons (and electrons) **c** Different numbers of neutrons **2** Has atoms with unstable nuclei **3** Removes electrons from atoms **4** Can kill or damage cells **5** Alpha, beta, gamma **6** Alpha **7** Gamma **8** Alpha **9** Small amount of radiation always present from natural sources

4.31
1 a Nucleus breaks up **b** Change in number of protons in nucleus **2** Takes 28 years for half strontium-90 nuclei to decay **3 a** 8 days **b** 24 days **4** So that activity quickly drops to safe level **5** neutrons from fission of uranium-235 causing further fission, and so on; nuclear reactor

4.32
1 a 24 hours **b** 365 days **c** 27 days **2** Less **3 b** Night **4** Apparent groupings of the brightest stars **5** Planets appear to move slowly across background of stars

4.33
1 Reflect Sun's light **2** Mercury, Venus, Mars, Pluto **3** Greater distance, longer orbit time **4 a** Mars, Jupiter, Saturn, Uranus, Neptune, Pluto **b** Further from Sun, so receive less heat **5** Venus hotter than Mercury, yet further from Sun **6 a** Star system with billions of stars **b** Distance travelled by light in one year **7 a** 2 million years **b** Take too long as nothing can travel faster than light

4.34
1 a Passes over whole of Earth's surface **b** Communications, space research, navigation **2** Earth turns at same rate as satellite orbits **3 a** B **b** B **c** A

4.35
1 Nuclear fusion **2 a** Huge cloud of gas and dust **b** Gravity **c** 4500 million years **d** Heavier elements present could only have been made in supernova **3 a** Expanded, cooled star **b** Hot core exposed after red giant loses outer layer **4** Collapse of very massive star's core after supernova **5** Galaxies moving apart, background microwave radiation

Answers to exam-style questions (pages 210–214)

1 a (i) Prevent loss of blood (ii) Supply blood with oxygen; maintain blood circulation **b** 100 beats per minute

2 a Making food (glucose) **b** Light needed for food-making can pass through cell walls and be absorbed by chlorophyll in chloroplasts **c** (i) 1st space – water; 2nd space – oxygen (ii) Through stomata **d** Microbes are decomposers which digest carbohydrate and release carbon dioxide as a product **e** (i) Carbon, hydrogen, oxygen (ii) Sodium chloride has strong ionic bonds, sugars have much weaker covalent bonds

3 a See p. 54 **b** (i) h; both F_1 offspring spotted despite one parent with black coat, so black coat must be recessive (ii) Parents are (from left to right) HH hh, F_1 are Hh Hh, F_2 are HH Hh Hh hh (iii) 1:1 (iv) Black better camouflage in forest than in grassland, so survival more likely in forest

4 a Rows within periodic table **b** (i) Doesn't contain H, Cu, Ag (ii) Elements in old Group don't all have same properties (iii) Alkali metals **c** (i) Halogens (ii) Mn is a metal **d** Elements not found or properties not investigated **e** Next group (0) contains unreactive (noble) gases, so no reactions observed

5 a Speeds up reactions **b** Harmful – sulphur dioxide (acid rain); useful – argon (lighting), carbon dioxide (coolant) **c** (i) Hydrogen, nitrogen (ii) Conditions must be right to drive reaction in direction required **d** (i) $2NH_3(g) + H_2SO_4(aq) \rightarrow (NH_4)2SO_4(aq)$ (ii) M_rs are 17 for NH_3 and 98 for H_2SO_4. From equation, $2 \times 17 = 34$ tonnes ammonia are neutralized by 98 tonnes sulphuric acid. Therefore 340 tonnes ammonia requires 980 tonnes sulphuric acid

6 a Metamorphic **b** (i) Igneous (ii) Liquid magma cools to form solid made up of crystals

7 a

```
H       H
 \     /
  C=C
 /     \
H       H
```

b

```
    H  H
    |  |
H—C—C—H
    |  |
   Br Br
```

c Small molecules add to form larger one **d** Liquid loses colour when added **e** (i) 2264 kJ (ii) 193 kJ (iii) 2579 kJ **f** 122 kJ **g** Endothermic; net input of energy required

8 a Semicircle round Sun **b** V is between S and E; M is beyond E

9 a (i) Switch next to battery (ii) One switch next to buzzer, other next to lamp **b** (i) Circuit with battery, lamp, and variable resistor in series (ii) No, because value of voltage divided by current changes (iii) 5.5 Ω when bright, 4.0 Ω when dim; resistance increases with temperature

10 a Gravitational potential **b** Gravitational potential + kinetic **c** (i) Generator wastes some energy (ii) Lost as heat **d** No fuel to buy, no fuel to tranport, no polluting gases

11 a Hot water rises by convection; if element were at top, bottom would stay cool **b** Insulation (lagging) round tank, using glass wool or plastic foam

12 a Absorbs energy carried by microwaves **b** 78 000 J **c** 0.12 m (12 cm) **d** (i) Infrared (ii) longer **e** X-rays, gamma rays **f** Radio waves, light

13 a (i) 1.4 m/s^2 (ii) 52.5 m **b** (i) 126 N (ii) 2205 J (iii) 2205 J (iv) 735 W

Units and symbols

Quantity	*Unit*	*Symbol*
mass	kilogram	kg
length	metre	m
time	second	s
force	newton	N
weight	newton	N
pressure	pascal	Pa
energy	joule	J
work	joule	J
power	watt	W
voltage	volt	V
current	ampere	A
resistance	ohm	Ω
charge	coulomb	C
temperature	degree Celsius	°C
frequency	hertz	Hz

Bigger or smaller

To make units bigger or smaller, prefixes are put in front of them.

micro (μ)	= 1 millionth	= 0.000 001	= 10^{-6}
milli (m)	= 1 thousandth	= 0.001	= 10^{-3}
kilo (k)	= 1 thousand	= 1000	= 10^3
mega (M)	= 1 million	= 1 000 000	= 10^6

For example

1 micrometre	= 1 μm	= 0.000 001 m
1 millisecond	= 1 ms	= 0.001 s
1 milliampere	= 1 mA	= 0.001 A
1 kilometre	= 1 km	= 1000 m
1 kilojoule	= 1 kJ	= 1000 J
1 megatonne	= 1 Mt	= 1 000 000 t

Index

Periodic table

1 H hydrogen 1

Period	Group 1	Group 2											Group 3	Group 4	Group 5	Group 6	Group 7	0
																		He helium 2
2	Li lithium 3	Be beryllium 4											B boron 5	C carbon 6	N nitrogen 7	O oxygen 8	F fluorine 9	Ne neon 10
3	Na sodium 11	Mg magnesium 12	transition metals										Al aluminium 13	Si silicon 14	P phosphorus 15	S sulphur 16	Cl chlorine 17	Ar argon 18
4	K potassium 19	Ca calcium 20	Sc scandium 21	Ti titanium 22	V vanadium 23	Cr chromium 24	Mn manganese 25	Fe iron 26	Co cobalt 27	Ni nickel 28	Cu copper 29	Zn zinc 30	Ga gallium 31	Ge germanium 32	As arsenic 33	Se selenium 34	Br bromine 35	Kr krypton 36
5	Rb rubidium 37	Sr strontium 38	Y yttrium 39	Zr zirconium 40	Nb niobium 41	Mo molybdenum 42	Tc technetium 43	Ru ruthenium 44	Rh rhodium 45	Pd palladium 46	Ag silver 47	Cd cadmium 48	In indium 49	Sn tin 50	Sb antimony 51	Te tellurium 52	I iodine 53	Xe xenon 54
6	Cs caesium 55	Ba barium 56	La lanthanum 57	Hf hafnium 72	Ta tantalum 73	W tungsten 74	Re rhenium 75	Os osmium 76	Ir iridium 77	Pt platinum 78	Au gold 79	Hg mercury 80	Tl thallium 81	Pb lead 82	Bi bismuth 83	Po polonium 84	At astatine 85	Rn radon 86
7	Fr francium 87	Ra radium 88	Ac actinium 89															

Ce cerium 58	Pr praseodymium 59	Nd neodymium 60	Pm promethium 61	Sm samarium 62	Eu europium 63	Gd gadolinium 64	Tb terbium 65	Dy dysprosium 66	Ho holmium 67	Er erbium 68	Tm thulium 69	Yb ytterbium 70	Lu lutetium 71
Th thorium 90	Pa protactinium 91	U uranium 92	Np neptunium 93	Pu plutonium 94	Am americium 95	Cm curium 96	Bk berkelium 97	Cf californium 98	Es einsteinium 99	Fm fermium 100	Md mendelevium 101	No nobelium 102	Lr lawrencium 103

Table of approximate relative atomic masses (A_r) for calculations

Element	Symbol	A_r	Element	Symbol	A_r	Element	Symbol	A_r
aluminium	Al	27	iodine	I	127	oxygen	O	16
bromine	Br	80	iron	Fe	56	phosphorus	P	31
calcium	Ca	40	lead	Pb	207	potassium	K	39
carbon	C	12	lithium	Li	7	silicon	Si	28
chlorine	Cl	35.5	magnesium	Mg	24	silver	Ag	108
copper	Cu	64	manganese	Mn	55	sodium	Na	23
helium	He	4	neon	Ne	20	sulphur	S	32
hydrogen	H	1	nitrogen	N	14	zinc	Zn	65